Advances in Industrial Control

Springer-Verlag London Ltd.

Other titles published in this Series:

Feedback Control Theory for Dynamic Traffic Assignment
Pusjkin Kachroo and Kaan Özbay

Robust Aeroservoelastic Stability Analysis
Rick Lind & Marty Brenner

Performance Assessment of Control Loops: Theory and Applications
Biao Huang & Sirish L. Shah

Advances in PID Control
Tan Kok Kiong, Wang Qing-Guo & Hang Chang Chieh with Tore J. Hägglund

Advanced Control with Recurrent High-order Neural Networks: Theory and Industrial Applications
George A. Rovithakis & Manolis A. Christodoulou

Structure and Synthesis of PID Controllers
Aniruddha Datta, Ming-Tzu Ho and Shankar P. Bhattacharyya

Data-driven Techniques for Fault Detection and Diagnosis in Chemical Processes
Evan L. Russell, Leo H. Chiang and Richard D. Braatz

Bounded Dynamic Stochastic Systems: Modelling and Control
Hong Wang

Non-linear Model-based Process Control
Rashid M. Ansari and Moses O. Tadé

Identification and Control of Sheet and Film Processes
Andrew P. Featherstone, Jeremy G. VanAntwerp and Richard D. Braatz

Precision Motion Control
Tan Kok Kiong, Lee Tong Heng, Dou Huifang and Huang Sunan

Nonlinear Identification and Control: A Neural Network Approach
Guoping Liu

Digital Controller Implementation and Fragility: A Modern Perspective
Robert S.H. Istepanian and James F. Whidborne

Optimisation of Industrial Processes at Supervisory Level
Doris Sáez, Aldo Cipriano and Andrzej W. Ordys

Applied Predictive Control
Huang Sunan, Tan Kok Kiong and Lee Tong Heng

Hard Disk Drive Servo Systems
Ben M. Chen, Tong H. Lee and Venkatakrishnan Venkataramanan

Robust Control of Diesel Ship Propulsion
Nikolaos Xiros

Hydraulic Servo-Systems
Mohieddine Jelali and Andreas Kroll

Silvio Simani, Cesare Fantuzzi
and Ron J. Patton

Model-based Fault Diagnosis in Dynamic Systems Using Identification Techniques

Springer

Silvio Simani, PhD
Dipartimento di Ingegneria, Università di Ferrara, Via Saragat, 1. 44100 Ferrara, Italia

Cesare Fantuzzi, PhD
Dipartimento di Scienze per l'Ingegneria, Università di Modena e Reggio Emilia, Italia

Ronald Jon Patton, Beng, MEng, PhD
School of Engineering, The University of Hull, Cottingham Road, Kingston-Upon-Hull, HU6 7RX, UK

ISBN 978-1-84996-895-9 ISBN 978-1-4471-3829-7 (eBook)
DOI 10.1007/978-1-4471-3829-7

British Library Cataloguing in Publication Data
Simani, Silvio
Model-based fault diagnosis in dynamic systems using identification techniques. - (Advances in industrial control)
1.Feedback control systems - Mathematical models 2.Fault location (Engineering) - mathematical models 3.System identification
I.Title II.Fantuzzi, Cesare III.Patton, Ron, 1949-
629.8'015118

Library of Congress Cataloging-in-Publication Data
A catalog record for this book is available from the Library of Congress.

http://www.springer.co.uk

Originally published by Springer-Verlag London Limited in 2003.
Softocover reprint of the hardcover 1st edition 2003

Typesetting: Electronic text files prepared by authors

69/3830-543210 Printed on acid-free paper SPIN 10830790

Advances in Industrial Control

Professor Emeritus O.P. Malik
Department of Electrical and Computer Engineering
University of Calgary
2500, University Drive, NW
Calgary
Alberta
T2N 1N4
Canada

Professor K.-F. Man
Electronic Engineering Department
City University of Hong Kong
Tat Chee Avenue
Kowloon
Hong Kong

Professor G. Olsson
Department of Industrial Electrical Engineering and Automation
Lund Institute of Technology
Box 118
S-221 00 Lund
Sweden

Professor A. Ray
Pennsylvania State University
Department of Mechanical Engineering
0329 Reber Building
University Park
PA 16802
USA

Professor D.E. Seborg
Chemical Engineering
3335 Engineering II
University of California Santa Barbara
Santa Barbara
CA 93106
USA

Doctor I. Yamamoto
Technical Headquarters
Nagasaki Research & Development Center
Mitsubishi Heavy Industries Ltd
5-717-1, Fukahori-Machi
Nagasaki 851-0392
Japan

SERIES EDITORS' FOREWORD

The series *Advances in Industrial Control* aims to report and encourage technology transfer in control engineering. The rapid development of control technology has an impact on all areas of the control discipline. New theory, new controllers, actuators, sensors, new industrial processes, computer methods, new applications, new philosophies..., new challenges. Much of this development work resides in industrial reports, feasibility study papers and the reports of advanced collaborative projects. The series offers an opportunity for researchers to present an extended exposition of such new work in all aspects of industrial control for wider and rapid dissemination.

As many technological systems become more complex, widespread and integrated, the effects of system failure can be simply devastating to the infrastructure of modern society. Feedback control is just one important component of total system supervision. Fault detection and isolation (FDI) is a second component with extensive commercial, industrial and societal implications if only we could work out how to do it in a reliable and inexpensive manner. Hardware or physical redundancy is the usual solution of the practical FDI problem, but as the authors Simani, Fantuzzi and Patton note in this monograph it is capital and maintenance costly. The search for reliable and inexpensive FDI methods has been active since the early 1970s. Since 1991, the International Federation of Automatic Control (IFAC) has created the SAFEPROCESS Steering Committee to promote research, developments and applications in the FDI field. The last decade has seen the formalisation of several theoretical approaches accompanied by some attempts to standardise nomenclature in the field.

The monograph series *Advances in Industrial Control* does not have many entries from this important research area but the monograph by Mangoubi, *Robust Estimation and Failure Detection* (ISBN 3-540-76251-5; 1998) and that by Russell, Chiang and Braatz, *Data-driven Techniques for Fault Detection and Diagnosis in Chemical Processes* (ISBN 1-85233-285-1; 2000) both make contributions even if they use quite different ideas. To these we can now add this monograph by S. Simani, C. Fantuzzi and R.J. Patton. Key features of this text include useful survey material (Chapter 2), a new approach based on model identification and an extended application study using a single shaft industrial gas turbine (Chapter 5). Different groups of readers ranging from industrial engineers wishing to gain insight into the applications potential of new FDI methods, to the academic control community looking for new problems to tackle (Chapter 6) will find much to learn from this monograph.

M.J. Grimble and M.A. Johnson
Industrial Control Centre
Glasgow, Scotland, U.K.

PREFACE

Control devices, which are nowadays exploited to improve the overall performance of industrial processes, involve both sophisticated digital system design techniques and complex hardware (input-output sensors, actuators, components and processing units). Such complexity results in an increased probability of failure. As a direct consequence of this, control systems must include automatic supervision of the closed-loop operation to detect and isolate malfunctions as early as possible.

Since the early 1970s, the problem of fault detection and isolation (FDI) in dynamic processes has received great attention, and a large variety of methodologies have been studied and developed based upon both physical and analytical redundancy. In the first case, the system is equipped with redundant physical devices, like sensors and actuators, so that if a fault occurs, the redundant device replaces the functionality of the faulty one.

The analytical redundancy approach is based on a completely different principle. The basic idea consists of using an accurate model of the system to mimic the real process behaviour. If a fault occurs, the residual signal (*i.e.* the difference between real system and model behaviours) can be used to diagnose and isolate the malfunction. This approach has some advantages with respect to physical (hardware-software) redundancy, mainly in economical and practical aspects. The analytical redundancy approach does not require additional equipment, but also suffers from some potential disadvantages, which are principally related to the need of an accurate model of the real system.

Model-based method reliability, which also includes false alarm rejection, is strictly related to the "quality" of the model and measurements exploited for fault diagnosis, as model uncertainty and noisy data can prevent an effective application of analytical redundancy methods.

This is not a simple problem. As model-based fault diagnosis methods are designed to detect any discrepancy between real system and model behaviours, it is assumed that a discrepancy signal is related to (has a response from) a fault. However, the same difference signal can respond to model mismatch or noise in real measurements, which can be (erroneously) detected as a fault, giving rise to a "false alarm" in detection. These considerations have led to research in the field of "robust" methods, in which particular attention is paid to the discrimination between actual faults and errors due to model mismatch. On the other hand, the availability of a "good" model of the monitored system can significantly improve the performance of diagnostic tools, minimising the probability of false alarms.

This monograph focuses on the explanation of what is a "good" model suitable for robust diagnosis of system performance and operation. The book also describes carefully how "accurate models" can be obtained from real data. A large amount of attention is paid to the "real system modelling problem", with reference to either linear-non-linear model structures. Special treatment is given to the case in which noise affects the acquired data. The mathematical description of the monitored system is obtained by means of a system identification scheme based on equation error and *errors-in-variables* models. This is a system identification approach that produces a reliable model of the plant under investigation as well as the variances of the input-output noises affecting the data.

After the discussion of identification procedures given in the first two chapters, the monograph focuses on the residual generation problem and fault diagnosis and identification for several cases, namely sensors, actuators and system faults.

The purpose of the monograph is to provide guidelines for the modelling and identification of real processes for fault diagnosis. Hence, significant attention is paid to practical application of the methods described to real system studies, as reported in the last chapters.

Both theoretical and practical arguments have been presented and discussed in a homogenous manner and the book targets both professional engineers working in industry and researchers in academic and scientific institutions.

Dr. S. Simani, Università di Ferrara
Dr. C. Fantuzzi, Università di Modena e Reggio Emilia
Prof. R.J. Patton, Department of Engineering, The University of Hull
Spring, 2002

CONTENTS

Symbols and Abbreviations xv

1. Introduction 1
1.1 Nomenclature 3
1.2 Fault Detection and Identification Methods based on Analytical Redundancy 5
1.3 Model-based Fault Detection Methods 7
1.4 Model Uncertainty and Fault Detection 8
1.5 The Robustness Problem in Fault Detection 9
1.6 System Identification for Robust FDI 11
1.7 Fault Identification Methods 12
1.8 Report on FDI Applications 13
1.9 Outline of the Book 16
1.10 Summary 18

2. Model-based Fault Diagnosis Techniques 19
2.1 Introduction 19
2.2 Model-based FDI Techniques 20
2.3 Modelling of Faulty Systems 21
2.4 Residual Generator General Structure 28
2.5 Residual Generation Techniques 31
2.5.1 Residual Generation via Parameter Estimation 32
2.5.2 Observer-based Approaches 35
2.5.3 Fault Detection with Parity Equations 40
2.6 Change Detection and Symptom Evaluation 44
2.7 The Residual Generation Problem 45
2.8 Fault Diagnosis Technique Integration 51
2.8.1 Fuzzy Logic for Residual Generation 51
2.8.2 Neural Networks in Fault Diagnosis 53
2.8.3 Neuro-fuzzy Approaches to FDI 54
2.8.4 Structure Identification of NF Models 56
2.8.5 NF Residual Generation Scheme for FDI 57
2.9 Summary 59

3. System Identification for Fault Diagnosis 61
3.1 Introduction 61
3.2 Models for Linear Systems 62
3.3 Parameter Estimation Methods 64
3.3.1 System Identification in Noiseless Environment 65
3.3.2 System Identification in Noisy Environment 68
3.3.3 The Frisch Scheme in the MIMO Case 73
3.4 Models for Non-linear Dynamic Systems 75
3.4.1 Piecewise Affine Model 75
3.4.2 Model Continuity and Domain Partitioning 79
3.4.3 Local Affine Model Identification 82
3.4.4 Multiple-Model Identification 85
3.5 Fuzzy Modelling and Identification 89
3.5.1 Fuzzy Multiple Inference Identification 90
3.5.2 Takagi-Sugeno Multiple-Model Paradigm 92
3.5.3 Fuzzy Clustering for Fuzzy Identification 95
3.5.4 Product Space Clustering and Fuzzy Model Identification 100
3.5.5 Non-linear Regression Problem and Black-Box Models 103
3.5.6 Fuzzy Model Identification From Clusters 107
3.6 Conclusion 112

4. Residual Generation, Fault Diagnosis and Identification .. 115
4.1 Introduction 115
4.2 Output Observers for Robust Residual Generation 116
4.3 Unknown Input Observer 119
4.3.1 UIO Mathematical Description 120
4.3.2 UIO Design Procedure 122
4.4 FDI Schemes Based on UIO and Output Observers 122
4.5 Sliding Mode Observers for FDI 127
4.5.1 Sliding Mode Observers 128
4.6 Kalman Filtering and FDI from Noisy Measurements 130
4.7 Residual Robustness to Disturbances 131
4.7.1 Disturbance Distribution Matrix Estimation 132
4.7.2 Additive Non-linear Disturbance and Noise 133
4.7.3 Model Complexity Reduction 133
4.7.4 Parameter Uncertainty 134
4.7.5 Distribution Matrix Low Rank Approximation 135
4.7.6 Model Estimation with Bounded Uncertainty 135
4.7.7 Disturbance Vector and Disturbance Matrix Estimation 136
4.7.8 Distribution Matrix Optimisation 139
4.7.9 Disturbance Distribution Matrix Identification 139
4.8 Residual Generation via Parameter Estimation 141
4.9 Residual Generation via Fuzzy Models 142
4.10 FDI Using Neural Networks 143

4.10.1 Neural Network Basics 145
4.11 Fault Diagnosis of an Industrial Plant at Different Operating Points Using Neural Networks 147
4.11.1 Operating Point Detection and Fault Diagnosis 147
4.11.2 FDI Method Development 149
4.12 Neuro-fuzzy in FDI 150
4.12.1 Methods of Neuro-fuzzy Integration 151
4.12.2 Neuro-fuzzy Networks 152
4.12.3 Residual Generation Using Neuro-fuzzy Models 154
4.12.4 Neuro-fuzzy-based Residual Evaluation 155
4.13 Summary 156

5. Fault Diagnosis Application Studies 157
5.1 Introduction 157
5.2 Physical Background and Modelling Aspects of an Industrial Gas Turbine 158
5.2.1 Gas Turbine Model Description 158
5.3 Identification and FDI of a Single Shaft Industrial Gas Turbine 168
5.3.1 System Identification 169
5.3.2 FDI Using Dynamic Observers 176
5.3.3 FDI Using Kalman Filters 183
5.3.4 Fuzzy System Identification and FDI 189
5.3.5 Sensor Fault Identification Using Neural Networks 191
5.3.6 Multiple Working Conditions FDI Using Neural Networks 196
5.3.7 FDI Method Development 196
5.3.8 Multiple Operating Point Simulation Results 197
5.4 Identification and FDI of Double Shaft Industrial Gas Turbine 199
5.4.1 Process Description 199
5.4.2 System Identification 201
5.4.3 FDI Using Unknown Input Observers 203
5.4.4 FDI Using Kalman Filters 208
5.4.5 Disturbance Decoupled Observers for Sensor FDI 209
5.4.6 Fuzzy Models for Fault Diagnosis. 210
5.5 Modelling and FDI of a Turbine Prototype 214
5.5.1 System Modelling and Identification 215
5.6 Turbine FDI Using Output Observers 220
5.6.1 Case 1: Compressor Failure (*Component Fault*) 221
5.6.2 Case 2: Fault Diagnosis of the Output Sensor 223
5.6.3 Case 3: Turbine Damage (*Turbine Component Fault*) . 227
5.6.4 Case 4: Actuator Fault (*Controller Malfunctioning*) ... 228
5.6.5 FDI in Noisy Environment Using Kalman Filters 233
5.6.6 Fault Isolation 235
5.6.7 Minimal Detectable Faults 239
5.7 FDI with Eigenstructure Assignment 242

5.7.1 Robust Fault Diagnosis of the Industrial Process 243
5.8 Robust Residual Generation Problem 247
5.9 Summary 249

6. Concluding Remarks 251
6.1 Suggestions for Future Work 253
6.1.1 Frequency Domain Residual Generation 253
6.1.2 Adaptive Residual Generators 255
6.1.3 Integration of Identification, FDI and Control 256
6.1.4 Fault Identification 256
6.1.5 Fault Diagnosis of Non–Linear Dynamic Systems 258

References 261

Index 279

SYMBOLS AND ABBREVIATIONS

The symbols and abbreviations listed here are used unless otherwise stated.

ARMAX	autoregressive moving average exogenous
ARX	autoregressive exogenous
BFDF	Beard fault detection filter
DOS	dedicated observer scheme
EE	equation error
EIV	errors-in-variables
FDD	fault detection and diagnosis
FDI	fault detection and isolation
FFT	fast Fourier transform
GK	Gustafson-Kessel
GOS	generalized observer scheme
IGV	inlet guided vane
KF	Kalman filter
LS	least-squares
MIMO	multiple-input multiple-output
MISO	multiple-input single-output
MLP	multiLayer perceptron
NN	neural network
OO	output observer
OLS	ordinary least-squares
RBF	radial basis function
RLS	recursive least-squares
SISO	single-Input single-Output
TS	Takagi-Sugeno
UIKF	unknown input Kalman filter
UIO	unknown input observer

CHAPTER 1
INTRODUCTION

There is an increasing interest in theory and applications of model-based fault detection and fault diagnosis methods, because of economical and safety related matters. In particular, well–established theoretical developments can be seen in many contributions published in the IFAC (International Federation of Automatic Control) Congresses and IFAC Symposium SAFEPROCESS (Fault Detection, Supervision and Safety of Technical Processes) [Isermann and Ballé, 1997, Isermann, 1997, Patton, 1999, Frank *et al.*, 2000].

The developments began at various places in the early 1970s. Beard [Beard, 1971] and Jones [Jones, 1973] reported, for example, the well–known "failure detection filter" approach for linear systems.

A summary of this early development is given by Willsky [Willsky, 1976]. Then Rault and his staff [Rault *et al.*, 1971] have considered the application of identification methods to the fault detection of jet engines. Correlation methods were applied to leak detection [Siebert and Isermann, 1976].

The first book on model-based methods for fault detection and diagnosis with specific application to chemical processes was published by Himmelblau [Himmelblau, 1978]. Sensor failure detection based on the inherent analytical redundancy of multiple observers was shown by Clark [Clark, 1978].

The use of parameter estimation techniques for fault detection of technical systems was demonstrated by Hohmann [Hohmann, 1977], Bakiotis [Bakiotis *et al.*, 1979], Geiger [Geiger, 1982], Filbert and Metzger [Filbert and Metzger, 1982].

The development of process fault detection methods based on modelling, parameter and state estimation was then summarised by Isermann [Isermann, 1984] and [Isermann, 1997]

Parity equation-based methods were treated early [Chow and Willsky, 1984], and then further developed by Patton and Chen [Patton and Chen, 1994b], Gertler [Gertler, 1991], Höfling and Pfeufer [Höfling and Pfeufer, 1994].

Frequency domain methods are typically applied when the effects of faults as well as disturbances have frequency characteristics which differ from each other and thus the frequency spectra serve as criterion to distinguish the faults [Massoumnia *et al.*, 1989, Frank *et al.*, 2000, Ding *et al.*, 2000].

The developments of fault detection and isolation methods to the present time is summarised in the books of Pau [Pau, 1981], then Patton *et al.* [Patton *et al.*, 2000], Basseville and Nikiforov [Basseville and Nikiforov, 1993], Chen and Patton [Chen and Patton, 1999], Gertler [Gertler, 1998], Isermann [Isermann, 1994b] and in survey papers by Gertler [Gertler, 1988], Frank [Frank, 1990] and Isermann [Isermann, 1994a].

Within IFAC, the increasing interest in this field was taken into account by creating first in 1991 a SAFEPROCESS (Fault Detection Supervision and Safety for Technical Processes) Steering Committee which then became a Technical Committee in 1993.

The first IFAC SAFEPROCESS Symposium was held in Baden–Baden, Germany in 1991 [Isermann and Freyermuth, 1992], and the second in Espo, Finland in 1994. The third symposium was scheduled at Hull, UK in 1997 and the fourth one was held in Budapest, Hungary in June 2000. The fifth is expected at Washington DC in July 2003.

Another tri–ennial series of IFAC Workshop exist for "Fault detection and supervision in the chemical process industries". Workshops were held in Newark, Delaware, Newcastle UK, Lyon and Korea between 1992 and 2001.

Most contributions in fault diagnosis rely on the analytical redundancy principle. The basic idea consists of using an accurate model of the system to mimic the real process behaviour. If a fault occurs, the residual signal (*i.e.* the difference between real system and model behaviour) can be used to diagnose and isolate the malfunction.

Model-based method reliability, which also includes false alarm rejection, is strictly related to the "quality" of the model and measurements exploited for fault diagnosis, as model uncertainty and noisy data can prevent an effective application of analytical redundancy methods.

This is not a simple problem, because model-based fault diagnosis methods are designed to detect any discrepancy between real system and model behaviours. It is assumed that this discrepancy signal is related to (has a response from) a fault. However, the same difference signal can respond to model mismatch or noise in real measurements, which are erroneously detected as a fault. These considerations have led to research in the field of "robust" methods, in which particular attention is paid to the discrimination between actual faults and errors due to model mismatch.

On the other hand, the availability of a "good" model of the monitored system can significantly improve the performance of diagnostic tools, minimising the probability of false alarms.

This monograph is devoted to the explanation of what is a "good" model suitable for robust diagnosis of system performance and operation. The book also explains how "robust models" can be obtained from real data. A large amount of attention is paid to the "real system modelling problem", with reference to either linear and non–linear model structures. Special treatment is given to the case in which noise affects the acquired data. The mathematical description of the monitored system is obtained by means of a system identification scheme based on equation error and errors–in–variables models. This is an identification approach which leads to a reliable model of the plant under investigation, as well as the estimation of the variances of the input–output noises affecting the data.

The purpose of the monograph is to provide guidelines for the modelling and identification of real processes for fault diagnosis. Hence, significant attention is paid to practical application of the methods described to real system studies, as reported in the last chapters.

In particular, this first chapter of the book outlines a new a common terminology in the fault diagnosis framework and gives some discussion and summary of developments in the field of fault detection and diagnosis based on papers selected during 1991–2001.

1.1 Nomenclature

By going through the literature, one recognises immediately that the terminology in this field is not consistent. This makes it difficult to understand the goals of the contributions and to compare the different approaches.

The SAFEPROCESS Technical Committee therefore discussed this matter and tried to find commonly accepted definitions. Some basic definitions can be found, for example, in the RAM (Reliability, Availability and Maintainability) dictionary [RAM, 1988], in contributions to IFIP (International Federation for Information Processing) [IFI, 1983].

Some of the terminology used in this book is given below. These are based on information obtained from the SAFEPROCESS Technical Committee and are considered “on–going” in the sense that new definitions and updates are being made.

1. ***States and Signals***

   Fault

   An unpermitted deviation of at least one characteristic property or parameter of the system from the acceptable, usual or standard condition.

   Failure

   A permanent interruption of a system’s ability to perform a required function under specified operating conditions.

   Malfunction

   An intermittent irregularity in the fulfilment of a system’s desired function.

   Error

   A deviation between a measured or computed value of an output variable and its true or theoretically correct one.

Disturbance
An unknown and uncontrolled input acting on a system.

Residual
A fault indicator, based on a deviation between measurements and model-equation-based computations.

Symptom
A change of an observable quantity from normal behaviour.

2. ***Functions***

Fault detection
Determination of faults present in a system and the time of detection.

Fault isolation
Determination of the kind, location and time of detection of a fault. Follows fault detection.

Fault identification
Determination of the size and time-variant behaviour of a fault. Follows fault isolation.

Fault diagnosis
Determination of the kind, size, location and time of detection of a fault. Follows fault detection. Includes fault detection and identification.

Monitoring
A continuous real-time task of determining the conditions of a physical system, by recording information, recognising and indication anomalies in the behaviour.

Supervision
Monitoring a physical and taking appropriate actions to maintain the operation in the case of fault.

3. ***Models***

Quantitative model
Use of static and dynamic relations among system variables and parameters in order to describe a system's behaviour in quantitative mathematical terms.

Qualitative model
Use of static and dynamic relations among system variables in order to describe a system's behaviour in qualitative terms such as causalities and IF–THEN rules.

Diagnostic model
A set of static or dynamic relations which link specific input variables, *the symptoms*, to specific output variables, the faults.

Analytical redundancy
Use of more (not necessarily identical) ways to determine a variable, where one way uses a mathematical process model in analytical form.

4. ***System properties***
   Reliability
   Ability of a system to perform a required function under stated conditions, within a given scope, during a given period of time.
   Safety
   Ability of a system not to cause danger to persons or equipment or the environment.
   Availability
   Probability that a system or equipment will operate satisfactorily and effectively at any point of time.

5. ***Time dependency of faults***
   Abrupt fault
   Fault modelled as stepwise function. It represents bias in the monitored signal.
   Incipient fault
   Fault modelled by using ramp signals. It represents drift of the monitored signal.
   Intermittent fault
   Combination of impulses with different amplitudes.

6. ***Fault terminology***
   Additive fault
   Influences a variable by an addition of the fault itself. They may represent, *e.g.*, offsets of sensors.
   Multiplicative fault
   Are represented by the product of a variable with the fault itself. They can appear as parameter changes within a process.

1.2 Fault Detection and Identification Methods based on Analytical Redundancy

A traditional approach to fault diagnosis in the wider application context is based on *hardware or physical redundancy* methods which use multiple sensors, actuators, components to measure and control a particular variable. Typically, a voting technique is applied to the hardware redundant system to decide if a fault has occurred and its location among all the redundant system components. The major problems encountered with hardware redundancy are the extra equipment and maintenance cost, as well as the additional space required to accommodate the equipment [Isermann and Ballé, 1997, Isermann, 1997].

In view of the conflict between reliability and the cost of adding more hardware, it is possible to use the dissimilar measured values together to cross-compare each other, rather than replicating each hardware individually. This is the meaning of ***analytical or functional redundancy.*** It exploits redundant analytical relationships among various measured variables of the monitored process [Patton *et al.*, 1989, Chen and Patton, 1999].

In the analytical redundancy scheme, the resulting difference generated from the comparison of different variables is called a ***residual or symptom signal.*** The residual should be zero when the system is in normal operation and should be different from zero when a fault has occurred. This property of the residual is used to determine whether or not faults have occurred [Patton *et al.*, 1989, Chen and Patton, 1999].

Consistency checking in analytical redundancy is normally achieved through a comparison between a measured signal with estimated values. The estimation is generated by a mathematical model of the considered plant. The comparison is done using the residual quantities which are computed as differences between the measured signals and the corresponding signals generated by the mathematical model [Patton *et al.*, 1989, Chen and Patton, 1999].

Figure 1.1 illustrates the concepts of hardware and analytical redundancy.

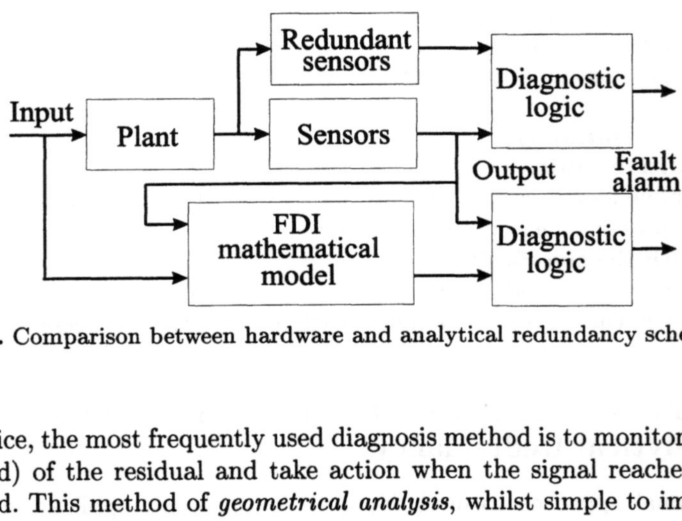

Fig. 1.1. Comparison between hardware and analytical redundancy schemes.

In practice, the most frequently used diagnosis method is to monitor the level (or trend) of the residual and take action when the signal reaches a given threshold. This method of ***geometrical analysis***, whilst simple to implement, has a few drawbacks. The most serious is that, in the presence of noise, input variations and change of operating point of the monitored process, false alarms are possible.

The major advantage of the model-based approach is that no additional hardware components are required in order to realize a Fault Detection and Isolation (FDI) algorithm. A model–based FDI algorithm can be implemented via software on a process control computer. In many cases, the measurements necessary to control the process are also sufficient for the FDI algorithm so that no additional sensors have to be installed [Patton *et al.*, 1989, Chen and Patton, 1999, Basseville and Nikiforov, 1993].

Analytical redundancy makes use of a mathematical model of the system under investigation and it is therefore often referred to as the *model–based approach* to fault diagnosis.

1.3 Model-based Fault Detection Methods

The task consists of the detection of faults on the technical process including actuators, components and sensors by measuring the available input and output variables $\boldsymbol{u}(t)$ and $\boldsymbol{y}(t)$. The principle of the model–based fault detection is depicted in Figure 1.2.

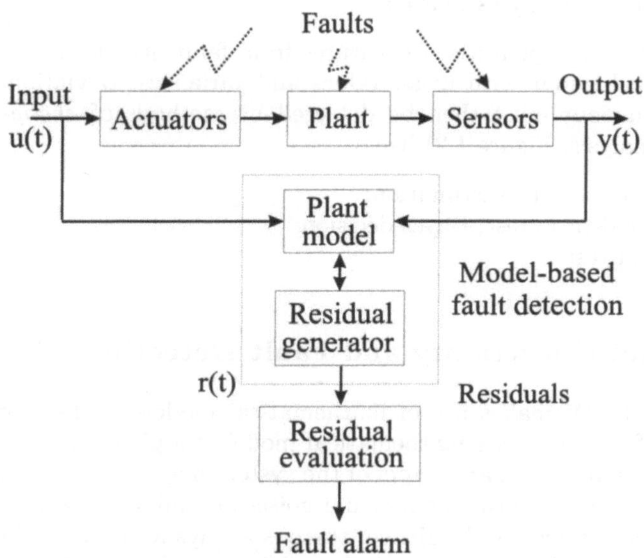

Fig. 1.2. Scheme for the model–based fault detection.

Basic process model–based FDI methods have been described by Patton *et al.* [Patton *et al.*, 1989], Basseville and Nikiforov [Basseville and Nikiforov, 1993], Gertler [Gertler, 1998] and Patton *et al.* [Chen and Patton, 1999, Patton *et al.*, 2000]:

1. Output observers (OO, estimators, filters);
2. Parity equations;
3. Identification and parameter estimation.

They generate residuals for output variables with fixed parametric models under method 1, fixed parametric or nonparametric models under method 2 and adaptive nonparametric or parametric models under method 3.

An important aspect of these methods is the kind of fault to be detected. As noted above, one can distinguish between *additive faults* which influence the variables of the process by a summation and *multiplicative faults* which are products of the process variables. The basic methods show different results, depending on these types of faults.

If only output signals $\boldsymbol{y}(t)$ can be measured, *signal model–based methods* can be applied, *e.g.* vibrations can be detected, which are related to rotating machinery or electrical circuits. Typical signal model–based methods of fault detection are:

1. Bandpass filters;
2. Spectral analysis (FFT);
3. Maximum-entropy estimation.

The characteristic quantities or features from fault detection methods show stochastic behaviour with mean values and variances. Deviations from the normal behaviour must then be detected by methods of *change detection* (residual analysis, Figure 1.2) like:

1. Mean and variance estimation;
2. Likelihood-ratio test, Bayes decision;
3. Run-sum test.

1.4 Model Uncertainty and Fault Detection

Model-based FDI makes use of mathematical models of the system. However, a perfectly accurate mathematical model of a physical system is never available. Usually, the parameters of the system may vary with time and the characteristics of the disturbances and noises are unknown so that they cannot be modelled accurately. Hence, there is always a mismatch between the actual process and its mathematical model even under no fault conditions. Such discrepancies cause difficulties in FDI applications, in particular, since they act as sources of false alarms and missed alarms. The effect of modelling uncertainties, disturbances and noise is therefore the most crucial point in the model–based FDI concept and the solution to this problem is the key for its practical applicability [Chen and Patton, 1999].

To overcome these problems, a model–based FDI scheme has to be insensitive to modelling uncertainty. Sometimes, a reduction of the sensitivity to modelling uncertainty does not solve the problem since the sensitivity reduction may be associated with a reduction of the sensitivity to faults [Chen and Patton, 1999, Gertler, 1998]. A more meaningful formulation of the FDI problem is to increase insensitivity to modelling uncertainty in order to provide increasing fault sensitivity.

The difficulties introduced by model uncertainties, disturbances and noises in model–based FDI have been widely considered during the last 10 years by both academia and industry [Gertler, 1998]. A number of methods have been proposed to tackle this problem, for example the Unknown Input Observer (UIO), eigenstructure assignment and parity relation methods.

An important task of the model–based FDI scheme is to be able to diagnose *incipient faults* in a system. With respect to *abrupt faults*, incipient faults may have a small effect on residuals and they can be hidden by disturbances. On the other hand, hard faults can be detected more easily because their effects are usually larger than modelling uncertainties and a simple fixed threshold is usually enough to diagnose their occurrence by residual analysis.

The presence of incipient faults may not necessarily degrade the performance of the plant, however, they may indicate that the component should be replaced before the probability of more serious malfunctions increases. The successful detection and diagnosis of incipient faults can therefore be considered a challenge for the design and evaluation of FDI algorithms.

1.5 The Robustness Problem in Fault Detection

In this monograph, observer–based approaches to robust FDI in industrial dynamic systems are summarised and applied to simulated and real plants. In the context of automatic control, the term robustness is used to describe the insensitivity or invariance of the performance of control systems with respect to disturbances, model–plant mismatches or parameter variations. Fault diagnosis schemes, on the other hand, must of course also be robust to the mentioned disturbances, but, in contrast to automatic control systems, they must not be robust to actual faults. On the contrary, while generating robustness to disturbances, the designer must maintain or even enhance the sensitivity of fault diagnosis schemes to faults. Furthermore, the robustness as well as the sensitivity properties must be independent of the particular fault and disturbance mode. Generally, the problem of robust FDI can be divided into the tasks of *robust residual generation* followed by *robust residual evaluation.*

In many cases, the disturbances and model–plant mismatches to which robustness must be generated, are due to the use of linear models for describing dynamic behaviour of non–linear processes. In this contribution, modelling errors are avoided from the very beginning by focusing on robust residual generation methods using linear and non–linear process models. This in turn simplifies the problem of residual evaluation without reducing the sensitivity to actual faults.

Effective tools for robust residual generation and even complete decoupling from external disturbances and unknown system parameters can be provided, *e.g.*, by unknown input observers which are introduced and applied to industrial processes. It is shown that the proposed solution to the disturbance de–coupling problem provides, in addition, the solution to both the fault detection and fault isolation problems.

On the other hand, many dynamic processes can only be described effectively using non-linear mathematical models. Most of the existing observer-based FDI techniques, however, are limited to the use of linear process models. The methods that can be found in the literature are based on the assumption that the system under supervision stays, during normal operation, in a neighbourhood of a certain known operating point [Chen and Patton, 1999, Patton *et al.*, 2000]

It is clear that, as almost every process system is non-linear, the modelling errors almost always reduce the accuracy of the linear model and therefore the performance of the FDI algorithm is compromised. Various methods for generating robustness to linearisation have been proposed in the literature and the reader is referred to [Patton *et al.*, 2000, Chap. 7] for a comprehensive treatment of this subject.

This monograph also surveys the state of the art of robustness methods and it presents some important ideas concerning the development of the use of non-linear models and predictors for FDI. In Chapter 4 observer-based approaches to robust FDI for dynamic systems are considered in more detail. In this contribution, the available model-based approaches are generalised, and thus extended to a wider class of dynamic systems.

In order to accommodate the application of robust FDI concepts, disturbances and parameter uncertainties of the monitored plants as well as faults are modelled in the form of unknown input signals. It is shown that, provided certain conditions can be met, complete decoupling of the residual from disturbances as well as from the parameter uncertainties of the process model can be achieved, whilst the sensitivity of the residual to faults is maintained. As the faults are also modelled in the form of external signals, this method additionally provides tools for the purpose of fault isolation. Fault isolation requires the de-coupling of the effects of different faults on the residual [Chen and Patton, 1999] and this, in turn, allows for decisions on which fault or faults out of a given set of possible faults has actually occurred.

These residual properties must be completely independent of the magnitude or frequency of the unknown inputs and the faults. This is crucial, in cases where no *a priori* knowledge about these properties is available. For systems, where the complete decoupling of the remaining unknown inputs or faults from the residual proves impossible, a threshold selection method, employing functional analytic methods and appropriate vector and operator norms can be exploited. This technique provides a tool for the robust evaluation of the residuals which have been generated by unknown input observers. Using the same functional analysis methods as employed for threshold selection, a performance index can be defined which allows for performance evaluation and, to a certain degree, also allows for optimal residual generator design [Patton *et al.*, 2000].

1.6 System Identification for Robust FDI

In earlier sections of this monograph, we have seen that model–based FDI methods formally require a high accuracy mathematic model of the monitored system. The better the model is as a representation of the dynamic behaviour of the system, the better will be the FDI performance. It is difficult to develop a highly accurate model of a complex system and hence the interesting question is: "what is a reasonable model to enable good performance in FDI to be guaranteed?".

It would be attractive to develop a robust FDI technique which is insensitive to modelling uncertainty, *i.e.*, so that a highly accurate mathematical model is no longer required. However, in order to design a robust FDI scheme, we should have a description (*i.e.*, some information) about the uncertainty, *e.g.*, its ***distribution matrix*** and spectral bandwidth, etc. Furthermore, this description should provide assistance for robust FDI design, *i.e.*, it can be handled in a systematic manner. Chapters 2 and 4 show how a typical uncertainty description makes use of the concept of "unknown inputs" acting upon a nominal linear model of the system. These unknown disturbances describe the uncertainties acting upon the system but disturbance distribution matrices are assumed known since they can be estimated by identification schemes.

It is clear that disturbances and faults act on the system in the same way, and thus we cannot easily discriminate between these excitations unless we know the structure of the disturbance distribution matrix. Once the disturbance distribution matrix is known, we can generate the residual with the disturbance de-coupling (robust) property, *i.e.*, the residual is de-coupled from the disturbance (uncertainty). The robust residual can then be used to achieve reliable FDI.

The theories underlying robust FDI approaches have been very well developed, but for real applications the following problems remain unsolved:

- estimation of reliable model for the monitored process;
- modelling accuracy of the real uncertainty by means of identified disturbance terms when no knowledge of the uncertainty is available;
- estimation of the disturbance terms and the structure of distribution matrices.

This book seeks to answer the above questions. Some simulation and real examples are given to test some of the theoretical results. These problems have to be addressed, otherwise the application domain of the disturbance de–coupling approach for robust FDI is very limited. In fact, few researchers and contributions have presented the application results of robust fault diagnosis to real processes.

As mentioned above, a primary requirement for model–based and disturbance de-coupling approaches to robust FDI is that both the system model and disturbance distribution matrices must be known. It is interesting that, within the framework of international research on this subject, there have been few attempts to address the problem by means of the ***identification approach***. This lack of information has obstructed the application of robust FDI in real engineering systems. Chapters 3 and 4 present the research developments surrounding the joint estimation of system and disturbance matrices in order to solve the robust fault diagnosis problem.

Concerning the identification schemes developed and exploited in Chapters 3, 4 and 5, when all observed variables of a dynamic process are affected by uncertainties, the parameter estimation task can be performed by the so–called *errors–in–variables* methods. On the other hand, *equation error* methods can be developed in the case of exactly known plant variables [Simani *et al.*, 2000a]. It is worthwhile noting that less attention has been paid to errors–in–variables schemes.

Under these considerations, Chapters 3, 4 and 5 present the robust FDI results concerning the description of monitored plants by means of equation error and error–in–variables identified models in the presence variable uncertainties. Moreover, for the examples presented, estimates obtained by the errors–in–variables approach and equation error estimates are computed and compared in Chapter 5.

1.7 Fault Identification Methods

If several symptoms change differently for certain faults, a first way of determining them is to use classification methods which indicate changes of symptom vectors.

Some classification methods are [Patton *et al.*, 1989, Basseville and Nikiforov, 1993, Gertler, 1998, Babuška, 1998, Chen and Patton, 1999]:

1. Geometrical distance and probabilistic methods;
2. Artificial neural networks;
3. Fuzzy clustering.

When more information about the relations between symptoms and faults is available in the form of diagnostic models, methods of reasoning can be applied. Diagnostic models then exist in the form of symptom–fault causalities, *e.g.* in the form of symptom-fault tree. The causalities can be expressed as IF–THEN rules. Then analytical as well as heuristic symptoms (from operators) can be processed. By considering these symptoms as vague facts, probabilistic or fuzzy set descriptions lead to a unified symptom representation. By using forward and backward reasoning, probabilities or possibilities of faults are obtained as a result of diagnosis. Typical approximate reasoning methods are [Basseville and Nikiforov, 1993, Chen and Patton, 1999]:

1. Probabilistic reasoning;
2. Possibilistic reasoning with fuzzy logic;
3. Reasoning with artificial neural networks.

This very short consideration shows that many different methods have been developed during the last 20 years. It is also clear that many combinations of them are possible.

Based on more than 100 publications during the last 5 years, it can be stated that parameter estimation and observer-based methods are the most frequently applied techniques for fault detection, especially for the detection of sensor and process faults. Nevertheless, the importance of neural network-based and combined methods for fault detection is steadily growing. In most applications, fault detection is supported by simple threshold logic or hypothesis testing. Fault isolation is often carried out using classification methods. For this task, neural networks are being more and more widely used.

The number of applications using non–linear models is growing, while the trend of using linearised models is diminishing. It seems that analytical redundancy-based methods have their best application areas in mechanical systems where the models of the processes are relatively precise. Most non–linear processes under investigation belong to the group of thermal and fluid dynamic processes. The field of applications to chemical processes has few developments, but the number of applications is growing. The favourite linear process under investigation is the DC motor. In general, the trend is changing from applications to safety-related processes with many measurements, as in nuclear reactors or aerospace systems, to applications in common technical processes with only a few sensors. For diagnosis, classification and rule-based reasoning methods are the most important and the use of neural network classification as well as fuzzy logic-based reasoning is growing.

1.8 Report on FDI Applications

Because of the many publications and increasing number of applications (IFAC Congress and IFAC Symposia SAFEPROCESS) between 1991–2000, it is of interest to show some trends [Patton *et al.*, 1989, Basseville and Nikiforov, 1993, Gertler, 1998, Chen and Patton, 1999, Frank *et al.*, 2000]. Therefore, a literature study of IFAC FDI–related Conferences is briefly presented in the following. Contributions taking into account the applications reported in Table 1.1 were considered. The type of faults considered are distinguished according to Table 1.2. Among all contributions, the fault detection methods were classified as in Table 1.3. The change detection and fault classification methods are indicated by Table 1.4. The reasoning strategies for fault diagnosis are reported in Table 1.5. The contributions considered are summarised in Table

1.6. The evaluation has been limited to the Fault Detection and Diagnosis (FDD) of laboratory, pilot and industrial processes.

Table 1.1. FDI applications and number of contributions.

Application	Number of contributions
Simulation of real processes	55
Large-scale pilot processes	44
Small-scale laboratory processes	18
Full-scale industrial processes	48

Table 1.2. Fault type and number of contributions.

Fault type	Number of contributions
Sensor faults	69
Actuator faults	51
Process faults	83
Control loop or controller faults	8

Table 1.3. FDI methods and number of contributions.

Method type	Number of contributions
Observer	53
Parity space	14
Parameter estimation	51
Frequency spectral analysis	7
Neural networks	9

Table 1.4. Residual evaluation methods and number of contributions.

Evaluation method	Number of contributions
Neural networks	19
Fuzzy logic	5
Bayes classification	4
Hypothesis testing	8

Table 1.6 shows that among mechanical and electrical processes, DC motor applications are mostly investigated. Parameter estimation and observer-based methods are used in the majority of applications on these kind of

Table 1.5. Reasoning strategies and number of contributions.

Reasoning strategy	Number of contributions
Rule based	10
Sign directed graph	3
Fault symptom tree	2
Fuzzy logic	6

Table 1.6. Applications of model–based fault detection.

FDD	Number of contributions
Milling and grinding processes	41
Power plants and thermal processes	46
Fluid dynamic processes	17
Combustion engine and turbines	36
Automotive	8
Inverted pendulum	33
Miscellaneous	42
DC motors	61
Stirred tank reactor	27
Navigation system	25
Nuclear process	10

processes, followed by parity space and combined methods. Thermal and chemical processes are investigated less frequently.

Table 1.3 shows that parameter estimation and observer-based methods are used in nearly 70% of all application considered. Neural networks, parity space and combined methods are significantly less often applied.

More than 50% of sensor faults are detected using observer-based methods, while parameter estimation and parity space and combined methods play a less important role. For the detection of actuator faults, observer-based methods are mostly used, followed by parameter estimation and neural networks methods.

Parity space and combined methods are rarely applied. In general, there are fewer applications for actuator faults than for sensor or process faults. The detection of process faults is mostly carried out with parameter estimation methods. Nearly 50% of all the applications considered use parameter estimation-based methods for detection of process faults. Observer-based, parity space and neural networks-based methods are used less often for this class of faults.

Among all the described processes, linear models have been used much more than non–linear ones. On processes with non–linear models, observer-based methods are mostly applied, but parity equations and neural networks also play an important role. On processes with linear or linearised models, parameter estimation and observer-based methods are mostly used. Parity space and combined methods are also used in several applications, but not to the same extent as observer-based and parameter estimation methods.

Taking into account the system considered, the number of non–linear process applications using non–linear models are decreasing. For linear processes, no significant change can be stated.

The use of neural networks and combinations seems to be increasing.

Concerning the fault diagnosis methods, in recent years, the field of classification approaches, especially with neural networks and fuzzy logic has steadily been growing. Also, rule–based reasoning methods are increasingly being based on fault diagnosis. A growing application of fuzzy rule-based reasoning can be stated. Applications using neural networks for classification are increasing and the trends are analogous to the increasing number of non–linear process investigations. Nevertheless, the classification of generated residuals seems to remain the most important application area for neural networks.

1.9 Outline of the Book

To detect and isolate faults in a dynamic system, based on the use of an analytical model, a residual signal has to be used. It is derived from a comparison between real measurements and the relative estimates (generated by the model). The modelling uncertainty problem can be tackled by designing a FDI scheme, whose residuals are insensitive to uncertainties whilst sensitive to faults. On the other hand, a model with satisfactory accuracy can be estimated using identification procedures [Norton, 1986, Söderström and Stoica, 1987, Ljung, 1999].

The aim of the design of a FDI scheme is to reduce the effects of uncertainties on the residuals and to enhance the effects of faults acting on the residuals. The *main aim of this monograph* is to develop a residual generator for model–based fault diagnosis of a process by means of input and output signals. An accurate model of the process under investigation will be estimated using identification procedures from data affected by noises and acquired from simulated and/or actual plants. The monograph consists of 6 chapters and the main contributions are presented in Chapters 3, 4 and 5. Chapters are devoted to the particular problem in residual generation and the are organised as follows.

Chapter 2 reviews the state of the art of the model–based FDI. The FDI problem is formalised in an uniform framework by presenting the mathematical description and definitions. The fundamental issue of model–based methods is the generation of residuals using the mathematical model of the monitored system. By analysing residuals, fault diagnosis can be performed. Some structures of the residual generator are presented in this Chapter in order to give ideas how to implement the residual generation. A residual generator can be designed for achieving the required diagnosis performances, *e.g.* fault isolation and disturbance decoupling.

In order to design the residual generator, some assumptions about the modelling uncertainties need to be made. The most frequently used hypothesis is that the modelling uncertainty is expressed as a disturbance term in the system dynamic equation. The disturbance vector is unknown whilst its distribution matrix can be estimated by using identification procedures. Based on this assumption, the disturbance decoupling residual generator can be design by using unknown input observer methods [Chen and Patton, 1999, Liu and Patton, 1998].

Chapter 3 demonstrates how to apply dynamic system identification methods in order to estimate an accurate model of the monitored system.

The FDI methods presented require, in fact, a linear mathematical model of the process under investigation, either in state space or input-output form.

In particular, since state space descriptions provide general and mathematically rigorous tools for system modelling, they may be used in the residual generator design, both for the deterministic case (UIO and OO) [Chen and Patton, 1999, Frank, 1990, Luenberger, 1979, Watanabe and Himmelblau, 1982] and the stochastic case (Kalman filters (KF) and unknown input Kalman filters (UIKF)) [Jazwinski, 1970, Xie *et al.*, 1994, Xie and Soh, 1994].

In such a manner, the suggested FDI tool does not require any physical knowledge of the process under observation since the linear models are obtained by means of an identification scheme which exploits equation error (EE) and errors–in–variables (EIV) models. In this situation, the identification technique is based on the rules of the Frisch scheme [Frisch, 1934], traditionally exploited to analyse economic systems. This approach, modified to be applied to dynamic system identification [Kalman, 1982b, Kalman, 1990, Beghelli *et al.*, 1990], gives a reliable model of the plant under investigation, as well as the variances of the input–output noises affecting the data.

For the non–linear case, piecewise affine and fuzzy models will be used as prototypes for the identification. In particular, the multiple-model approach, using several local affine submodels each describing a different operating condition of the process, is exploited.

Chapter 4 aims to define a comprehensive methodology for actuator, process component and sensor fault detection. It is based on an output estimation approach, in conjunction with residual processing schemes, which include a simple threshold detection, in deterministic case, as well as statistical analysis when data are affected by noise. The final result consists of a strategy based on fault diagnosis methods well–known in the literature for generating redundant residuals.

In particular, this Chapter studies the approach to residual generation with the aid of OO, UIO, KF and UIKF. The residual is defined as the *output estimation error*, obtained by difference between the measurement of one output and the relative estimate. This Chapter also presents the design of such estimators both in the deterministic and stochastic environment.

The diagnosis procedure may be further specialised for actuators, input or output sensors and process components. In fact, the fault diagnosis of input sensors and actuators uses a bank of UIO in high signal to noise ratio conditions or a bank of UIKF, otherwise. The i–th UIO or UIKF is designed to be insensitive to the i–th input of the system. On the other hand, output sensor and process component faults affecting a single residual can be detected by means of a OO or a classical KF, driven by a single output and all the inputs of the system.

Chapter 5 shows how the proposed algorithms can be applied to the FDI of actuators, process components and input-output sensors of industrial plants.

In particular, the FDI techniques presented in this book have been tested on time series of data acquired from different simulated and real industrial gas turbine working in parallel with electrical mains, whose linear mathematical description is obtained by using identification procedures.

Results from simulation show that minimum detectable faults are perfectly compatible with the industrial target of this application.

Chapter 6 summarises the contributions and achievements of the monograph providing some suggestions for possible further research topics as an extension of this work.

1.10 Summary

Chapter 1 has provided a common terminology in the fault diagnosis framework in order to comment on some developments in the field of fault detection and diagnosis based on papers selected during the last 10 years.

The structure of the six chapters of this monograph and the main contributions presented have also been outlined briefly.

CHAPTER 2
MODEL-BASED FAULT DIAGNOSIS TECHNIQUES

2.1 Introduction

The model–based approach to fault detection in dynamic systems has been receiving more and more attention over the last two decades, in the contexts of both research and real plant application.

Stemming from this activity, a great variety of methods are found in current literature, based on the use of mathematical models of the process under investigation and exploiting modern control theory.

Model–based fault detection methods use residuals which indicate changes between the process and the model. One general assumption is that the residuals are changed significantly so that a detection is possible. This means that the residual size after the appearance of a fault is large and long enough to be detectable.

This chapter provides an overview on the various fault detection methods, with particular attention to the FDI techniques related to the applications described in this book.

All the methods considered require that the process can be described by a mathematical model. As there is almost never an exact agreement between the model used to represent the process and the process itself, the model–reality discrepancy is of primary interest.

Hence, the most important issue in model–based fault detection is concerned with the accuracy of the model describing the behaviour of the monitored system. This issue has become a central research theme over recent years, as modelling uncertainty arises from the impossibility of obtaining complete knowledge and understanding of the monitored process.

The main focus of this Chapter is the modelling aspects of the process whose faults are to be detected and isolated. The Chapter also studies the general structure of a controlled system, its possible fault locations and modes. Residual generation is then identified as an essential problem in model–based FDI, since, if it is not performed correctly, some fault information could be lost. A general framework for the residual generation is also recalled.

Residual generators based on different methods, such as state and output observers, parity relations and parameter estimations, are just special cases in this general framework. In the following, some commonly used residual generation and evaluation methods are discussed and their mathematical formulation presented.

Finally, the chapter presents and summarises special features and problems regarding the different methods.

2.2 Model-based FDI Techniques

According to the definitions given in Section 1.1, model–based FDI can be defined as the ***detection***, ***isolation*** and ***identification*** of faults on a system by means of methods which extract features from measured signals and use *a priori* information on the process available in term of a mathematical models.

Faults are thus detected by setting fixed or variable thresholds on residual signals generated from the difference between actual measurements and their estimates obtained by using the process model.

A number of residuals can be designed with each having sensitivity to individual faults occurring in different locations of the system. The analysis of each residual, once the threshold is exceeded, then leads to fault isolation.

Figure 2.1 shows the general and logic block diagram of model–based FDI system.

It comprises two main stages of residual generation and residual evaluation. This structure was first suggested by Chow and Willsky in [Chow and Willsky, 1980] and now is widely accepted by the fault diagnosis community.

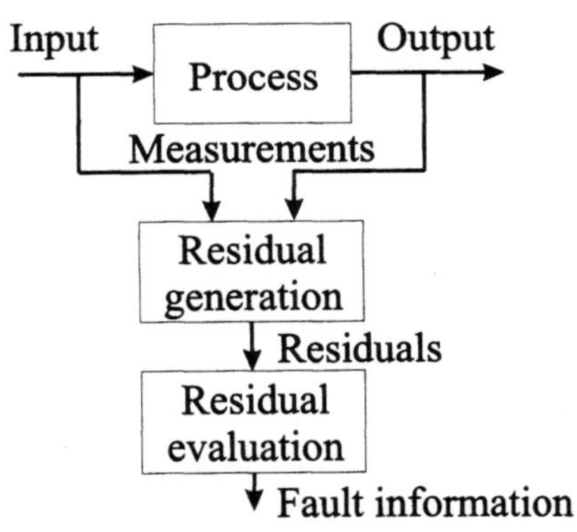

Fig. 2.1. Structure of model-based FDI system.

The two main blocks are described as follows:

1. **Residual generation:** this block generates residual signals using available inputs and outputs from the monitored system. This residual (or fault symptom) should indicate that a fault has occurred. It should normally be zero or close to zero under no fault condition, whilst distinguishably different from zero when a fault occurs. This means that the residual is characteristically independent of process inputs and outputs, in ideal conditions. Referring to Figure 2.1, this block is called *residual generation.*
2. **Residual evaluation:** This block examines residuals for the likelihood of faults and a decision rule is then applied to determine if any faults have occurred. The *residual evaluation* block, shown in Figure 2.1, may perform a simple threshold test (geometrical methods) on the instantaneous values or moving averages of the residuals. On the other hand, it may consist of statistical methods, *e.g.*, generalised likelihood ratio testing or sequential probability ratio testing [Isermann, 1997, Willsky, 1976, Basseville, 1988, Patton *et al.*, 2000].

Most contributions in the field of quantitative model–based FDI focus on the residual generation problem, since the decision–making problem can be considered relatively straightforward if residuals are well–designed.

Section 2.3 presents a number of different strategies for solving the quantitative residual generation problem.

2.3 Modelling of Faulty Systems

This book is concerned with Multi–Input Single–Output (MISO) and Multi–Input Multi–Output (MIMO) dynamic systems.

The first step in FDI model–based approach consists of providing a mathematical description of the system under investigation which shows all the possible fault cases, as well.

The detailed scheme for FDI techniques here presented is depicted by Figure 2.2.

The main components are the *Plant* under investigation, the *Actuators* and *Sensors*, which can be further sub–divided as *input* and *output* sensors, and finally the *Controller.*

In the following, the system working conditions will be monitored by means of its input $\boldsymbol{u}(t)$ and output $\boldsymbol{y}(t)$ measurements and signals from the controller $\boldsymbol{u}_R(t)$ which are supposed completely available for FDI purposes. Also, as shown in Figure 2.3, the behaviour of any controller that drives the system is inherently taken into consideration.

It is worth noting that, when the signals $\boldsymbol{u}_R(t)$ from the controller or measurements of plant inputs $\boldsymbol{u}(t)$ are not available, the controller plays an important role in the design of the FDI scheme, as a robust controller may desensitise faults effects and make diagnosis difficult.

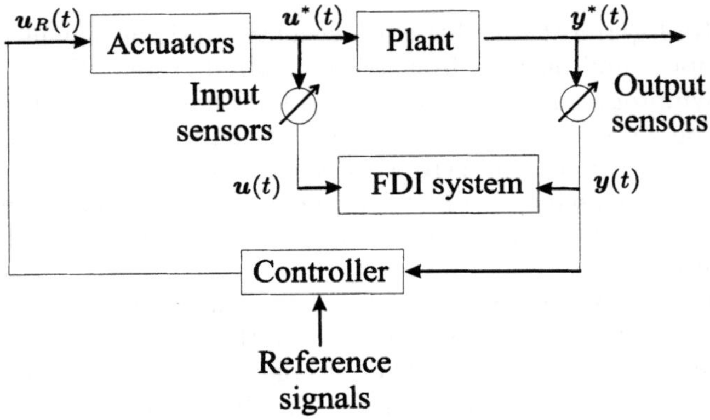

Fig. 2.2. Fault diagnosis in a closed-loop system.

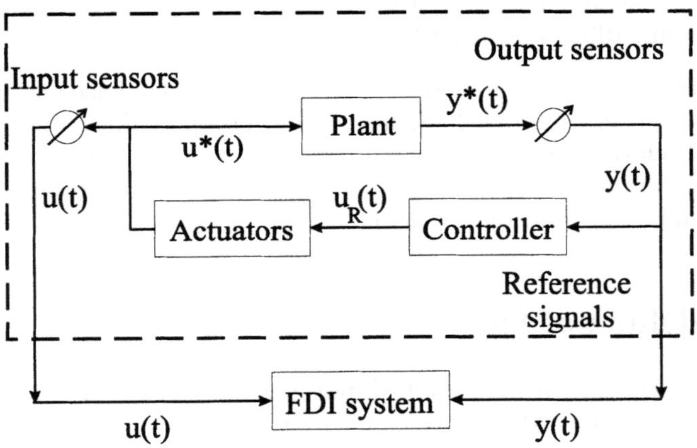

Fig. 2.3. The rearranged fault diagnosis scheme.

Once the actual process inputs and outputs $\boldsymbol{u}^*(t)$ and $\boldsymbol{y}^*(t)$ (usually not available) are measured by the input and output sensors, FDI theory can be treated as an observation problem of $\boldsymbol{u}(t)$ and $\boldsymbol{y}(t)$. The monitored system considered for FDI purpose can be therefore rearranged as illustrated in Figure 2.3.

Concerning the occurrence of malfunctions, the *location of faults* and their modelling, the system under diagnosis can be separated into the following different parts which can be affected by faults:

- Actuators,
- Process or system components,
- Input sensors,

- Output sensors,
- Controller.

With respect to previous work (see, *e.g.*, in the References [Patton *et al.*, 1989, Gertler, 1998, Patton *et al.*, 2000]), it is necessary to distinguish between input and output sensors.

Figure 2.3 shows that the input and output signals $\boldsymbol{u}^*(t)$ and $\boldsymbol{y}^*(t)$ are acquired in order to obtain the measurements $\boldsymbol{u}(t)$ and $\boldsymbol{y}(t)$ from the sensors.

This fault scenario can be summarised by the diagram shown in Figure 2.4.

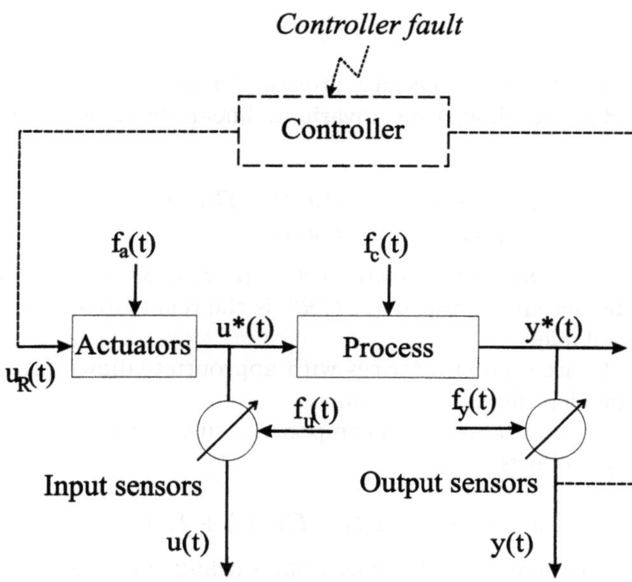

Fig. 2.4. The controlled system and fault topology.

Figure 2.4 also shows the situation where the controller can be affected by faults, since the monitored process consists of a closed-loop system. However, because of technological reasons (*e.g.*, the control action is performed by a digital computer), when the actuator is considered as a part or a component of the whole controller device, the former can be treated as subsystem where faults are likelier to occur whilst the latter remains free from faults.

Under these assumptions, as depicted in Figure 2.5 when system is considered in view of fault location, since input and output measurements are supposed completely available for FDI purposes, hence the controller behaviour in the design of a fault diagnosis scheme can be neglected as well as the interconnection between control system and the process.

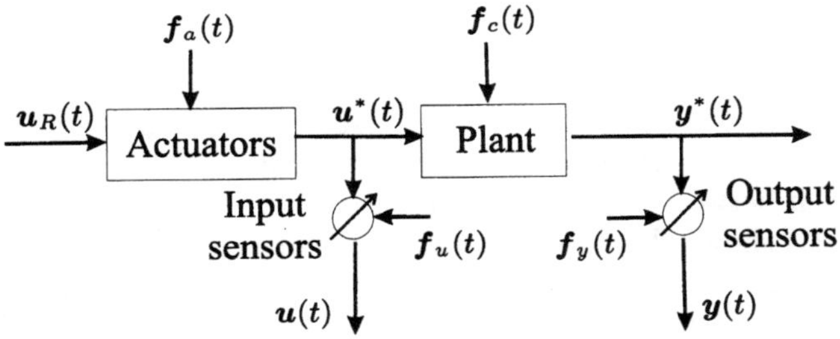

Fig. 2.5. The monitored system and fault topology.

Under the hypothesis of linearity, process dynamics can be described by the following discrete–time, time–invariant, linear dynamic system in the state–space form

$$\begin{cases} \boldsymbol{x}(t+1) &= \boldsymbol{A}\boldsymbol{x}(t) + \boldsymbol{B}\boldsymbol{u}^*(t) \\ \boldsymbol{y}^*(t) &= \boldsymbol{C}\boldsymbol{x}(t) \end{cases} \tag{2.1}$$

where $\boldsymbol{x}(t) \in \Re^n$ is the system state vector, $\boldsymbol{u}^*(t) \in \Re^r$ is the input signal vector driven by actuators, and $\boldsymbol{y}^*(t) \in \Re^m$ is the real system output vector, not directly available.

$\boldsymbol{A}$, $\boldsymbol{B}$, and $\boldsymbol{C}$ are system matrices with appropriate dimensions obtained by modelling or identification procedure.

With reference to Figure 2.5, a component fault vector $\boldsymbol{f}_c(t)$ affects process dynamics as follows:

$$\boldsymbol{x}(t+1) = \boldsymbol{A}\boldsymbol{x}(t) + \boldsymbol{B}\boldsymbol{u}^*(t) + \boldsymbol{f}_c(t) \tag{2.2}$$

In some cases, component faults come from a change in the system parameters, *e.g.*, a change in entries of the $\boldsymbol{A}$ matrix. For example, a change in the i-th row and the j-th column of the $\boldsymbol{A}$ matrix, leads to a fault vector $\boldsymbol{f}_c(t)$ described as

$$\boldsymbol{f}_c(t) = I_i \Delta a_{ij} x_j(t) \tag{2.3}$$

where $x_j(t)$ in the j-th element of the vector $\boldsymbol{x}(t)$ and I_i is a n-dimensional vector with all zero except a "1" in the i-th element.

As stated previously, as the actual process output $\boldsymbol{y}^*(t)$ is not directly available, a sensor is used to acquire a measure of the system outputs.

Moreover, generally speaking, a sensor can be also used to measure the system inputs $\boldsymbol{u}^*(t)$ (*e.g.*, for uncontrolled system).

By neglecting sensor dynamics, faults on input and output sensors are modelled with additive signals, respectively, as

$$\begin{cases} \boldsymbol{u}(t) &= \boldsymbol{u}^*(t) + \boldsymbol{f}_u(t) \\ \boldsymbol{y}(t) &= \boldsymbol{y}^*(t) + \boldsymbol{f}_y(t) \end{cases} \tag{2.4}$$

where the vectors $\boldsymbol{f}_u(t) = [f_{u_1}(t) \dots f_{u_r}(t)]^T$ and $\boldsymbol{f}_y(t) = [f_{y_1}(t) \dots f_{y_m}(t)]^T$ are chosen to describe a fault situation.

For example, if the sensor outputs are stuck at a fixed value $\bar{\boldsymbol{u}}$ because of a malfunction, the measurement vector is $\boldsymbol{u}(t) = \bar{\boldsymbol{u}}$ and the fault can be written as $\boldsymbol{f}_u(t) = -\boldsymbol{u}^*(t) + \bar{\boldsymbol{u}}$.

On the other hand, when the sensors are affected by a multiplicative fault δ, the measurements become $\boldsymbol{u}(t) = (1+\delta)\boldsymbol{u}^*(t)$, and the fault vector can be written as $\boldsymbol{f}_u(t) = \delta\boldsymbol{u}^*(t)$.

Usually, as shown in the following, fault modes can be described by step and ramp signals in order to model abrupt and incipient (hard to detect) faults, representing bias and drift, respectively.

Moreover, for technical reasons, sensor output signals are generally affected by measurement noise. Fault–free sensor signals $\boldsymbol{u}(t)$ and $\boldsymbol{y}(t)$, with additive noise can be modelled as:

$$\begin{cases} \boldsymbol{u}(t) &= \boldsymbol{u}^*(t) + \tilde{\boldsymbol{u}}(t) \\ \boldsymbol{y}(t) &= \boldsymbol{y}^*(t) + \tilde{\boldsymbol{y}}(t) \end{cases} \tag{2.5}$$

in which the sequences $\tilde{\boldsymbol{u}}(t)$ and $\tilde{\boldsymbol{y}}(t)$ are usually described as white, zero–mean, uncorrelated Gaussian processes.

In this case, taking into account the effects of faults and noise, 2.4 has to be replaced by:

$$\begin{cases} \boldsymbol{u}(t) &= \boldsymbol{u}^*(t) + \tilde{\boldsymbol{u}}(t) + \boldsymbol{f}_u(t) \\ \boldsymbol{y}(t) &= \boldsymbol{y}^*(t) + \tilde{\boldsymbol{y}}(t) + \boldsymbol{f}_y(t) \end{cases} \tag{2.6}$$

By neglecting the actuator block, Figure 2.6 shows the structure of the measurement process.

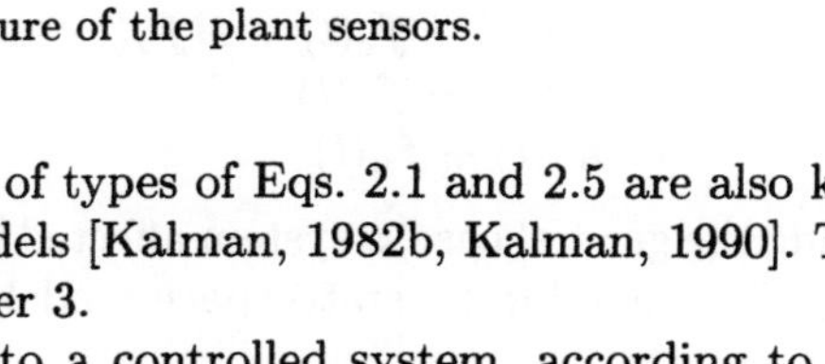

Fig. 2.6. The structure of the plant sensors.

Model descriptions of types of Eqs. 2.1 and 2.5 are also known as Error–In–Variable (EIV) models [Kalman, 1982b, Kalman, 1990]. They will be briefly presented in Chapter 3.

With reference to a controlled system, according to Figure 2.5, signals $\boldsymbol{u}^*(t)$ are the actuator response to the command signals $\boldsymbol{u}_R(t)$.

A purely algebraic actuator (*i.e.* with gain equal to 1) can be described by:

$$\boldsymbol{u}^*(t) = \boldsymbol{u}_R(t) + \boldsymbol{f}_a(t) \tag{2.7}$$

where, similarly to input-output sensor fault situation, $\boldsymbol{f}_a(t) \in \Re^r$ is the actuator fault vector.

In general, as shown in Figure 2.5, if the the actuation signals $\boldsymbol{u}^*(t)$ are assumed to be measurable, by neglecting input and output sensor noises, the process model with fault can be described by the following system equation:

$$\begin{cases} \boldsymbol{x}(t+1) &= \boldsymbol{A}\boldsymbol{x}(t) + \boldsymbol{f}_c(t) + \boldsymbol{B}\boldsymbol{u}^*(t) \\ \boldsymbol{y}(t) &= \boldsymbol{C}\boldsymbol{x}(t) + \boldsymbol{f}_y(t) \\ \boldsymbol{u}(t) &= \boldsymbol{u}^*(t) + \boldsymbol{f}_u(t) \end{cases} \tag{2.8}$$

On the other hand, Figure 2.7 represents the case where the $\boldsymbol{u}_R$ signals are measured only by the input sensors.

Such a configuration represents a critical situation with respect to the input sensor connection depicted in Figure 2.5.

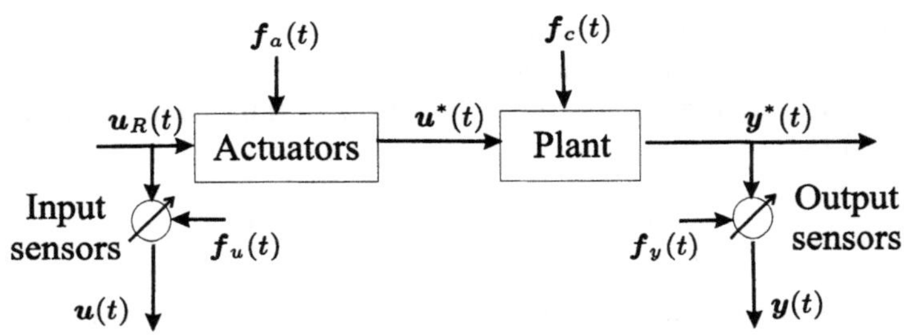

Fig. 2.7. Fault topology with actuator input signal measurement.

In this situation, actuator faults cannot be directly related to the input measurements $\boldsymbol{u}(t)$ but their effects can only be detected by means of output signals $\boldsymbol{y}(t)$.

By taking into account also actuator faults $\boldsymbol{f}_a(t)$, the description below is obtained:

$$\begin{cases} \boldsymbol{x}(t+1) &= \boldsymbol{A}\boldsymbol{x}(t) + \boldsymbol{f}_c(t) + \boldsymbol{B}\boldsymbol{f}_a(t) + \boldsymbol{B}\boldsymbol{u}^*(t) \\ \boldsymbol{y}(t) &= \boldsymbol{C}\boldsymbol{x}(t) + \boldsymbol{f}_y(t) \\ \boldsymbol{u}(t) &= \boldsymbol{u}^*(t) + \boldsymbol{f}_u(t) \end{cases} \tag{2.9}$$

Moreover, considering the general case, a system affected by all possible faults can be described by the the following state–space model:

$$\begin{cases} \boldsymbol{x}(t+1) &= \boldsymbol{A}\boldsymbol{x}(t) + \boldsymbol{B}\boldsymbol{u}^*(t) + \boldsymbol{L}_1\boldsymbol{f}(t) \\ \boldsymbol{y}(t) &= \boldsymbol{C}\boldsymbol{x}(t) + \boldsymbol{L}_2\boldsymbol{f}(t) \\ \boldsymbol{u}(t) &= \boldsymbol{u}^*(t) + \boldsymbol{L}_3\boldsymbol{f}(t) \end{cases} \tag{2.10}$$

where entries of the vector $\boldsymbol{f}(t) = [\boldsymbol{f}_a^T, \ \boldsymbol{f}_u^T, \ \boldsymbol{f}_c^T, \ \boldsymbol{f}_y^T]^T \in \Re^k$ correspond to specific faults.

In practice, it is reasonable to assume that the fault signals are described by *unknown* time functions. The matrices $\boldsymbol{L}_1, \boldsymbol{L}_2, \boldsymbol{L}_3$ are known as faulty entry matrices which describe how the faults enter the system.

The vectors $\boldsymbol{u}(t)$ and $\boldsymbol{y}(t)$ are the available and measurable inputs and outputs, respectively. Both vectors are supposed known for FDI purpose.

The distribution of the fault in the system depicted in Figure 2.5 can be described as input–output transfer matrix representation in the following form:

$$\boldsymbol{y}(z) = \boldsymbol{G}_{yu^*}(z)\boldsymbol{u}^*(z) + \boldsymbol{G}_{yf}(z)\boldsymbol{f}(z) \tag{2.11}$$

z being the unitary advance operator whilst the transfer matrices $\boldsymbol{G}_{yu^*}(z)$ and $\boldsymbol{G}_{yf}(z)$ are defined as:

$$\begin{cases} \boldsymbol{G}_{yu^*}(z) & = & \boldsymbol{C}(z\boldsymbol{I} - \boldsymbol{A})^{-1}\boldsymbol{B} \\ \boldsymbol{G}_{yf}(z) & = & \boldsymbol{C}(z\boldsymbol{I} - \boldsymbol{A})^{-1}\boldsymbol{L}_1 + \boldsymbol{L}_2 \end{cases} \tag{2.12}$$

Both the general models for FDI described by Equations 2.10 and 2.11 in the time and frequency domain, respectively, have been widely accepted in the fault diagnosis literature [Patton *et al.*, 1989, Patton *et al.*, 2000, Chen and Patton, 1999, Gertler, 1998].

Under these assumptions, the general model–based FDI problem here treated can be performed on the basis of the knowledge only of the measured sequences $\boldsymbol{u}(t)$ and $\boldsymbol{y}(t)$.

Frequency domain descriptions are typically applied when the effects of faults as well as the disturbances have frequency characteristics which differ from each other and thus information in the frequency spectra serve as criteria to distinguish the faults [Ding and Frank, 1990, Massoumnia *et al.*, 1989].

On the other hand, since state–space descriptions provide general and mathematically rigorous tools for system modelling and robust residual generation, for both the deterministic (noise free measurements) and the stochastic case (measurements affected by noises), the system matrices $\boldsymbol{A}$, $\boldsymbol{B}$ and $\boldsymbol{C}$, 2.10, in canonical forms can be obtained by multivariable identification procedures [Guidorzi, 1975, Norton, 1986, Söderström and Stoica, 1987, Ljung, 1999].

Moreover, in the case of a MIMO system, the choice of state–space representations in canonical form [Guidorzi, 1975] instead of parity space methods [Gertler, 1995] may avoid unexpected false alarm problems [Delmaire *et al.*, 1999].

As shown in Chapter 3, the FDI methods proposed here do not require any physical knowledge of the processes under observation, since the mathematical description of the monitored system is obtained by means of a system identification scheme based on Equation Error (EE) and EIV models.

It is worthy to note how this approach represents a novel point of view of the model–based fault diagnosis. The new aspect consists of exploiting linear system identification procedures, presented in Chapter 3, in connection with the model–based residual generation problem, shown in Chapter 4.

Although most systems to be monitored are actually non–linear, linear system modelling and identification methods are described here to avoid the complexities that would otherwise be inevitable when non–linear models are used.

There is certainly an increasing interest in the use of non–linear methods (non–linear observers, extended Kalman filters, fuzzy-logic methods, etc). However, as the feature of system supervision is to monitor the operation and performance of the system with respect to an expected point of operation, linear system methods are still very valid. Deviations from expected behaviour can be used to monitor system performance changes as well as component malfunctions.

2.4 Residual Generator General Structure

In this section, a review is given on fault detection methods based on *process models* and *signal models*. The basic methods are described briefly whilst their presentation and application are shown in Chapter 4 and 5, respectively.

The most frequently used FDI methods exploit the *a priori* knowledge of characteristics of certain signals. As an example, the spectrum, the dynamic range of the signal and its variations may be checked.

However, the necessity of *a priori* information concerning the monitored signals and the dependence of the signal characteristics on unknown working conditions of the system under diagnosis are main drawbacks of such a class of methods.

The most significant contribution in modern model–based approaches is the introduction of the *symptom or residual signals*, which depend on faults and are independent of system operating states.

They represent the inconsistency between the actual system measurements and the corresponding signals of the mathematical model.

The residual generator block introduced in Figure 2.1 can be interpreted as illustrated in Figure 2.8 [Basseville, 1988].

In the above structure, the auxiliary redundant signal $\boldsymbol{z}(t)$ is generated by the function $W_1\big(\boldsymbol{u}(\cdot), \boldsymbol{y}(\cdot)\big)$ and, together with the measurement $\boldsymbol{y}(t)$, the symptom signal $\boldsymbol{r}(t)$ is computed by means of $W_2\big(\boldsymbol{z}(\cdot), \boldsymbol{y}(\cdot)\big)$.

In the fault–free case, the following relations are satisfied

$$\left\{ \begin{array}{lcll} \boldsymbol{z}(t) & = & W_1\big(\boldsymbol{u}(\cdot), \boldsymbol{y}(\cdot)\big) & \\ \boldsymbol{r}(t) & = & W_2\big(\boldsymbol{z}(\cdot), \boldsymbol{y}(\cdot)\big) & = \ \boldsymbol{0}. \end{array} \right. \tag{2.13}$$

When a fault occurs in the plant, the residual $\boldsymbol{r}(t)$ will be different from zero.

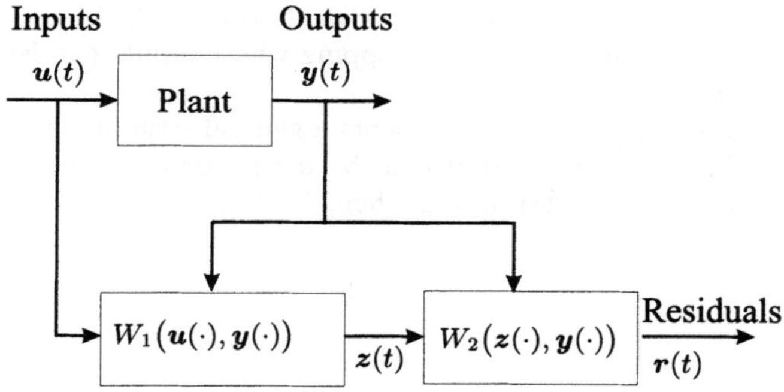

Fig. 2.8. Residual generator general structure.

The simplest residual generator is depicted in Figure 2.9 and it is obtained when the system W_1 is a plant identical model $\boldsymbol{z}(t) = W_1\big(\boldsymbol{u}(\cdot)\big)$ or it is an input–output description for the actual process obtained from system identification procedure (*e.g.*, an Auto Regressive eXogenous (ARX) model, see Chapter 3).

In the former case, the measurement $\boldsymbol{y}(t)$ is not required in W_1 because it is a *system simulator*. The signal $\boldsymbol{z}(t)$ represents the simulated output and the residual is computed as $\boldsymbol{r}(t) = \boldsymbol{z}(t) - \boldsymbol{y}(t)$. Since it is an open–loop system, the process simulation may become unstable.

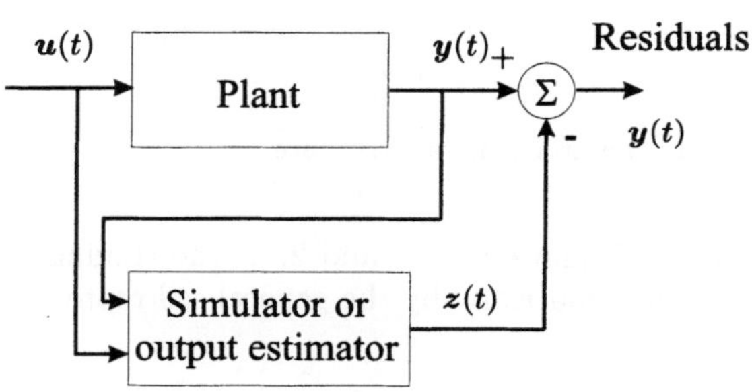

Fig. 2.9. Residual generation via system simulator.

An extension to the model–based residual generation is to replace $W_1\big(\boldsymbol{u}(\cdot)\big)$ by $W_1\big(\boldsymbol{u}(\cdot), \boldsymbol{y}(\cdot)\big)$, *i.e.* an *output estimator* fed by both system input and output.

In such a case, function W_1 generates an estimation of a linear function of the output $W_1\big(\boldsymbol{u}(\cdot), \boldsymbol{y}(\cdot)\big) = \boldsymbol{M}\boldsymbol{y}(t)$ whilst function W_2 can be defined as $W_2\big(\boldsymbol{z}(\cdot), \boldsymbol{y}(\cdot)\big) = \boldsymbol{W}\big(\boldsymbol{z}(t) - \boldsymbol{M}\boldsymbol{y}(t)\big)$, $\boldsymbol{W}$ being a weighting matrix.

Concluding, no matter which type of method is used, the residual generation process is nothing but a liner mapping whose inputs consist of process inputs and outputs.

As an example, Figure 2.10 represents a general structure for all residual generators using the input–output transfer matrix description was presented by Patton and Chen in [Patton and Chen, 1991a].

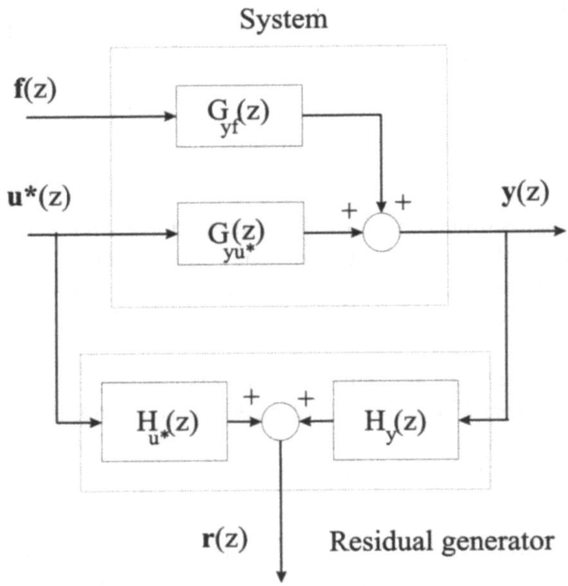

Fig. 2.10. Residual generator general structure.

With reference to Equations 2.11 and 2.12, the residual generator structure is expressed mathematically by the generalised representation:

$$\boldsymbol{r}(z) = \begin{bmatrix} \boldsymbol{H}_{u^*}(z) & \boldsymbol{H}_y(z) \end{bmatrix} \begin{bmatrix} \boldsymbol{u}^*(z) \\ \boldsymbol{y}(z) \end{bmatrix} = H_{u^*}(z)\boldsymbol{u}^*(z) + \boldsymbol{H}_y(z)\boldsymbol{y}(z) \quad (2.14)$$

where $\boldsymbol{H}_{u^*}(z)$ and $\boldsymbol{H}_y(z)$ are discrete transfer matrices which can be designed using stable discrete–time linear systems. The functions $\boldsymbol{u}^*(z)$, $\boldsymbol{y}(z)$, $\boldsymbol{r}(z)$ and $\boldsymbol{f}(z)$ are the Z-transform of the corresponding discrete–time signals.

According to the definition, the residual $\boldsymbol{r}(t)$ has to be designed to become zero for for fault–free case and different from zero in case of failures. This means that

$$\boldsymbol{r}(t) = \mathbf{0} \text{ if and only if } \boldsymbol{f}(t) = \mathbf{0} \quad (2.15)$$

In order to satisfy the Equation 2.15, the design of the transfer matrices $\boldsymbol{H}_{u^*}(z)$ and $\boldsymbol{H}_y(z)$ must satisfy to the constraint conditions

$$\boldsymbol{H}_{u^*}(z) + \boldsymbol{H}_y(z)\boldsymbol{G}_{yu^*} = \mathbf{0} \tag{2.16}$$

It is worth noting that different residual generators can be obtained by using different parametrisations of $\boldsymbol{H}_{u^*}(z)$ and $\boldsymbol{H}_y(z)$ [Patton and Chen, 1991a, Chen and Patton, 1999].

After generating the residual, the simplest and most widely used way to fault detection is achieved by directly comparing residual signal $\boldsymbol{r}(t)$ or a residual function $J(\boldsymbol{r}(t))$ with a fixed threshold ϵ or a threshold function $\varepsilon(t)$ as follows

$$\begin{cases} J(\boldsymbol{r}(t)) \leq \varepsilon(t) & \text{for} \quad \boldsymbol{f}(t) = \mathbf{0} \\ J(\boldsymbol{r}(t)) > \varepsilon(t) & \text{for} \quad \boldsymbol{f}(t) \neq \mathbf{0} \end{cases} \tag{2.17}$$

where $\boldsymbol{f}(t)$ is the general fault vector defined in Equation 2.10. If the residual exceeds the threshold, a fault may be occurred.

This test works especially well with fixed thresholds ε if the process operates approximately in a steady state and it reacts after relatively large feature, *i.e.* after either a large sudden or a long-lasting gradually increasing fault.

On the other hand, adaptive thresholds $\varepsilon(t)$ can be exploited which depend on plant operating conditions, for example when $\varepsilon(t)$ is expressed as a function of plant inputs [Clark, 1989, Chen and Patton, 1999].

2.5 Residual Generation Techniques

The generation of symptoms is the main issue in model–based fault diagnosis.

A variety of methods are available in literature for residual generation and this section presents briefly some of the most common methods.

Most of the residual generation techniques are based on both continuous and discrete system models, however, in this book, the attention is focused only on discrete–time dynamic linear models.

The following process model–based fault detection schemes will be considered and summarised [Isermann and Ballé, 1997, Patton *et al.*, 2000]:

1. Fault detection via parameter estimation [Isermann, 1984, Isermann and Freyermuth, 1992, Isermann, 1993, Isermann and Ballé, 1997, Patton *et al.*, 2000].
2. Observer–based approaches [Beard, 1971, Frank, 1993, Frank and Ding, 1997, Patton and Chen, 1997, Willsky, 1976, Basseville, 1988],

3. Parity vector (relation) methods [Chow and Willsky, 1984, Gertler and Singer, 1990, Patton and Chen, 1991a, Gertler and Monajemy, 1993, Delmaire *et al.*, 1999].

2.5.1 Residual Generation via Parameter Estimation

In most practical cases, the process parameters are not known at all, or they are not known exactly enough. Then, they can be determined with parameter estimation methods, by measuring input and output signals, $\boldsymbol{u}(t)$ and $\boldsymbol{y}(t)$, if the basic structure of the model is known [Isermann, 1997, Patton *et al.*, 2000].

This approach is based on the assumption that the faults are reflected in the physical system parameters and the basic idea is that the parameters of the actual process are estimated on–line using well–known parameter estimations methods.

The results are thus compared with the parameters of the reference model; obtained initially under fault–free assumptions. Any discrepancy can indicate that a fault may have occurred.

Now we compare two different approaches for modelling the input–output behaviour of the monitored system.

Equation Error Methods. The SISO process discrete–time model of order n is written in the vector form

$$y(t) = \boldsymbol{\Psi}^T \boldsymbol{\Theta} \tag{2.18}$$

where

$$\boldsymbol{\Theta}^T = [a_1 \dots a_n,\ b_1 \dots b_n] \tag{2.19}$$

is the parameter vector and

$$\boldsymbol{\Psi}^T = [y(t-1) \dots y(t-n) \ \ u(t-1) \dots u(t-n)] \tag{2.20}$$

the discrete–time data vector.

According to Figure 2.11, for parameter estimation, the equation error $e(t)$ is introduced

$$e(t) = y(t) - \boldsymbol{\Psi}^T \boldsymbol{\Theta} \tag{2.21}$$

or, if

$$\frac{y(t)}{u(t)} = \frac{B(z)}{A(z)} \tag{2.22}$$

is the transfer function of the process, the equation error via Z-transformation becomes

$$e(t) = \hat{B}(z)u(t) - \hat{A}(z)y(t). \tag{2.23}$$

in which $\hat{A}(z)$ and $\hat{B}(z)$ correspond to the estimates of $A(z)$ and $B(z)$.
The least–squares (LS) estimate

$$\hat{\boldsymbol{\Theta}} = [\boldsymbol{\Psi}^T\, \boldsymbol{\Psi}]^{-1}\boldsymbol{\Psi}^T y \tag{2.24}$$

is obtained if the minimisation of the sum of least–squares is computed

$$\begin{cases} J(\boldsymbol{\Theta}) & = & \sum_t e^2(t) = \boldsymbol{e}^T\boldsymbol{e} \\ \frac{d\ J(\boldsymbol{\Theta})}{d\ \boldsymbol{\Theta}} & = & \mathbf{0}. \end{cases} \tag{2.25}$$

As described in *e.g.*, [Patton *et al.*, 2000, Isermann, 1992], the least–squares estimate can be also expressed in recursive form (RLS) with respect to the estimates at the instant t, with $t = 0,\ 1,\ 2, \cdots$

$$\hat{\boldsymbol{\Theta}}(t+1) = \hat{\boldsymbol{\Theta}}(t) + \gamma(t)\left[y(t+1) - \boldsymbol{\Psi}^T(t+1)\hat{\boldsymbol{\Theta}}(t+1)\right] \tag{2.26}$$

where

$$\begin{cases} \gamma(t) & = & \frac{1}{\boldsymbol{\Psi}^T(t+1)\boldsymbol{P}(t)\boldsymbol{\Psi}(t+1)+1}\boldsymbol{P}(t)\boldsymbol{\Psi}(t+1) \\ \boldsymbol{P}(t+1) & = & \left[I - \gamma(t)\boldsymbol{\Psi}^T(t+1)\right]\boldsymbol{P}(t). \end{cases} \tag{2.27}$$

For improved estimates, filtering methods can be exploited. In particular, as shown in Section 4.8, when measurements are affected by noise, a Kalman filter can be used for the parameter estimation [Jazwinski, 1970].

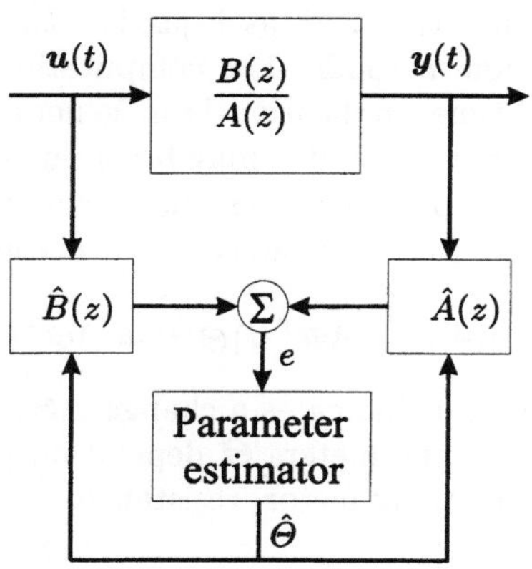

Fig. 2.11. Parameter estimation equation error.

Output Error Methods. Instead of the equation error computed in Equation 2.21, the output error

$$e(t) = y(t) - \hat{y}(\boldsymbol{\Theta}, t) \tag{2.28}$$

where

$$\hat{y}(\boldsymbol{\Theta}, z) = \frac{\hat{B}(z)}{\hat{A}(z)} u(z) \tag{2.29}$$

is the model output, can also be used, as depicted in Figure 2.12.

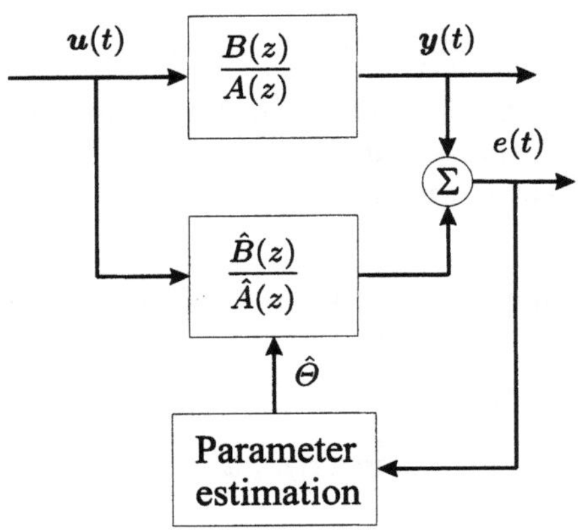

Fig. 2.12. Parameter estimation output error.

Unfortunately, direct calculation of the parameter estimate $\boldsymbol{\Theta}$ is not possible, because $e(t)$ is non–linear in the parameters.

Therefore, the loss function 2.28 as Equation 2.21 has to be minimised by numerical optimisation methods. The computational effort is then much larger and on-line real-time application is in general impossible. However, relatively precise parameter estimates may be obtained.

If a fault within the process changes one or several parameters by $\Delta\boldsymbol{\Theta}$, the output signal changes for small deviations according to

$$\Delta y(t) = \boldsymbol{\Psi}^T(t)\Delta\boldsymbol{\Theta}(t) + \Delta\boldsymbol{\Psi}^T(t)\boldsymbol{\Theta}(t) + \Delta\boldsymbol{\Psi}^T(t)\Delta\boldsymbol{\Theta}(t) \tag{2.30}$$

and the parameter estimator indicates a change $\Delta\boldsymbol{\Theta}$.

Generally, the process parameters $\boldsymbol{\Theta}$ depend on physical process coefficients $\boldsymbol{p}$ (like stiffness, damping factor, resistance, ...)

$$\boldsymbol{\Theta} = f(\boldsymbol{p}) \tag{2.31}$$

via non–linear algebraic equations. If the inversion of the relationship

$$p = f^{-1}(\boldsymbol{\Theta}) \tag{2.32}$$

exists [Patton *et al.*, 2000, Isermann, 1992], changes $\boldsymbol{\Delta p}$ of the process coefficients can be calculated. These changes in the coefficients are in many cases directly related to faults.

Thus, although the knowledge of $\boldsymbol{\Delta p}$ facilitates the fault diagnosis problem, it is not necessary for fault detection only. Parameter estimation can also be applied to non–linear static process models [Isermann, 1993].

2.5.2 Observer-based Approaches

The basic idea behind the observer or filter–based techniques is to estimate the outputs of the system from the measurements by using either Luenberger observers in a deterministic setting or Kalman filters in a noisy environment. The output estimation error (or its weighted value) is therefore used as residual.

It is worth noting that when an observer is exploited for FDI purpose, the estimation of the outputs is necessary, whilst the estimation of the state vector is usually not needed [Chen and Patton, 1999]. Moreover, the advantage of using the observer is the flexibility in the selection of its gains which leads to a rich variety of FDI schemes [Frank, 1994b, Frank and Ding, 1997, Chen *et al.*, 1996b, Liu and Patton, 1998].

In order to obtain the structure of a (generalised) observer, the discrete-time, time-invariant, linear dynamic model for the process under consideration in a state-space form is considered

$$\begin{cases} \boldsymbol{x}(t+1) &= \boldsymbol{A}\boldsymbol{x}(t) + \boldsymbol{B}\boldsymbol{u}(t) \\ \boldsymbol{y}(t) &= \boldsymbol{C}\boldsymbol{x}(t). \end{cases} \tag{2.33}$$

being $\boldsymbol{u}(t) \in \Re^r$, $\boldsymbol{x}(t) \in \Re^n$ and $\boldsymbol{y}(t) \in \Re^m$.

Assuming that all matrices $\boldsymbol{A}$, $\boldsymbol{B}$ and $\boldsymbol{C}$ are perfectly known, an observer is used to reconstruct the system variables based on the measured inputs and outputs $\boldsymbol{u}(t)$ and $\boldsymbol{y}(t)$

$$\begin{cases} \hat{\boldsymbol{x}}(t+1) &= \boldsymbol{A}\hat{\boldsymbol{x}}(t) + \boldsymbol{B}\boldsymbol{u}(t) + \boldsymbol{H}\boldsymbol{e}(t) \\ \boldsymbol{e}(t) &= \boldsymbol{y}(t) - \boldsymbol{C}\hat{\boldsymbol{x}}(t). \end{cases} \tag{2.34}$$

The observer scheme described by Equation 2.34 is depicted in Figure 2.13.

For the state estimation error $\boldsymbol{e}_x(t)$, it follows from Equations 2.34 that

$$\begin{cases} \boldsymbol{e}_x(t) &= \boldsymbol{x}(t) - \hat{\boldsymbol{x}}(t) \\ \boldsymbol{e}_x(t+1) &= (\boldsymbol{A} - \boldsymbol{H}\boldsymbol{C})\boldsymbol{e}_x(t). \end{cases} \tag{2.35}$$

The state error $\boldsymbol{e}_x(t)$ (and the error $\boldsymbol{e}(t)$) vanishes asymptotically

$$\lim_{t \to \infty} \boldsymbol{e}_x(t) = \boldsymbol{0} \tag{2.36}$$

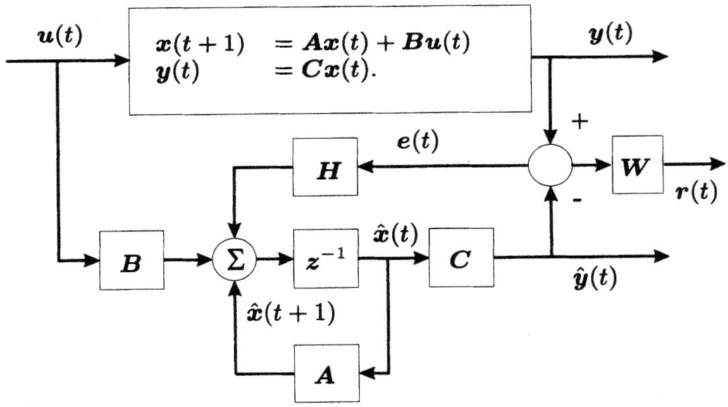

Fig. 2.13. Process and state observer.

if the observer is stable, which can be achieved by proper design of the observer feedback $\boldsymbol{H}$.

If the process is influenced by disturbance and faults, by comparing Figure 2.14) and Equations 2.10 it is described by the following system

$$
\begin{cases} \boldsymbol{x}(t+1) &= \boldsymbol{A}\boldsymbol{x}(t) + \boldsymbol{B}\boldsymbol{u}(t) + \boldsymbol{Q}\boldsymbol{v}(t) + \boldsymbol{L}_1\boldsymbol{f}(t) \\ \boldsymbol{y}(t) &= \boldsymbol{C}\boldsymbol{x}(t) + \boldsymbol{R}\boldsymbol{w}(t) + \boldsymbol{L}_2\boldsymbol{f}(t) \end{cases} \tag{2.37}
$$

where $\boldsymbol{v}(t)$ is the non–measurable disturbance vector at the input, $\boldsymbol{w}(t)$ the non–measurable disturbance vector at the output, $\boldsymbol{f}(t)$ fault signals at the input and output acting through L_1 and L_2, respectively.

They can represent actuator, process, input and output sensor additive faults.

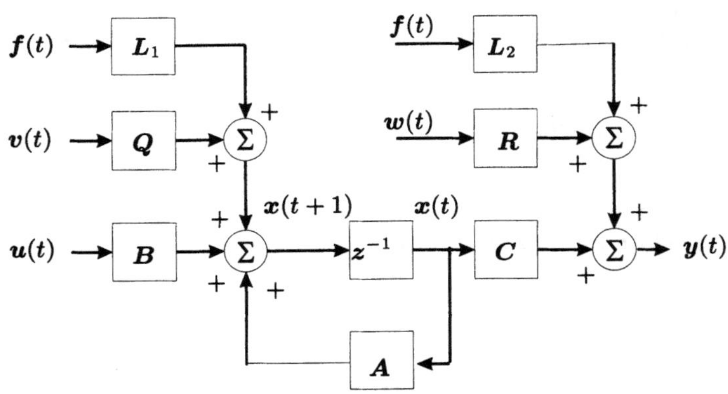

Fig. 2.14. MIMO process with faults and noises.

For the state estimation error, the following equations hold if the disturbances $\boldsymbol{v}(t) = \mathbf{0}$ and $\boldsymbol{w}(t) = \mathbf{0}$

$$\boldsymbol{e}_x(t+1) = (\boldsymbol{A} - \boldsymbol{HC})\boldsymbol{e}_x(t) + \boldsymbol{L}_1\boldsymbol{f}(t) - \boldsymbol{HL}_2\boldsymbol{f}(t) \tag{2.38}$$

and the output error $\boldsymbol{e}(t)$ becomes

$$\boldsymbol{e}(t) = \boldsymbol{C}\boldsymbol{e}_x(t) + \boldsymbol{L}_2\boldsymbol{f}(t). \tag{2.39}$$

The vector $\boldsymbol{f}(t)$ represents *additive faults* because they influence $\boldsymbol{e}(t)$ and $\boldsymbol{x}(t)$ by a summation.

When sudden and permanent faults $\boldsymbol{f}(t)$ occur, the state estimation error will deviate from zero.

$\boldsymbol{e}_x(t)$ as well as $\boldsymbol{e}(t)$ show dynamic behaviour which are different for $\boldsymbol{L}_1\boldsymbol{f}(t)$ and $\boldsymbol{L}_2\boldsymbol{f}(t)$. Both $\boldsymbol{e}_x(t)$ or $\boldsymbol{e}(t)$ can be taken as residuals.

In particular, the residual $\boldsymbol{e}(t)$ is the basis for different fault detection methods based on output estimation.

For the generation of residual with special properties, the design of the observer feedback matrix $\boldsymbol{H}$ is of interest [Chen and Patton, 1999, Liu and Patton, 1998].

Limiting conditions are the stability and the sensitivity against disturbances $\boldsymbol{v}(t)$ and $\boldsymbol{w}(t)$. If the signals are affected by noise, the Kalman filter must be used instead of classical observers [Jazwinski, 1970].

If faults appear as changes $\Delta\boldsymbol{A}$ or $\Delta\boldsymbol{B}$ of the parameters, the process behaviour becomes

$$\begin{cases} \boldsymbol{x}(t+1) &= (\boldsymbol{A} + \Delta\boldsymbol{A})\boldsymbol{x}(t) + (\boldsymbol{B} + \Delta\boldsymbol{B})\boldsymbol{u}(t) \\ \boldsymbol{y}(t) &= \boldsymbol{C}\boldsymbol{x}(t) \end{cases} \tag{2.40}$$

while the state $\boldsymbol{e}_x(t)$ and the output estimation $\boldsymbol{e}(t)$ errors

$$\begin{cases} \boldsymbol{e}_x(t+1) &= (\boldsymbol{A} - \boldsymbol{HC})\boldsymbol{e}_x(t) + \Delta\boldsymbol{A}\boldsymbol{x}(t) + \Delta\boldsymbol{B}\boldsymbol{u}(t) \\ \boldsymbol{e}(t) &= \boldsymbol{C}\boldsymbol{e}_x(t). \end{cases} \tag{2.41}$$

The changes $\Delta\boldsymbol{A}$ and $\Delta\boldsymbol{B}$ are then *multiplicative faults* [Isermann, 1997, Patton *et al.*, 2000].

In this case, the changes in the residuals depend on the parameter changes, as well as input and state variable changes. Hence, the influence of parameter changes on the residuals is not as straightforward as in the case of the additive faults $\boldsymbol{f}(t)$.

The following observer–based fault detection schemes and configurations are briefly summarised and recalled [Isermann, 1997, Willsky, 1976, Patton *et al.*, 1989, Chen and Patton, 1999, Patton *et al.*, 2000].

1. **Dedicated observers for MIMO processes**
   - *Observer excited by one output*: one observer is driven by one sensor output. The other outputs $\hat{\boldsymbol{y}}(t)$ are reconstructed and compared with measured outputs $\boldsymbol{y}(t)$. This allows the detection of single output sensor faults [Clark, 1978].
   - *Bank of observers, excited by all outputs*: several observers are designed for a definite fault signal and detected by hypothesis test [Willsky, 1976].
   - *Bank of observers, excited by single outputs*: several observers for single sensors outputs are used. The estimated outputs $\hat{\boldsymbol{y}}(t)$ are compared with the measured outputs $\boldsymbol{y}(t)$. This allows the detection of multiple sensor fault (DOS, Dedicated Observer Scheme) [Clark, 1978].
   - *Bank of observers, excited by all outputs except one*: as before, but each observer is excited by all outputs except one sensor output, which is supervised (GOS, Generalised Observer Scheme) [Wünnenberg and Frank, 1987, Frank, 1993].

2. **Fault detection filters for MIMO processes**
   - The feedback $\boldsymbol{H}$ of the state observer in Equation 2.34 is chosen so that particular fault signals $\boldsymbol{L}_1\boldsymbol{f}(t)$ change in a definite direction and fault signals $\boldsymbol{L}_2\boldsymbol{f}(t)$ in a definite plane [Beard, 1971, Jones, 1973, Speyer, 1999].

With directional residual vectors, the fault isolation problem consists of determining which of the known fault signature directions the residual vector lies the closest to. The original form of the "failure detection filter" was proposed by Beard [Beard, 1971] and Jones [Jones, 1973] to generate directional residual vectors. Many more straightforward methods have followed, including methods to achieve "robust fault detection filter" [Chen *et al.*, 1996b].
The fault (or failure) detection is a class of Luenberger observers with a specially designed feedback gain matrix. It allows output estimation errors having directional characteristics associated with some known fault directions, to be obtained.
These fault detection methods mostly require several measurable output signals and make use of internal analytical redundancy of multivariable systems. Recently it was proposed to improve their robustness with respect to process parameter changes and unknown input signals $\boldsymbol{v}(t)$ and $\boldsymbol{w}(t)$ [Patton and Chen, 1994a, Chen *et al.*, 1996b, Chung and Speyer, 1998, Speyer, 1999].
This can be reached, for example, through filtering the output error of the observer by

$$\boldsymbol{r}(t) = \boldsymbol{W}\boldsymbol{e}(t) \tag{2.42}$$

together with a special design of the observer feedback matrix $\boldsymbol{H}$.

3. **Output observers**
   Another possibility is the use of output observers (or UIO, see Section 4.3) in the reconstruction of the output signals, if the estimate of the state variable $\hat{\boldsymbol{x}}(t)$ is not of primary interest.
   In this context, it is worthy to mention the paper by Chen, Patton and Zhang [Chen *et al.*, 1996b] concerning the design of output observers for robust FDI using eigenstructure assignment method.
   Through a linear transformation

$$\boldsymbol{z}(t) = \boldsymbol{T}\boldsymbol{x}(t) \tag{2.43}$$

the state–space representation of the observer becomes

$$\hat{\boldsymbol{z}}(t+1) = \boldsymbol{F}\hat{\boldsymbol{z}}(t) + \boldsymbol{J}\boldsymbol{u}(t) + \boldsymbol{G}\boldsymbol{y}(t) \tag{2.44}$$

and the residual is determined by

$$\boldsymbol{r}(t) = \boldsymbol{W}_z\hat{\boldsymbol{z}}(t) + \boldsymbol{W}_y\boldsymbol{y}(t). \tag{2.45}$$

This situation is depicted in Figure 2.15.

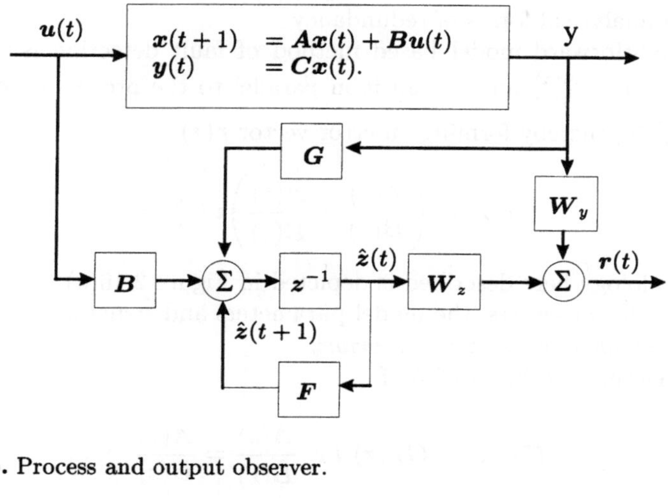

Fig. 2.15. Process and output observer.

The state estimation error

$$\boldsymbol{e}_x(t) = \hat{\boldsymbol{z}}(t) - \boldsymbol{z}(t) = \hat{\boldsymbol{z}}(t) - \boldsymbol{T}\boldsymbol{x}(t) \tag{2.46}$$

and the residuals $\boldsymbol{r}(t)$ are then designed, such that they are independent of the process states $\boldsymbol{x}(t)$, the known input $\boldsymbol{u}(t)$ and the unknown inputs $\boldsymbol{v}(t)$ and $\boldsymbol{w}(t)$, as depicted in Figure 2.14.

In this way, the residuals are dependent only on fault signals $\boldsymbol{f}(t)$ [Patton and Chen, 1994a, Chen *et al.*, 1996b, Gertler, 1998, Patton *et al.*, 2000].

2.5.3 Fault Detection with Parity Equations

The basic idea of the parity relations approach is to provide a proper check of the parity (consistency) of the measurements acquired from the monitored system.

In the early development of fault diagnosis, the parity vector (relation) approach was applied to static or parallel redundancy schemes [Potter and Suman, 1977] which may be obtained directly from measurements (hardware redundancy) or from analytical relations (analytical redundancy). A survey of these methods can be found in [Ray and Luck, 1991].

In the case of hardware redundancy, two methods can be exploited to obtain redundant relations. The first requires the use of several sensors having identical or similar functions to measure the same variable. The second approach consists of dissimilar sensors to measure different variables but with their outputs being relative to each other.

Even if these techniques have been successfully applied for fault diagnosis [Potter and Suman, 1977, Daly *et al.*, 1979], the attention of this section is focused on analytical forms of redundancy.

A straightforward model–based method of fault detection is to take a model $G_M(z) = \frac{\hat{\boldsymbol{A}}(z)}{\hat{\boldsymbol{B}}(z)}$ and to run it in parallel to the process described by $G_P(z) = \frac{\boldsymbol{A}(z)}{\boldsymbol{B}(z)}$, thereby forming an error vector $\boldsymbol{r}(z)$

$$\boldsymbol{r}(z) = \left(\frac{\boldsymbol{A}(z)}{\boldsymbol{B}(z)} - \frac{\hat{\boldsymbol{A}}(z)}{\hat{\boldsymbol{B}}(z)} \right) \boldsymbol{u}(z). \tag{2.47}$$

The methodology here described is depicted in Figure 2.16(a).
However, as for observers, the model parameters and structure of the monitored process have to be known *a priori*.

With reference to Figure 2.5, if

$$\boldsymbol{G}_M(z) = \boldsymbol{G}_P(z) \ i.e. \ \frac{\hat{\boldsymbol{A}}(z)}{\hat{\boldsymbol{B}}(z)} = \frac{\boldsymbol{A}(z)}{\boldsymbol{B}(z)} \tag{2.48}$$

for additive input $\boldsymbol{f}_u(z)$ and output $\boldsymbol{f}_y(z)$ faults, the $\boldsymbol{r}(z)$ error then becomes

$$\boldsymbol{r}(z) = \frac{\boldsymbol{A}(z)}{\boldsymbol{B}(z)} \boldsymbol{f}_u(z) + \boldsymbol{f}_y(z). \tag{2.49}$$

According to Figure 2.16(b), another possibility is to generate a polynomial error

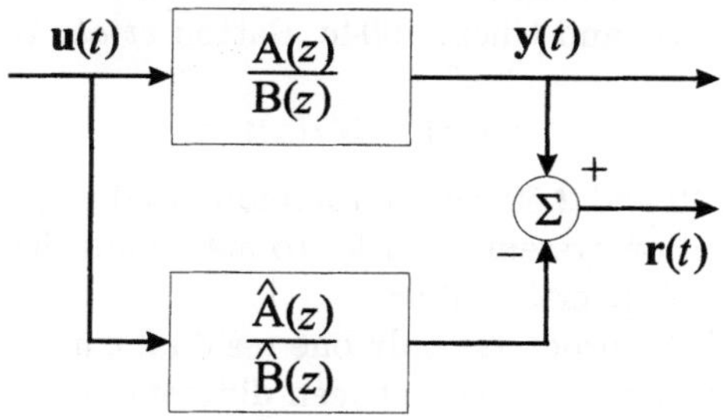

(a) Output error

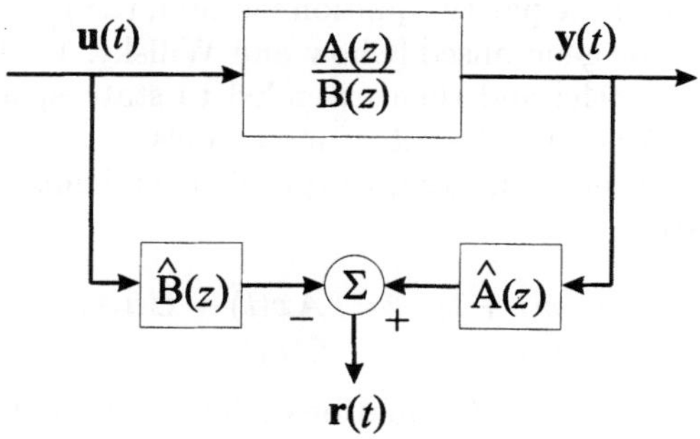

(b) Equation error

Fig. 2.16. Parity equation methods.

$$\begin{aligned} \boldsymbol{r}(z) &= \hat{\boldsymbol{A}}(z)\boldsymbol{y}(z) - \hat{\boldsymbol{B}}(z)\boldsymbol{u}(z) \\ &= \boldsymbol{B}(z)\boldsymbol{f}_u(z) + \boldsymbol{A}(z)\boldsymbol{f}_y(z). \end{aligned} \tag{2.50}$$

In both cases, different time responses are obtained for an additive input or output fault.

Moreover, the error vector $\boldsymbol{r}(z)$ computed by Equation 2.49 corresponds to the output error of parameter estimation method computed by Equation 2.28.

On the other hand, $\boldsymbol{r}(z)$ in Equation 2.50 concerns the equation error of Equation 2.21.

Equations 2.49 and 2.50 generate residuals and are called *parity equations* [Gertler, 1991] under the assumptions of fault occurrence and of exact agreement between process and model.

However, within the parity equations, the model parameters are assumed to be known and constant, whereas the parameter estimations can vary the parameters of $\hat{\boldsymbol{A}}(z)$ and $\hat{\boldsymbol{B}}(z)$ in order to minimise the residuals.

Moreover, for the generation of specific characteristics of the parity vector $\boldsymbol{r}(z)$ and for obtaining fault detection and isolation properties, the residuals can be filtered according to matrix $\boldsymbol{G}_f(z)$ to compute the vector $\boldsymbol{r}_f(z)$ [Gertler, 1991, Patton and Chen, 1994c, Patton *et al.*, 2000]:

$$\boldsymbol{r}_f(z) = \boldsymbol{G}_f(z)\boldsymbol{r}(z). \tag{2.51}$$

Equations 2.51, 2.49 and 2.50 can be therefore used to implement and design the residual generation system, in order to meet fault detection and isolation specifications, as well [Gertler, 1998].

However, for SISO processes only one residual can be generated and it is therefore not easy to distinguish between different faults.

On the other hand, more freedom in the design of parity equations can be obtained when for SISO processes intermediate signals can be measured (see Figure 2.5), or for MIMO systems.

As an extension of the parity equation method, the parity relation concept presented here can be generalised [Chow and Willsky, 1984, Lou *et al.*, 1986, Patton and Chen, 1994c] and then extended to state–space descriptions, as shown in [Gertler, 1998] for discrete–time models.

The redundancy relations are now specified mathematically as follow.

Given the system

$$\begin{cases} \boldsymbol{x}(t+1) & = & \boldsymbol{A}\boldsymbol{x}(t) + \boldsymbol{B}\boldsymbol{u}(t) \\ \boldsymbol{y}(t) & = & \boldsymbol{C}\boldsymbol{x}(t) \end{cases} \tag{2.52}$$

by substituting the second of Equations 2.52 in the first one and delaying several times, the following system is obtained

$$\begin{bmatrix} \boldsymbol{y}(t) \\ \boldsymbol{y}(t+1) \\ \boldsymbol{y}(t+2) \\ \vdots \end{bmatrix} = \begin{bmatrix} \boldsymbol{C} \\ \boldsymbol{CA} \\ \boldsymbol{CA}^2 \\ \vdots \end{bmatrix} \boldsymbol{x}(t) + \begin{bmatrix} \boldsymbol{0} & \boldsymbol{0} & \boldsymbol{0} & \dots \\ \boldsymbol{CB} & \boldsymbol{0} & \boldsymbol{0} & \dots \\ \boldsymbol{CAB} & \boldsymbol{CB} & \boldsymbol{0} & \dots \\ \vdots & \vdots & \vdots & \ddots \end{bmatrix} \begin{bmatrix} \boldsymbol{u}(t) \\ \boldsymbol{u}(t+1) \\ \boldsymbol{u}(t+2) \\ \vdots \end{bmatrix} \tag{2.53}$$

$$\boldsymbol{Y}_f(t) = \boldsymbol{T}\boldsymbol{x}(t) + \boldsymbol{Q}\boldsymbol{U}_f(t). \tag{2.54}$$

In order to remove the non–measurable states $\boldsymbol{x}(t)$, and to obtain a parity vector useful for FDI, Equation 2.53 is multiplied by $\boldsymbol{W}$, such that

$$\boldsymbol{W}\boldsymbol{T} = \boldsymbol{0}. \tag{2.55}$$

This leads to residuals

$$\boldsymbol{r}(t) = \boldsymbol{W}\boldsymbol{Y}_f - \boldsymbol{W}\boldsymbol{Q}\boldsymbol{U}_f(t) \tag{2.56}$$

as shown in Figure 2.17.

The filtered input and output vectors $\boldsymbol{U}_f$ and $\boldsymbol{Y}_f$ are obtained by delaying the corresponding signals.

The design of the matrix $\boldsymbol{W}$ gives some freedom to generate a structured set of residuals.

One possibility is to select the elements of $\boldsymbol{W}$ such that one measured variable has no impact on a specific residual. Then, this residual remains small in the case of an additive fault on this variable, and the other residuals increase [Patton and Chen, 1994c, Chen and Patton, 1999].

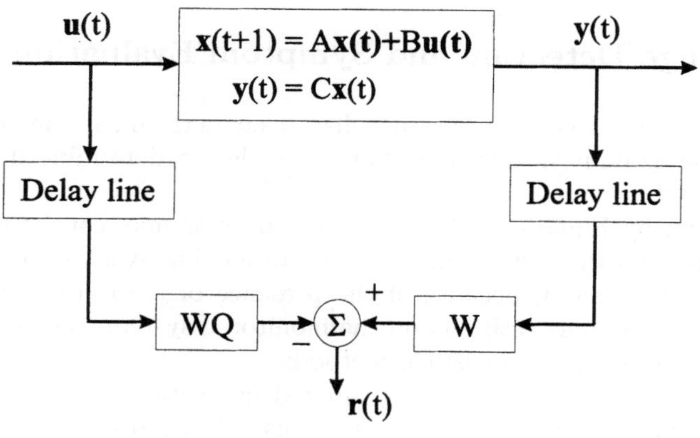

Fig. 2.17. Parity equation methods for a MIMO model.

Finally, because of the previous results, it is clear therefore that some correspondence exists between parity relation and observer-based methods. This aspect was firstly pointed out by Massoumnia [Massoumnia, 1986] and later was demonstrated by Frank and Wunnenberg [Wünnenberg, 1990, Patton *et al.*, 1989].

The problem was re–examined in detail by Chen and Patton [Patton and Chen, 1994c] and the equivalence under different conditions and in different meanings was discussed. It was shown that the parity relation approach is equivalent to the use of a dead–beat observer.

This implies that the parity relation scheme provides less design flexibility when compared with methods which are based on observers without any restriction.

More recently, a comparison between observer–based and parity space techniques was proposed [Delmaire *et al.*, 1999]. Both the methods were first explored for SISO systems and therefore extended the comparison to MIMO systems. The comparison was performed using linear discrete-time models.

In particular, considering MIMO systems described by estimated input–output discrete–time forms (*e.g.*, ARX or Auto Regressive Moving Average eXogenous (ARMAX) models) of Equations 2.49 and 2.50 leads to a representation in which parameters redundancy can not be avoided. To overcome this drawback Delmaire *et al.* proposed in [Delmaire *et al.*, 1999] to use observers designed from identified canonical state–space forms [Guidorzi, 1975]. Moreover, in the case of parameters redundancy, multiple identification of some parameters may occur, leading to inconsistent estimations which might produce inconsistent FDI decisions [Delmaire *et al.*, 1999].

This states again the FDI capabilities of the observer–based methods with respect to parity relation schemes.

2.6 Change Detection and Symptom Evaluation

When the residual generation stage has been performed, the second step requires the examination of symptoms in order to determine if any faults have occurred.

As shown by Equation 2.17, a decision process may consist of a simple threshold test on the instantaneous values of moving averages of residuals.

On the other hand, because of the presence of noise, disturbances and other unknown signals acting upon the monitored system, the decision making process can exploits statistical methods.

In this case, the measured or estimated quantities, such as signals, parameters, state variables or residuals are usually represented by stochastic variables $\boldsymbol{r}(t) = \{r_i(t)\}_i^q$, with mean value and variance [Willsky, 1976]

$$\bar{r}_i = E\{r_i(t)\}; \qquad \bar{\sigma}_i^2 = E\{[r_i(t) - \bar{r}_i]^2\} \tag{2.57}$$

as normal values for the fault-free process.

Analytic symptoms are then obtained as changes

$$\Delta r_i = E\{r_i(t) - \bar{r}_i\}; \qquad \Delta\sigma_i = E\{\sigma_i(t) - \bar{\sigma}_i\} \tag{2.58}$$

with reference to the normal values. Usually, the time instant $t > t_f$ represents the unknown instant of the fault occurrence.

In order to separate normal from faulty behaviour, usually a fixed threshold Δr_{tol} defined as

$$\Delta r_{tol} = \epsilon\bar{\sigma}_r, \qquad \epsilon \geq 2 \tag{2.59}$$

has to be selected.

By a proper choice of ϵ, a compromise has to be made between the detection of small faults and false alarms.

Another class of methods can be exploited for detecting residual changes due to faults. Therefore, techniques of change detection, *e.g.*, as a likelihood–ratio–test or Bayes decision, a run–sum test are commonly used [Isermann, 1984, Basseville and Benveniste, 1986, Basseville and Nikiforov, 1993].

Moreover, fuzzy or adaptive thresholds may improve the binary decision [Chen and Patton, 1999, Patton *et al.*, 2000].

Finally, when several variables change, classification methods are used. In a multidimensional space, the symptom vector

$$\Delta \boldsymbol{r} = [\Delta r_1 \ \Delta r_2 \ \cdots \ \Delta r_q] \tag{2.60}$$

belongs to a q–dimensional space and its direction depends on the fault occurrence.

In this case, the process of residual evaluation consists of determining the direction as well as the distance of $\Delta \boldsymbol{r}$ from the origin. Geometrical distance methods [Carpenter and Grossberg, 1987, Tou and Gonzalez, 1974] or artificial neural networks [Himmelblau *et al.*, 1991, Meneganti *et al.*, 1998] can be hence applied.

The generation and evaluation of analytic symptoms concludes the task of fault–detection within the framework of model–based fault diagnosis of Figure 2.8.

2.7 The Residual Generation Problem

Although the analytical redundancy method for residual generation has been recognised as an effective technique for detecting and isolating faults, the critical problem of unavoidable modelling uncertainty has not been fully solved.

The main problem regarding the reliability of FDI schemes is the modelling uncertainty which is due, for example, to process noise, parameter variations and non–linearities.

On the other hand, all model–based methods use a model of the monitored system to produce the symptom generator. If the system is not complex and can be described accurately by the mathematical model, FDI is directly performed by using a simple geometrical analysis of residuals. In real industrial systems however, the modelling uncertainty is unavoidable.

The design of an effective and reliable FDI scheme for residual generation should take into account of the modelling uncertainty with respect to the sensitivity of the faults. Therefore, the task of the design of an FDI system is thus to generate residuals which are *robust* [Chow and Willsky, 1984, Ding and Frank, 1990, Frank, 1994b, Frank and Ding, 1997, Patton and Chen, 1994c].

Several papers addressed this problem. For example, optimal robust parity relations were proposed [Chow and Willsky, 1984, Chung and Speyer, 1998, Speyer, 1999, Lou *et al.*, 1986] and the threshold selector concept was introduced [Emami-Naeini *et al.*, 1988]. Robust FDI using the disturbance decoupling technique was also used [Patton and Chen, 1994c, Chen *et al.*, 1996b]. The Patton and Chen approach is an interesting contrast to the Chow and Willsky method which seems to minimise the modelling uncertainty over several points of operation. Patton and Chen deal directly with this problem by estimating the optimum unknown input distribution matrix over a range of operating points and exploiting the eigenstructure assignment approach [Patton and Chen, 1994c, Chen and Patton, 1999].

The model–based FDI technique requires a high accuracy mathematical description of the monitored system. The better the model represents the dynamic behaviour of the system, the better will be the FDI precision. If a FDI method can be developed which is insensitive to modelling uncertainty, a very accurate model is not necessarily needed.

All uncertainties can be are summarised as disturbances acting on the system. Although the disturbance vector is unknown, its distribution matrix can be obtained by an identification procedure. Under this assumption, the "disturbance de–coupling" principle can be exploited to design a robust FDI scheme.

In order to summarise the approach to the robustness problem, the state–space model of the monitored system should be considered [Patton and Chen, 1993]:

$$\begin{cases} \boldsymbol{x}(t+1) & = & (\boldsymbol{A}+\Delta\boldsymbol{A})\,\boldsymbol{x}(t)+(\boldsymbol{B}+\Delta\boldsymbol{B})\,\boldsymbol{u}(t)+\boldsymbol{E}_1\boldsymbol{\varepsilon}(t)+\boldsymbol{R}_1\boldsymbol{f}(t) \\ \boldsymbol{y}(t) & = & \boldsymbol{C}\boldsymbol{x}(t)+\boldsymbol{E}_2\boldsymbol{\varepsilon}(t)+\boldsymbol{R}_2\boldsymbol{f}(t) \end{cases} \tag{2.61}$$

where $\boldsymbol{\varepsilon}(t)$ is the disturbance vector, and $\boldsymbol{E}_1$ and $\boldsymbol{E}_2$ are the known or unknown input distribution matrices. The matrices $\Delta\boldsymbol{A}$ and $\Delta\boldsymbol{B}$ are the parameter errors or variations which represent modelling errors.

The discrete transfer matrix description between the output $\boldsymbol{y}(t)$ and input $\boldsymbol{u}(t)$ of the system 2.61 is then

$$\boldsymbol{y}(z)=(\boldsymbol{G}_u(z)+\Delta\boldsymbol{G}_u(z))\,\boldsymbol{u}(z)+\boldsymbol{G}_\varepsilon(z)\boldsymbol{\varepsilon}(z)+\boldsymbol{G}_f(z)\boldsymbol{f}(z) \tag{2.62}$$

where $\Delta\boldsymbol{G}_u(z)$ is used to describe modelling errors, whilst both $\Delta\boldsymbol{G}_u(z)$ and $\boldsymbol{G}_\varepsilon(z)$ represent modelling uncertainty.

With reference to the residual generator of Figure 2.10 and described by Equation 2.14, the z–domain residual vector has to be rewritten as

$$\boldsymbol{r}(z)=\boldsymbol{H}_y(z)\boldsymbol{G}_f(z)\boldsymbol{f}(z)+\boldsymbol{H}_y(z)\boldsymbol{G}_\varepsilon(z)\boldsymbol{\varepsilon}(z)+\boldsymbol{H}_y(z)\Delta\boldsymbol{G}_u(z)\boldsymbol{u}(z). \tag{2.63}$$

With respect to Equation 2.14, the terms $\boldsymbol{H}_y(z)\boldsymbol{G}_\varepsilon(z)$ and $\boldsymbol{H}_y(z)\Delta\boldsymbol{G}_u(z)$ cannot be deleted.

Both faults and modelling uncertainty (disturbance and modelling error) affect the residual and hence discrimination between these two effects is difficult.

The principle of disturbance de–coupling for robust residual generation requires that the residual generator satisfies

$$\boldsymbol{H}_y(z)\boldsymbol{G}_\varepsilon(z)=\boldsymbol{0} \tag{2.64}$$

in order to achieve total de–coupling between residual $\boldsymbol{r}(z)$ and disturbance $\boldsymbol{\varepsilon}(z)$.

This property can be achieved by using the unknown input observer [Watanabe and Himmelblau, 1982, Wünnenberg and Frank, 1987, Chen *et al.*, 1996b, Frank *et al.*, 2000], optimal (robust) parity relations [Chow and Willsky, 1984, Lou *et al.*, 1986, Wünnenberg, 1990, Wünnenberg and Frank, 1990, Frank *et al.*, 2000] or alternatively the eigenstructure assignment approach [Patton *et al.*, 1986, Patton and Chen, 1991b, Liu and Patton, 1998, Patton and Chen, 2000, Duan *et al.*, 2002].

These approaches are presented in detail in Chapter 4 where the design of a robust residual generator is also achieved in connection with different identification tools summarised in Chapter 3.

Hence, for disturbance de–coupling approaches in FDI, the aim is to completely eliminate the disturbance effect from the residual. However, the complete elimination of disturbance effects may not be possible due to the lack of degree of freedom. Moreover, it may be problematic, in some cases, because the fault effect may also be eliminated. Hence, an appropriate criterion for robust residual design should take into account the effects of both modelling error and faults. There is a trade–off between sensitivity to faults and robustness to modelling uncertainty and hence robust residual generation can be considered as a *multi–objective optimisation problem* [Chen and Patton, 1999, chapt. 6]. It consists of the maximisation of fault effects and the minimisation of uncertainty effects [Wünnenberg, 1990, Frank *et al.*, 2000].

Therefore, the approach to the design of optimal residuals can require the satisfaction of a set of objectives. These objectives are essential for achieving robust diagnosis of incipient faults. If such joint optimisation problems, which can be also expressed in the frequency domain, were reformulated for satisfying a set of inequalities on the performance indices, Genetic Algorithms (GA) [Goldberg, 1989, Davis, 1991] and Linear Matrix Inequalities (LMI) [Boyd *et al.*, 1994] can be successfully exploited to search the optimal solution [Chen *et al.*, 1996a, Hou and Patton, 1997, Chen *et al.*, 1997], [Chen and Patton, 1999, Chen and Patton, 2001].

Disturbance de–coupling can also be achieved using frequency domain design techniques. As an example, the robust fault detection problem can be managed by using the standard H_∞ filtering formulation [Ding and Frank, 1990, Hou and Patton, 1996, Frank and Ding, 1997].

With this method, the minimisation of the disturbance effect on the residual is formulated as a standard H_∞ filtering problem [Chen and Patton, 2000, Frank *et al.*, 2000]. On the other hand, the so–called H_∞/H_- approach can be also exploited [Hou and Patton, 1996, Hou and Patton, 1997, Frank *et al.*, 2000, Chen and Patton, 2000].

Among the many ways for eliminating or minimising disturbance and modelling error effects on the residual and hence for achieving robustness in FDI [Patton *et al.*, 2000] H_∞ optimisation is a robust design method with the original motivation firmly rooted in the consideration of various uncertainties, especially the modelling errors. It is reasonable to seek an application of this technique in the robust design of FDI systems. Therefore, the H_∞ optimisation method can be successfully exploited for robust residual generation of FDI.

The early work of using H_∞ optimisation techniques for robust FDI was based on the use of factorisation approach [Ding and Frank, 1990, Ding *et al.*, 2000]. The factorisation–based H_∞ optimisation technique is useful in solving FDI problems. However, the more elegant and advanced H_∞ optimisation methods are based on the use of the Algebraic Riccati Equation (ARE) [Zhou *et al.*, 1996]. Mangoubi *et al.* [Mangoubi *et al.*, 1992] first solved the robust FDI estimation problem using the ARE approach via the use of H_∞ and μ robust estimator synthesis methods developed by Appleby *et al.* [Appleby *et al.*, 1991]. A direct formulation of the FDI problem as a robust H_∞ filter design problem with the solution of an ARE was given in Edelmayer *et al.* [Edelmayer *et al.*, 1997]. To deal with modelling errors as well as disturbances in robust FDI design, Niemann and Stoustrup [Niemann and Stoustrup, 1996] introduced modelling error blocks into the standard H_∞ observer design. The weighting factors are then introduced in the problem formulation for finding an optimal FDI solution. This is further extended to non–linear systems where the non–linearity is treated in the same way as a modelling error block [Stoustrup and Niemann, 1998, Stoustrup *et al.*, 1997].

The majority of studies discussed so far involve the use of a slightly modified H_∞ filter for the residual generation, *i.e.* the design objective is to minimise the effect of disturbances and modelling errors on the estimation error and subsequently on the residual. However, robust residual generation is different from the robust estimation because it does not only require the disturbance attenuation. The residual has to remain sensitive to faults whilst the effect of disturbance is minimised. Sauter et at. [Sauter *et al.*, 1997] studied this problem where the fault sensitivity is enhanced by applying an optimal post–filter to the "primary residual". The problem of enhancing fault sensitivity while increasing robustness against disturbances and modelling errors was studied extensively by Sadrnia *et at.* [Sadrnia *et al.*, 1997]. The essential idea is to reach an acceptable compromise between disturbance robustness and fault sensitivity. In the beginning, an observer with very small disturbance sensitivity bound is designed via an ARE. Then, the fault sensitivity is checked. If the fault sensitivity is too small, the disturbance robustness requirement should be relaxed, *i.e.* to design another optimal observer with an increased disturbance sensitivity bound. This procedure is likely to be repeated several times. The final goal is to find a design which provides the maximum ratio between fault sensitivity and disturbance sensitivity.

Recently, Chen and Patton [Chen and Patton, 1999, Chen and Patton, 2000] have formulated the robust residual generation problem within the standard H_∞ filtering framework, *i.e.* to generate the residual whose sensitivity to disturbances is minimised. To facilitate reliable FDI, the residual sensitivity to faults has to be maintained (or maximised) in addition to the minimisation of the disturbance sensitivity. This problem was solved via the minimisation of the difference between the residual and the fault against the disturbance and the fault, *i.e.* the objective is to replicate the fault using the residual. In this way, the residual sensitivity to the fault is indirectly maximised. The residual sensitivity to the modelling error can be minimised if the modelling error is approximately represented by the disturbance vector with the estimated distribution matrix [Chen and Patton, 1999]. However, the modelling error can be handled directly using standard H_∞. In [Chen and Patton, 1999, Chen and Patton, 2000] the way of including the modelling error in the robust residual design within the standard H_∞ framework was shown.

Generally speaking, the robust FDI approach can be approached in different ways. It is therefore important to mention the design principle of residual generators under a certain performance index [Basseville, 1997, Frank *et al.*, 2000]. This is indeed a reasonable extension of the unknown input residual generator design, in which, instead of full de–coupling, a compromise between the robustness and sensitivity is made.

It is worth focusing the attention to this scheme, due to its important role in theoretical studies and its relationship to the residual evaluation and integrated design of FDI systems. Since the goal of residual generation is to enhance the robustness of the residual to the model uncertainty without loss of its sensitivity to the faults, the minimisation of performance index [Frank *et al.*, 2000]

$$J = \frac{\|\frac{\partial r}{\partial d}\|}{\|\frac{\partial r}{\partial f}\|} \text{ or } J = \|\frac{\partial r}{\partial d}\| \text{ with } \|\frac{\partial r}{\partial f}\| > \alpha \tag{2.65}$$

is widely recognised as a suitable design objective. Associated to the norm used, the type of the residual generator and the mathematical tool adopted, a number of optimisation approaches have been developed [Frank *et al.*, 2000]. Recently, [Ding *et al.*, 2000] derived a unified solution for a number of optimisation problems and provided thus a satisfactory solution to the above–defined optimisation problem ten years after it was first proposed. In [Frank *et al.*, 2000] a briefly review the state of art of the solutions can be found whilst [Hou and Patton, 1996, Hou and Patton, 1997, Frank *et al.*, 2000] address the H_∞/H_- method.

According to the norm selected, by minimising the performance index 2.65 over a specified range, an approximate de–coupling design can be achieved [Ding and Frank, 1990, Patton and Hou, 1997, Frank and Ding, 1997, Ding *et al.*, 1999].

Moreover, the approximated design for optimal disturbance de–coupling can also be solved in the time domain [Wünnenberg, 1990, Chen *et al.*, 1993].

On the other hand, with reference to the modelling errors in Equation 2.63, represented by the term $\Delta \boldsymbol{G}_u(z)$ the robust problem is more difficult to solve.

Two main techniques have been described by Patton and Chen. In the first case, the uncertainty is taken into account at the residual design stage [Chen *et al.*, 1996b]; this is known as ***active robustness*** in fault diagnosis [Patton and Chen, 1994c].

The active way of achieving a robust solution is to approximate uncertainties, *i.e.* representing approximately modelling errors as disturbances [Chen and Patton, 1999]

$$\Delta \boldsymbol{G}_u(z)\boldsymbol{u}(z) \approx \boldsymbol{G}_d(z)\boldsymbol{d}(z) \tag{2.66}$$

where $\boldsymbol{d}(z)$ is an unknown vector and $\boldsymbol{G}_d(z)$ is an estimated transfer function. When this approximate structure is exploited to design disturbance de–coupling residual generators, robust FDI can be achieved. This disturbance approximation technique will be presented in Section 4.7.

The second approach called ***passive robustness*** makes use of a residual evaluator with adaptive threshold. As a simple example, it is assumed that the residual generation uncertainty 2.63 is only represented by modelling errors.

The fault–free residual $\boldsymbol{r}(z)$ is

$$\boldsymbol{r}(z) = \boldsymbol{H}_y(z)\Delta \boldsymbol{G}_u(z)\boldsymbol{u}(z). \tag{2.67}$$

Under the assumption that the modelling errors are bounded by a value δ, such that

$$\| \Delta \boldsymbol{G}_u(w) \| \leq \delta \tag{2.68}$$

an adaptive threshold $\varepsilon(t)$ can be generated by a linear system

$$\varepsilon(t) = \delta \boldsymbol{H}_y(z)\boldsymbol{u}(z) \tag{2.69}$$

In such case, the threshold $\varepsilon(t)$ is no longer fixed but depend on the input $\boldsymbol{u}(t)$, thus being adaptive to the system operating point. A fault is then detected if

$$\| \boldsymbol{r}(t) \| > \| \varepsilon(t) \| \tag{2.70}$$

A robust FDI technique with the threshold adaptor or selector is therefore briefly recalled [Clark, 1989], [Emami-Naeini *et al.*, 1988], [Ding and Frank, 1991]. This method represents a passive approach since no effort is made to design a robust residual.

Even if disturbance de–coupling methods for robust FDI has been studied extensively, their effectiveness regarding real problems has not been fully demonstrated.

The main difficulty arises as most of the disturbance only account for a small percentage of the uncertainty in the real system. The presented disturbance decoupling methods cannot be directly applied to the systems with other uncertainties such as modelling errors.

The estimation and approximate representation of modelling errors as well as other uncertain factors as the disturbance term provides a practical way to tackle the robustness issue for real plants.

Chapter 4 provides a study of a different approach for representing modelling errors and other uncertain factors via the disturbance term with an estimated distribution matrix. As presented in Chapter 3, this identified distribution matrix will be used for the design of the disturbance de–coupled residual in order to solve the robust FDI problem.

2.8 Fault Diagnosis Technique Integration

Several FDI techniques have been developed and their application shows different properties with respect of the diagnosis of different faults in a process. In order to achieve a reliable FDI technique, a good solution consists of a proper integration of several methods which take advantages of the different procedures [Isermann, 1994a, Isermann and Ballé, 1997].

Furthermore, a comprehensive approach to fault diagnosis should exploit a knowledge–based treatment of all available analytical and heuristic information. This successful approach can be performed by an integrated method to knowledge–based fault diagnosis.

2.8.1 Fuzzy Logic for Residual Generation

As stated in Section 2.2, model–based FDI consists of two stages, residual generation and decision making.

The first block is exploited to generate residuals by means of the available inputs and outputs from the monitored system.

For the first step, classical fault diagnosis model–based methods can exploit state–space of input–output dynamic models of the process under investigation. Within this framework, faults are supposed to appear as changes on the system state or output caused by malfunctions of the components as well as of the sensors. Such fault indices are often monitored using estimation techniques.

The main problem with these techniques is that the precision of the process model affects the accuracy of the detection and isolation system as well as the diagnostic sensibility.

On the other hand, the majority of real industrial processes are non–linear [Chen and Patton, 1999, Gertler, 1998, Patton and Chen, 1997] and cannot be modelled by using a single model for all operating conditions.

Since a mathematical model is a description of system behaviour, accurate modelling for a complex non–linear system is very difficult to achieve in practice. Sometimes for some non–linear systems, it can be impossible to describe them by analytical equations. Moreover, sometimes the system structure or parameters are not precisely known and if diagnosis has to be based primarily on heuristic information, no qualitative model can be set up

Because of these assumptions, fuzzy system theory seems to be a natural tool to handle complicated and uncertain conditions [Babuška, 1998].

Instead of exploiting complicated non–linear models obtained by modelling techniques, it is also possible to describe the plant by a collection of local affine fuzzy and non–fuzzy models [Leontaritis and Billings, 1985a, Leontaritis and Billings, 1985b, Takagi and Sugeno, 1985], whose parameters are obtained by identification procedures.

The second stage of model–based FDI consists of a logic decision process that transforms residual signal information (quantitative knowledge) into qualitative statements (faulty or normal working conditions). Therefore, the problem of decision–making can be treated in a novel way by means of fuzzy logic.

As noise contamination and uncertainty affect the residuals even in fault–free conditions, so that they fluctuate and become unequal to zero. This common situation, which may hide the fault effects, can be handled by means of the fuzzy logic framework.

The interesting feature of fuzzy logic is that it represents a powerful tool for describing vague and imprecise fact and is therefore suited for applications where complete information about fault and system is not available to the designer.

Even if much effort has been spent on trying to decrease the uncertainty associated with quantitative residual generation, it is impossible to fully eliminate the effect of uncertainty. On the basis of this limitation, the residual evaluation problem consists of making the correct decision with respect to uncertain information. Fuzzy logic can be a suitable tool for this task. For instance, a lot of processes can be managed by humans heuristically since an analytical description is impossible to use. Fuzzy logic can express expert knowledge in the form of a rule–based knowledge format.

The introduction of fuzzy logic can improve the decision making in order to provide reliable FDI methods which are applicable for real industrial systems.

As an example, fuzzy logic can be exploited for residual evaluation mainly in the decision making stage for releasing the final yes–no decision [Ulieru and Isermann, 1993, Frank, 1994a, Meneganti *et al.*, 1998].

Rule–based expert systems have therefore been investigated very intensively for fault detection and diagnosis problems [Rich and Venkatasubramanian, 1987, Kramer, 1987, Patton *et al.*, 1989, Patton *et al.*, 2000]. Fault diagnosis using rule–based system needs a database of rules and the accuracy of diagnosis depend on the rules. Moreover, creating a rich and detailed database of rules is usually a time-consuming task and many process experts are needed.

It should finally be pointed out how the fuzzy approach in FDI can solve the problem at two levels: first, fuzzy descriptions are used to generate symptoms and then, the fault detection and isolation is achieved using again fuzzy logic [Dexter and Benouarets, 1997, Isermann, 1998].

2.8.2 Neural Networks in Fault Diagnosis

Quantitative model–based fault diagnosis generates symptoms on the basis of the analytical knowledge of the process under investigation. In most cases however, this does not provide enough information to perform an efficient FDI, *i.e.*, to indicate the location and the mode of the fault.

A typical integrated fault diagnosis system uses both analytical and heuristic knowledge of the monitored system. The knowledge can be processed in terms of residual generation (analytical knowledge) and feature extraction (heuristic knowledge). The processed knowledge is then provided to an inference mechanism which can comprise residual evaluation, symptom observation and *pattern recognition.*

In particular, when the process model is only known to a certain extent of precision, pattern recognition method can provide a convenient approach to solve the fault identification problem, *i.e.* to determine the size of the fault [Himmelblau, 1978, Pau, 1981].

In recent years, neural networks (NN) have been used successfully in pattern recognition as well as system identification, and they have been proposed as a possible technique for fault diagnosis, too.

NN can handle non–linear behaviour and partially known process because they learn the diagnostic requirements by means of the information of the training data.

NN are noise tolerant and their ability to generalise the knowledge as well as to adapt during use are extremely interesting properties [Hoskins and Himmelblau, 1988, Dietz *et al.*, 1989, Venkatasubramanian and Chan, 1989, McDuff and Simpson, 1990, Chen *et al.*, 1990a]. Some example processes were considered in which FDI was performed by a NN using input and output measurements. In these works the NN is trained to identify the fault from measurement patterns, however the classification of individual measurement pattern is not always unique in dynamic situations, therefore the straightforward use of NN in fault diagnosis of dynamic plant is not practical and other approaches should be investigated.

A NN could be exploited in order to find a dynamic model of the monitored system or connections from faults to residuals. In the latter case, the NN is used as pattern classifier or non–linear function approximator. In fact, artificial neural networks are capable of approximating a large class of functions, for fault diagnosis of an industrial plant.

Under these considerations, in Chapter 4, the identification of fuzzy and non–fuzzy models for the system under diagnosis as well as the application of NN as function approximator will be shown.

Quantitative and qualitative approaches have a lot of complementary characteristics which can be suitably combined together to exploit their advantages and to increase the robustness of quantitative techniques. The suggested combination can also minimise the disadvantages of the two procedures; in particular, it is important that partial knowledge deriving from qualitative reasoning is reduced by quantitative methods. Hence, the main aim of further research on model–based fault diagnosis consists of finding the way to properly combine these two approaches together to provide highly reliable diagnostic information.

2.8.3 Neuro-fuzzy Approaches to FDI

Identification of multivariable processes can be interpreted as a problem of approximation to an input–output mapping. The mathematical model used in traditional methods is sensitive to modelling errors, parameter variation, noise and disturbance [Chen and Patton, 1999, Patton *et al.*, 2000]. Process modelling has limitations, especially when the system is complex and uncertain and the data are ambiguous and not information rich.

As previously stated, NN are known to approximate any non–linear even dynamic function, given suitable weighting factors and architecture. Moreover, on–line training makes it possible to change the FDI system easily in cases where changes are made in the physical process or the control system. NN can generalise when presented with inputs not appearing in the training data and make intelligent decisions in cases of noisy or corrupted data. They are also readily applicable to multivariable systems and have a highly parallel structure, which is expected to achieve a higher degree of fault tolerance. A NN can operate simultaneously on qualitative and quantitative data. NN can be very useful when no mathematical model of the system is available, *i.e.* analytical models cannot be applied. 152

Almost all the physical processes are dynamic in nature. Combining dynamic elements such as filters and delays yield a powerful modelling technique. But the NN operates as a "black box" with no qualitative/quantitative information available of the model it represents. Usually, engineers and operators want to visualise how the system is working and what rules govern its operation. There is also ambiguity about the performance of the NN in case of unexpected situation [Korbicz *et al.*, 1999].

Fuzzy logic systems, on the other hand, have the ability to mimic the sensing, generalising, processing, operating and learning abilities of a human operator. They offer a linguistic model of the system dynamics which can be easily understood by certain rules. They also have inherent abilities to deal with imprecise or noisy data.

Fuzzy logic can be used with neural networks [Chiang *et al.*, 2001, chapt. 12]. A fuzzy neuron has the same basic structure as the artificial neuron, except that some or all of its components and parameters may be described through fuzzy logic. A fuzzy neural network is built on fuzzy neurons or on standard neurons but dealing with fuzzy data. A fuzzy neural network is a connectionist model for the implementation and inference of fuzzy rules. There are many different ways to fuzzify an artificial neuron, which results in a variety of fuzzy neurons and fuzzy networks [Chiang *et al.*, 2001, chapt. 12], [Nelles, 2001].

Different neuro–fuzzy structures can be therefore designed to combine the advantages of both neural networks and fuzzy logic [Patton *et al.*, 1999b, Calado *et al.*, 2001]. These structures have been successfully applied to a wide range of applications from industrial processes to financial systems, because of the ease of rule base design, linguistic modelling, application to complex and uncertain systems, inherent non-linear nature, learning abilities, parallel processing and fault–tolerance abilities [Wu and Harris, 1996, Ayoubi, 1995]. However, successful implementation depends heavily on prior knowledge of the system and the training data. There are three common methods of combining neural networks with the fuzzy logic.

1. Fuzzification of the inputs or outputs of the neural networks.
2. Fuzzification of the interconnections of conventional neural networks.
3. Using neural networks in fuzzy models where neurons provide the necessary membership functions and rule base.

All of the Neuro–fuzzy (NF) modelling structures combine, in a single framework, both numerical and symbolic knowledge about the process. Automatic linguistic rule extraction is a useful aspect of NF especially when little or no prior knowledge about the process is available [Brown and Harris, 1994a, Jang and Sur, 1995]. For example, a NF model of a non–linear dynamical system can be identified from the empirical data. This modelling approach can give us some insight about the non–linearity and dynamical properties of the system.

The most common NF systems are based on two types of fuzzy models TSK [Takagi and Sugeno, 1985, Sugeno and Kang, 1988] and [Mamdani, 1976, Mamdani and Assilian, 1995] combined with NN learning algorithms. TSK models use local linear models in the consequents, which are easier to interpret and can be used for control and fault diagnosis [Füssel *et al.*, 1997, Isermann and Ballé, 1997]. Mamdani models use fuzzy sets or rules as consequents and therefore give a more qualitative description.

The B–spline neural network (with triangular basis functions) is the simplest of all of the Mamdani NF structures, but the large consequent rule set means that the method is not easy to use due to low transparency.

Many neuro–fuzzy structures have been successfully applied to a wide range of applications from industrial processes to financial systems, because of the ease of rule base design, linguistic modelling, application to complex and uncertain systems, inherent non-linear nature, learning abilities, parallel processing and fault-tolerance abilities. However, successful implementation depends heavily on prior knowledge of the system and the empirical data [Ayoubi, 1995].

NF networks by their intrinsic nature can handle a limited number of inputs and can usually be identified in a not very transparent way from the empirical data. Transparency corresponds here to a more meaningful description of the process *i.e.* less rules with appropriate membership functions. In ANFIS [Jang, 1993, Jang and Sur, 1995] a fixed structure with grid partition is used. Antecedent and consequent parameters are identified by a combination of least-squares estimates and gradient based methods, the so–called called ***hybrid learning rule***. This method is fast and easy to implement for low dimensional input spaces. It is more prone to losing the transparency and the local model accuracy because of the use of ***error back-propagation*** that is a global and not locally non–linear optimisation procedure. One possible method to overcome this problem can be to find the antecedents and rules separately *e.g.* clustering and constrain the antecedents, and then apply optimisation.

Hierarchical NF networks can be used to overcome the dimensionality problem by decomposing the system into a series of MISO and/or SISO systems called ***hierarchical systems*** [Tachibana and Furuhashi, 1994]. The local rules use subsets of input spaces and are activated by higher level rules.

The criteria on which to build a NF model are based on the requirements for fault diagnosis and the system characteristics. The function of the NF model in the FDI scheme is also important *i.e.* pre–processing data, identification (residual generation) or classification (decision making/fault isolation). For example, a NF model with high approximation capability and disturbance rejection is needed for identification so that the residuals are more accurate. Whereas, in the classification stage, a NF network with more transparency is required.

2.8.4 Structure Identification of NF Models

For complexity reduction and transparency, structure identification methods can be applied to find appropriate input partition, rules and membership functions (MFs). Methods like Evolutionary Algorithms (EA), Classification and Regression Trees CART [Jang, 1994], Clustering and unsupervised NN (*e.g.* like the Kohonen feature maps) can be used. Once the structure is determined *i.e.* the rules and input membership functions, the consequent parameters can be identified by optimisation techniques like Least–Squares Estimation. The Product Space Clustering approach can be used [Babuška, 1998] for structure identification of TSK and Mamdani fuzzy models. For a MISO non–linear dynamic system with p inputs, the Product Space $X \times Y \subset \Re^{p+1}$ is divided in subspaces in which linear models can approximate the non–linear system. The locally linear model tree LOLIMOT algorithm developed by Nelles and Isermann [Nelles and Isermann, 1996] can be used to identify a TSK fuzzy model with dynamic linear models as consequent. When using such structure identification techniques, a major issue is the sensitivity to uneven distribution of data. For example in most clustering algorithms, more clusters are created in regions with more data. A possible solution to this is problem may be to initialise the algorithm with large number of clusters.

Transparency of the NF models can be enhanced by tuning rules and MFs [Babuška, 1998]. This type of method is referred to as structure simplification/optimisation techniques. To find the optimal number of rules, different cluster validity measures and methods like Compatible Cluster Merging CCM [Krishnapuram and Freg, 1992] can be used. At the NF model level the rules are further simplified by merging similar fuzzy sets and removing fuzzy sets similar to the universal set. Setnes *et al.*, [Setnes and Kaymak, 1998] used a supervised fuzzy clustering algorithm that uses input–output data, orthogonal techniques and tuning for complexity reduction.

2.8.5 NF Residual Generation Scheme for FDI

Fig. 2.18 describes a FDI scheme in which several NF models are constructed to identify the faulty and the fault–free behaviour of the system.

$$r_i(t) = f\big(\boldsymbol{u}(t), \ldots, \boldsymbol{u}(t-n), \boldsymbol{y}(t), \ldots, \boldsymbol{y}(t-n)\big), \quad i = 1, \ldots, m \qquad (2.71)$$

Each residual $r_i(t)$ in 2.71 is ideally sensitive to one particular fault in the system. In practice however, as a consequence of noise and disturbances, residuals are sensitive to more than one fault.

To take into account the sensitivity of residuals to various faults and noise we apply a NF classifier. A linguistic style (Mamdani) NF network is used which processes the residuals to indicate the fault.

This NF model is constructed with following set of rules:

$$\textbf{If } r_1 \text{ is small } \ldots r_j \text{ is large } r_m \text{ is small } \textbf{then } \text{fault}_r \text{ is large} \qquad (2.72)$$

Fuzzy threshold evaluation in Equation 2.73 is employed to take into account the imprecision of the residual generator at different regions in the input space

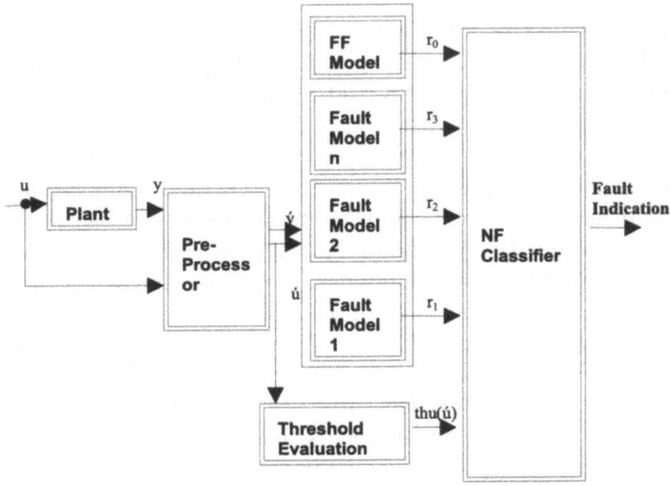

Fig. 2.18. Neuro–fuzzy based FDI scheme.

$$\mathrm{th}_\nu(\boldsymbol{u}) = \frac{\sum_{i=1}^{C} \mathrm{th}_i \; \eta_i(\boldsymbol{u})}{\sum_{i=1}^{C} \eta_i(\boldsymbol{u})} \tag{2.73}$$

where C is the total number of I/P regions with different sensitivity to faults and a multidimensional fuzzy set η_i defines the fuzzy boundary of i–th such region. This approach depends heavily on the availability of the faulty and fault–free data and it is more difficult to isolate faults that appear in the dynamics.

Residuals can also be generated by a non–linear dynamic model of the plant that approximates a non–linear dynamic system by local linear models. Such a model can be obtained by *Product space clustering* [Babuška, 1998], or tree–like algorithms (LOLIMOT algorithm by Nelles *et al.*, [Nelles and Isermann, 1996]). Each local model is a linear approximation of the process in an I/P subspace and the selection of the local model is fuzzy. The output of such a model can be described by:

$$y = \frac{\sum_{i=1}^{C} \alpha_i(\boldsymbol{u}_s) \; f_i}{\sum_{i=1}^{C} \alpha_i(\boldsymbol{u}_s)} \tag{2.74}$$

where f_i is the i–th local linear model given by:

$$f_i = \sum_{k=0}^{n} b_{i,k} \; \boldsymbol{u}(t-k) + \sum_{k=0}^{n} a_{i,k} \; \boldsymbol{y}(t-k) + c_i \tag{2.75}$$

$a_{i,k}$, $b_{i,k}$ and c_i are the parameters of the i–th model, $\boldsymbol{u}_s$ is the I/P subspace defining the operating point, α_i is the degree to which the i–th local model is valid at this operating point.

From $a_{i,k}$, $b_{i,k}$ and c_i, physical parameters like time constants, static gains, offsets, *etc.* [Füssel *et al.*, 1997] can be extracted for each operating point and can be compared with the parameters estimated online. This approach heavily depends on the accuracy of the non-linear dynamic model described above. Also the output error should be minimum when operated in parallel to the system. Moreover, this method requires that there is sufficient excitation at each operating point for online estimation of parameters. The TSK NF based FDI scheme is depicted in Figure 2.19.

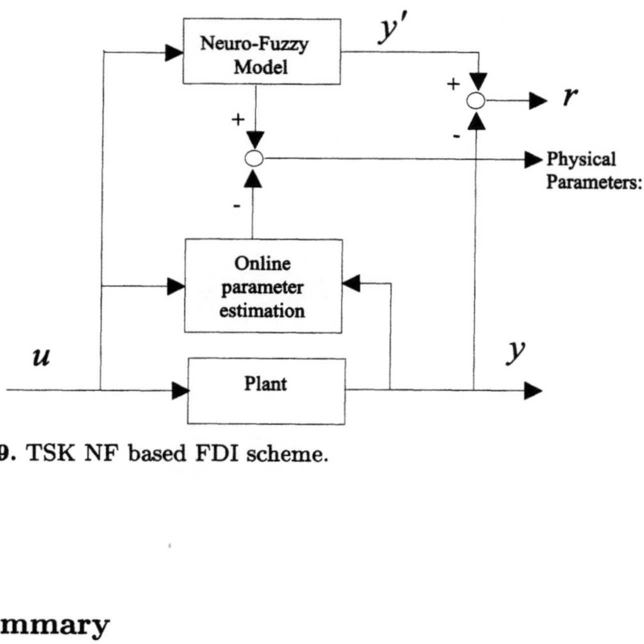

Fig. 2.19. TSK NF based FDI scheme.

2.9 Summary

This Chapter has presented a tutorial treatment on the basis principles of model-based FDI.

The FDI problem has been formalised in a uniform framework by presenting a mathematical description and definition. Within this framework, the residual generation has been identified as a central issue in model–based FDI. By choosing the proper design approach, the FDI task can be performed.

The residual generator has been summarised in different residual generation structures. The ways of designing residuals for isolation have also been discussed. The most commonly used residual generation techniques have been introduced by presenting related problems and discussing the applicability of model-based FDI methods.

It is worth noting that the success of fault diagnosis depends on the quality of the residuals. Successful diagnosis requires residual signals which should be robust with respect to modelling uncertainty. The robust FDI problem has been also discussed in this chapter and the implementation of a robust residual generator will be shown in the following chapters of the book.

Other FDI methods such as fuzzy logic, qualitative modelling and NN have been briefly discussed and the concept of integrated knowledge–based fault diagnosis, utilising both analytical and heuristic information has been presented.

CHAPTER 3
SYSTEM IDENTIFICATION FOR FAULT DIAGNOSIS

3.1 Introduction

The problem of identifying an unknown system given samples of its behaviour is well–known [Söderström and Stoica, 1987, Kalman, 1990, Ljung, 1999] to be ill-posed in the sense of Hadamard [Hadamard, 1964], as its solution is neither unique nor depends continuously on the given data.

When *a priori* knowledge on the characteristics of the unknown system is available, the identification procedure can be enhanced. This knowledge may act as a set of constraints shaping the space of possible models so that identification problem in this new space became more tractable. As an example, the regularity of the unknown system can be translated into smoothness constraints of some kind, transforming the identification problem into a minimisation problem [Tikhonov and Arsenin, 1977, Morozov, 1984].

This point of view can be successfully applied to estimate algebraic and dynamic affine systems from noisy samples, by assuming certain good properties of the noise and of the sampling process [Kalman, 1982b, Beghelli and Soverini, 1992].

The identification method described in this chapter starts from the results based on an algebraic case with the purpose of showing the possibility of extending the Frisch scheme [Frisch, 1934] to dynamic systems determining the whole family of models compatible with noisy sequences.

As often happens to new disciplines, Systems Theory borrowed some tools and viewpoints from existing and well–established fields. Thus the identification of static and dynamical systems, *i.e.* the determination of models from noisy data, has relied heavily on techniques developed by statisticians (*e.g.* R. Frisch) who have traditionally considered it mandatory to associate a unique model to every available set of data, whether or not contaminated by noise.

Kalman [Kalman, 1982b, Kalman, 1982a, Kalman, 1984] reconsidered this problem pointing out how the association of a single model to uncertain data if often based on the introduction of additional information, unrelated to the data, *i.e.* of prejudices. While the introduction of such prejudices can be convenient in some practical cases it is, of course, very important to evaluate the family of solutions that can be found without introducing prejudices or, at least, by introducing only mild ones.

Such an analysis has been performed by Kalman in the previously mentioned works with reference to the Frisch scheme and to noisy data generated by linear algebraic systems. In this context Kalman describes the class of *Errors–In–Variables* (*EIV*) models (see Section 3.2) compatible with the given noisy sequences; his solution is in accordance with the Uncertainty Principle stating that no uncertain data can lead, in general, to a unique model.

The research described in this chapter started from the algebraic results with the purpose of investigating the possibility of extending the Frisch scheme to dynamical systems determining the whole family of models compatible with noisy sequences. The results obtained differ from the expectations of the authors in that, as it is proved in the following sections, a single model is, in general, compatible with the data. This result is not in contrast with Kalman's; in fact, in the dynamic case, the additional information necessary to obtain a single model is carried by the correlations established among the samples by the dynamic nature of the generating process [Fantuzzi *et al.*, 2002].

A frequency approach for EIV models [Kalman, 1982b, Kalman, 1990] and its application to the dynamic Frisch scheme identification in still in development [Beghelli *et al.*, 1994a, Beghelli *et al.*, 1994b, Beghelli *et al.*, 1997].

This chapter also addresses the problem of the identification of both linear and non–linear dynamic systems. In the case of non–linear dynamic systems the identification will be performed by exploiting parametric non–linear models, such as affine, piecewise affine and fuzzy models [Fantuzzi *et al.*, 2002].

Model estimation from noisy data, *i.e.* model parameters, order and structure, is then achieved by using an extension of the so–called Frisch scheme procedure [Fantuzzi *et al.*, 2002].

3.2 Models for Linear Systems

In the following it is assumed that the monitored system can be described when working in some steady state operating condition by a linear, discrete-time, time-invariant, dynamic model of the type

$$\begin{cases} \boldsymbol{x}(t+1) &= \boldsymbol{A}\boldsymbol{x}(t) + \boldsymbol{B}\boldsymbol{u}^*(t) \\ \boldsymbol{y}^*(t) &= \boldsymbol{C}\boldsymbol{x}(t) \qquad\qquad t = 1, 2, \cdots \end{cases} \tag{3.1}$$

where $\boldsymbol{x}(t) \in \Re^n$ is the state vector, $\boldsymbol{y}^*(t) \in \Re^m$ the process output vector and $\boldsymbol{u}^* \in \Re^r$ the control input vector.

$\boldsymbol{A}$, $\boldsymbol{B}$, $\boldsymbol{C}$ are constant matrices of appropriate dimensions obtained by means of modelling techniques or identification procedures.

The input and output variables are measured through sensors for identification purposes. Generally, sensor measurements are affected by additive noise, that can be modelled as:

$$\begin{cases} \boldsymbol{u}(t) &= \boldsymbol{u}^*(t) + \tilde{\boldsymbol{u}}(t), \\ \boldsymbol{y}(t) &= \boldsymbol{y}^*(t) + \tilde{\boldsymbol{y}}(t). \end{cases} \tag{3.2}$$

variables $\tilde{\boldsymbol{u}}(t)$ and $\tilde{\boldsymbol{y}}(t)$ represent noises which affect sensor output, that are generally described as white, zero-mean, uncorrelated Gaussian noises.

It is assumed that $\boldsymbol{u}(t)$ and $\boldsymbol{y}(t)$ are the only available measurements from the real process. It is worthwhile noting that representations of types as Equations 3.1 and 3.2 are known as Errors–In–Variables (EIV) models.

Two conditions were considered regarding the identification of parametric models characterising system behaviour, namely, high and low signal to noise ratios.

In the case of high signal to noise ratios ($\tilde{\boldsymbol{u}}(t) \cong \mathbf{0}$ and $\tilde{\boldsymbol{y}}(t) \cong \mathbf{0}$), equation error identification can be exploited and, in particular, different equation error models can be extracted from the data. A specific discrete-time, time-invariant, linear dynamic model, e.g. ARX or ARMAX (Auto Regressive eXogenous or Auto Regressive Moving Average eXogenous) [Söderström and Stoica, 1987, Ljung, 1999] can be selected only inside an assumed family of models.

On the other hand, if the signal to noise ratios on the input and output of the process are low, the Frisch scheme [Frisch, 1934] can be applied to perform dynamic system identification. Such a scheme facilitates the determination of the linear discrete system which has generated the sequences as well as the variances of the noises $\tilde{\boldsymbol{u}}(t)$ and $\tilde{\boldsymbol{y}}(t)$ affecting the data [Beghelli *et al.*, 1990]. In the Frisch scheme these signals are assumed zero–mean white noises, mutually uncorrelated and uncorrelated with every component of $\boldsymbol{u}^*(t)$ and $\boldsymbol{y}^*(t)$.

The input to output discrete–time model behaviour can be mathematically described by a set of ARX Multi–Input Single–Output (MISO) models of the type

$$y_i^*(t) = \sum_{j=1}^{n} \alpha_{i,j} y_i^*(t-j) + \sum_{j=1}^{r} \sum_{k=1}^{n} \beta_{i,j,k} u_j^*(t-k) + \varepsilon_i(t) \ , \quad i = 1, \ldots, m \tag{3.3}$$

whose number is equal to the number m of the output variables. The order n and the parameters $\alpha_{i,j}$ and $\beta_{i,j,k}$, with $i = 1, \cdots, m$, of the model have to be determined by the identification approach. The term $\varepsilon_i(t)$ takes into account the modelling error, which is due to process noise, parameter variations, etc.

In either the case, the subsequent step consist of transforming the input–output discrete–time time–invariant linear models of the form of Eq. 3.3 into state space representations.

The state space systems obtained by the equation errors models are useful to design dynamic observers [Luenberger, 1971], whilst the ones coming from the Frisch scheme can be used in order to build Kalman filters [Jazwinski, 1970].

As presented by Söderström and Stoica [Söderström and Stoica, 1987], the input–output equation error model of Eq. 3.3 has a state space realisation as follows:

$$\begin{cases} \boldsymbol{x}_i(t+1) = \boldsymbol{A}_i\boldsymbol{x}_i(t) + \boldsymbol{B}_i\boldsymbol{u}^*(t) + \boldsymbol{B}_{\omega_i}\varepsilon_i & i = 1, \cdots, m \\ \boldsymbol{y}_i^*(t) = \boldsymbol{C}_i\boldsymbol{x}_i(t) + \boldsymbol{D}_{\omega_i}\varepsilon_i, & t = 1, 2, \cdots \end{cases} \tag{3.4}$$

where the matrices $\boldsymbol{A}_i$ $(n \times n)$, $\boldsymbol{B}_i$ $(n \times r)$, $\boldsymbol{B}_{\omega_i}$ $(n \times 1)$, $\boldsymbol{C}_i$ $(1 \times n)$ and $\boldsymbol{D}_{\omega_i}$ depend on system order and on $\alpha_{i,j}$ and $\beta_{i,j,k}$ parameters.

The matrices $(\boldsymbol{A}_i, \boldsymbol{B}_i, \boldsymbol{C}_i, \boldsymbol{B}_{\omega_i}, \boldsymbol{D}_{\omega_i})$ of a state space representation in canonical form of the n–th order system given by Eq. 3.4 can be defined as follows

$$\begin{aligned} \boldsymbol{A}_i &= \begin{bmatrix} 0 & 1 & 0 & \cdots & 0 \\ 0 & 0 & 1 & \cdots & 0 \\ \vdots & \vdots & \vdots & \ddots & \vdots \\ \alpha_{i,1} & \alpha_{i,2} & \alpha_{i,3} & \cdots & \alpha_{i,n} \end{bmatrix}, \\ \boldsymbol{C}_i &= \begin{bmatrix} 1 & 0 & \cdots & 0 \end{bmatrix}, \end{aligned} \tag{3.5}$$

and

$$\begin{bmatrix} \boldsymbol{B}_i & \boldsymbol{B}_{\omega_i} \\ \boldsymbol{0} & \boldsymbol{D}_{\omega_i} \end{bmatrix} = \boldsymbol{S}_i^{-1} \begin{bmatrix} \beta_{i,1,1} & \cdots & \beta_{i,r,1} & 0 \\ \vdots & \vdots & \vdots & \vdots \\ \beta_{i,1,n} & \cdots & \beta_{i,r,n} & 0 \\ 0 & \cdots & 0 & 1 \end{bmatrix}. \tag{3.6}$$

where the $\boldsymbol{S}_i$ $((n+1) \times (n+1))$ matrix is defined as:

$$\boldsymbol{S}_i = \begin{bmatrix} -\alpha_{i,1} & -\alpha_{i,2} & \cdots & -\alpha_{i,n} & 1 \\ -\alpha_{i,2} & -\alpha_{i,3} & \cdots & 1 & 0 \\ \vdots & \vdots & \vdots & \ddots & \vdots \\ -\alpha_{i,n} & 1 & 0 & 0 & 0 \\ 1 & 0 & 0 & 0 & 0 \end{bmatrix} \tag{3.7}$$

Section 3.3 discusses procedures for obtaining the needed parameters, based on the model structures outlined above.

3.3 Parameter Estimation Methods

Many classes of models can be considered from an identification standpoint; the most relevant description considered in this work are the discrete–time, parametric, deterministic and stochastic, input–output, MISO dynamic models.

Real systems are always affected by disturbances (noise entering into the system and/or affecting the measures, unknown inputs, etc.). These disturbances or their global effect can be described by means of noises acting on the input, state and output of the model which can be called, in this case, stochastic. Often, the global effect of disturbances can be modelled as the output of a filter driven by white noise, and is added to the output of the deterministic part of the model, which is thus decomposed into a deterministic and a stochastic part. Depending on the applications, it can be sufficient to identify the deterministic part of the model (high signal–to–noise ratio case, Section 3.3.1) or it can be necessary to identify both parts (low signal–to–noise ratio, Section 3.3.2).

3.3.1 System Identification in Noiseless Environment

In this section we consider a SISO (Single-Input Single Output, $m = 1$ and $r = 1$) ARX model with order n (see Equation 3.3), and the input-output sequences $\{\boldsymbol{u}(t), \boldsymbol{y}(t)\}$ observed in the time interval $[1, L]$. If the model 3.3 is used to compute predicted output values $\boldsymbol{y}^*(t)$ in the $N = L - n$ times, for a given set of parameters

$$\theta = \begin{bmatrix} \alpha_n & \cdots & \alpha_1 & \beta_n & \cdots & \beta_1 \end{bmatrix} \tag{3.8}$$

the mean–square prediction error $J(\theta)$ is given by

$$J(\theta) = \frac{1}{N}\sum_{t=n+1}^{L} \left(y^*(t) - y(t)\right)^2 . \tag{3.9}$$

By introducing now the following Hankel matrices $\boldsymbol{H}_u$ and $\boldsymbol{H}_y$

$$\begin{aligned} \boldsymbol{H}_n(u) &= \begin{bmatrix} u(1) & \cdots & u(n) \\ \vdots & \ddots & \vdots \\ u(L-n) & \cdots & u(L-1) \end{bmatrix} \text{ and} \\ \boldsymbol{H}_n(y) &= \begin{bmatrix} y(1) & \cdots & y(n) \\ \vdots & \ddots & \vdots \\ y(L-n) & \cdots & y(L-1) \end{bmatrix}, \end{aligned} \tag{3.10}$$

it follows that

$$\begin{bmatrix} y^*(n+1) \\ \vdots \\ y^*(L) \end{bmatrix} = \left[\boldsymbol{H}_n(y)\boldsymbol{H}_n(u)\right]\theta = \boldsymbol{H}_n\theta. \tag{3.11}$$

It can be proved that the parameter vector minimising the cost function 3.9 is given by

$$\hat{\theta} = \boldsymbol{H}_n^{+} \begin{bmatrix} y(n+1) \\ \vdots \\ y(L) \end{bmatrix} = \boldsymbol{H}_n^{+} \boldsymbol{y}_n^{o} \tag{3.12}$$

where $\boldsymbol{H}_n^{+}$ denotes the pseudo–inverse of the $\boldsymbol{H}_n$ matrix. The algorithm gives an estimate $\hat{\theta}$ of θ, which converges asymptotically to the real parameter of the process that has generated the data.

To estimate the order n of the ARX process, an integer $k > 0$ and the $(N \times 2k+1)$ matrix of input-output samples given by

$$\boldsymbol{H}_k^{*} = [\boldsymbol{H}_k(y)\boldsymbol{H}_k(u)y_k^{o}] \tag{3.13}$$

are considered.

If $\varepsilon_i(t) = 0$ in Equation 3.3 the following properties hold

$$\begin{array}{rcll} \text{rank}\,(\boldsymbol{H}_k^{*}) & = & 2k+1 & \text{for } 2k+1 < n \\ \text{rank}\,(\boldsymbol{H}_k^{*}) & = & 2k & \text{for } 2k \geq n \end{array} \tag{3.14}$$

It could be possible to consider the increasing sequence of matrices

$$\boldsymbol{S}_1 \ \boldsymbol{S}_2 \ \cdots \ \boldsymbol{S}_n \ \cdots \tag{3.15}$$

where $\boldsymbol{S}_k = \boldsymbol{H}_k^{*T} \boldsymbol{H}_k^{*}$ and to evaluate their singularity.

The first singular matrix $\boldsymbol{S}_k$ would define the correct order for the model $(k = n)$. Unfortunately, the presence of $\varepsilon_i(t) \neq 0$ in Equation 3.3 leads to the non–singularity of every matrix in Equation 3.15.

It can be proved that if N is large enough, an estimate of the standard deviation σ_ε of the process $\varepsilon(t)$ in Equation 3.3, is given by [Ljung, 1999]

$$\sigma_\varepsilon = \sqrt{\frac{\det(\boldsymbol{S}_n)}{N\det(\boldsymbol{H}_n^T \boldsymbol{H}_n)}}. \tag{3.16}$$

In the following, the quantity σ_{ε_k} is defined,

$$\sigma_{\varepsilon_k} = \sqrt{\frac{\det(\boldsymbol{S}_k)}{N\det(\boldsymbol{H}_k^T \boldsymbol{H}_k)}}. \tag{3.17}$$

it can be shown that $\sigma_{\varepsilon_h} > \sigma_{\varepsilon_k}$ for $h < k$, $\sigma_{\varepsilon_k} > \sigma_\varepsilon$ for $k < n$ and $\sigma_{\varepsilon_k} \cong \sigma_\varepsilon$ for $k \geq n$. In other words, if N is large enough, a sequence of decreasing values of σ_{ε_k} followed by a stabilisation once the correct order is reached, can be noted. The criterion can be used to evaluate a suitable order or, at least, an interval of admissible orders for the model before computing its parameters.

It can be shown that $\sigma_{\varepsilon_k}^2 = J_k(\theta)$ for an ARX model with order k and parameters θ given by Equation 3.12.

If the value in Equation 3.17 is expressed as a percentage of the standard deviation of the measured output, the well-established Predicted Per Cent Reconstruction Error criterion (PPCRE) is obtained [Guidorzi *et al.*, 1982].

The PPCRE(k) gives the prediction error of an ARX model of order k without requiring any computation of its parameters and predictions.

The application of the PPCRE criterion consists of computing an increasing sequence of PPCRE(k) (or $J_k(\theta)$) and in selecting the minimal order that, once increased, does not lead to a significantly better performance.

An example of the PPCRE(k) and $J_k(\theta)$ increasing sequences with ARX model order k for a simulated second–order model ($n = 2$) is reported in Figure 3.1(a) and Figure 3.1(b).

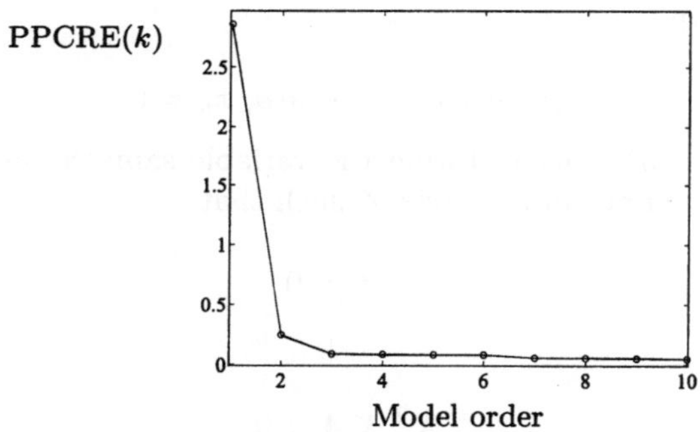

(a) PPCRE criterion.

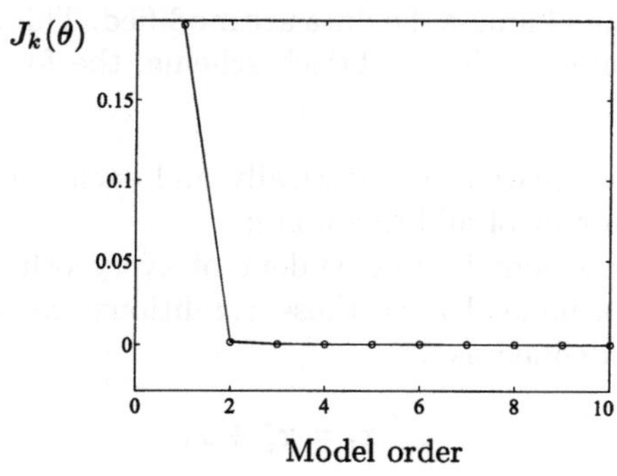

(b) Mean square error criterion.

Fig. 3.1. (a) Predicted reconstruction error and (b) mean square errors for different ARX model orders k.

It is clear the change in the slope of both the curves represented in Figure 3.1(a) and Figure 3.1(b) corresponding to the order $n = 2$.

Relation 3.17 can also be used in the application of the well-known FPE, AIC and MDL order estimation criteria [Söderström and Stoica, 1987].

3.3.2 System Identification in Noisy Environment

In this section the identification of linear models from noisy data is presented. In the interests of clarity, we consider the algebraic and dynamic cases separately.

The Frisch Scheme in the Algebraic Case. We consider the finite sequence of n variables $x_1, x_2, \cdots, x_n$ observed at N different times with $N > n$. If linear relations exist among these variables, they are described by models of the type

$$a_1x_1 + a_2x_2 + \cdots + a_nx_n = 0. \tag{3.18}$$

If $\boldsymbol{X}$ is the $(N \times n)$ matrix storing the variable samples, model 3.18 is described by the columns of a matrix $\boldsymbol{A}$ such that

$$\boldsymbol{XA} = 0 \tag{3.19}$$

or, equivalently

$$\boldsymbol{X}^T\boldsymbol{XA} = 0 \tag{3.20}$$

where $\Sigma = \boldsymbol{X}^T\boldsymbol{X}$ is the sample covariance matrix, under a zero-mean assumption for all variables.

When the data are corrupted by noise, then the rank$[\Sigma] = n$, so that no relation can be obtained unless the data are modified. This explains the reason for using the EIV models. In the Frisch scheme, the following assumptions are therefore added:

1. All variables are treated symmetrically and each variable is affected by an unknown amount of additive noise;
2. Each noise component is independent of every other noise component and of every variable. Under these conditions, each variable x_i, with $i = 1, \cdots, n$, is defined as

$$x_i = x_i^* + \tilde{x}_i \tag{3.21}$$

where the unknown terms x_i^* are the true value of the i-th variable whilst $\tilde{x}_i$, the additive noises on this variable.

The problem of determining the true data from the available noisy sequences can thus be formulated as follows:

Problem 3.3.1. Given an $(n \times n)$ symmetric positive-definite covariance matrix Σ, find all diagonal matrices $\tilde{\boldsymbol{\Sigma}}$ with non–negative elements elements such that $\boldsymbol{\Sigma}^* = \boldsymbol{\Sigma} - \tilde{\boldsymbol{\Sigma}} \geq 0$.

In this context, the solution of the problem is not unique.

The rank of $\boldsymbol{\Sigma}^*$ may change by varying $\tilde{\boldsymbol{\Sigma}}$, which models the noise and, consequently, the same set of data may be linked by different numbers of linear relations.

Even if it is not the case, the problem has infinite solutions; the rank of $\boldsymbol{\Sigma}^*$, for instance, is always equal to $(n-1)$ if and only if $\boldsymbol{\Sigma}^{-1}$ can be reduced to a matrix with strictly positive entries by the transformation $\boldsymbol{L}\boldsymbol{\Sigma}^{-1}\boldsymbol{L}$ with $\boldsymbol{L} = \text{diag}[\pm 1]$.

The solution set is the convex simplex the n vertices of which are the least-squares solutions which can be found assuming that one variable is noisy and all others are noise-free [Beghelli *et al.*, 1990].

Before solving Problem 3.3.1, the following theorem can be considered [Beghelli *et al.*, 1990].

Theorem 3.3.1. *Given the $(n \times n)$ symmetric positive-definite covariance matrix $\boldsymbol{\Sigma}$, the maximal variance of the additive noise on the i-th variable, when all others are noise-free, is computed by*

$$\tilde{\sigma}_i = \frac{det[\boldsymbol{\Sigma}]}{det[\boldsymbol{\Sigma}_i]} \tag{3.22}$$

where $\boldsymbol{\Sigma}_i$ is obtained from $\boldsymbol{\Sigma}$ by deleting its i-th row and column.

Every allowable noise covariance matrix $\tilde{\boldsymbol{\Sigma}}$ (*i.e.* such that $\boldsymbol{\Sigma}^* = \boldsymbol{\Sigma} - \tilde{\boldsymbol{\Sigma}} \geq 0$) defines a point $(\sigma_1, \cdots, \sigma_n)$ belonging to the first orthant of the noise space $\Re^n$, which is mapped into one and only one point $(a_1, \cdots, a_n)$ of the solution space $\Re^n$. Moreover, the following result can be proved [Beghelli *et al.*, 1990].

Theorem 3.3.2. *The solution set defined by all points $(\sigma_1, \cdots, \sigma_n)$ defined by the matrix set $\tilde{\boldsymbol{\Sigma}}$ is a convex hypersurface belonging to the first orthant of the noise space the section of which, with a plane parallel to a coordinate one, is a hyperbola segment.*

Note that if noise values, corresponding to a rank of $\boldsymbol{\Sigma}^*$ lower than $n-1$, exist, they belong to the hypersurface defined by Theorem 3.3.2.

In these conditions, in the parameter space, the solution set might be a collection of convex polyhedral sets lying in the orthants.

The hypersurface defined by Theorem 3.3.2 partitions the first orthant of the noise space into two regions. The points over the hypersurface correspond to non–definite matrices $\boldsymbol{\Sigma}^*$, those under the hypersurface to positive-definite matrices.

The Frisch Scheme in the Dynamic Case. Consider a finite sequence of the variables $u_1(\cdot), \cdots, u_r(\cdot), y(\cdot)$ observed with a constant sampling interval.

If dynamic linear relations exist among these variables, they can be described by models of the type

$$y^*(t+n) = \sum_{i=0}^{n-1} \alpha_i y^*(t+i) + \sum_{i=0}^{n-1}\sum_{j=1}^{r} \beta_{ij} u_j^*(t+i) \tag{3.23}$$

which represent linear MISO discrete-time systems whose order is n and whose parameters are α_i and β_{ij}.

We consider first the following problem:

Problem 3.3.2. Given a noiseless input–output sequence $u_1^*(\cdot), \cdots, u_r^*(\cdot)$, $y^*(\cdot)$ generated by a system in the form given by Eq. 3.23, determine the order n and the parameters α_i, β_{ij} of the system.

The following vectors and matrices are thus defined as:

$$\begin{aligned}
u_j^{*N}(t+k) &= \left[\begin{array}{ccc} u_j^*(t+k) & \ldots & u_j^*(t+k+N-1) \end{array}\right]^T & (3.24)\\
y^{*N}(t+k) &= \left[\begin{array}{ccc} y^*(t+k) & \ldots & y^*(t+k+N-1) \end{array}\right]^T & (3.25)\\
X_k(u_j^*) &= \left[\begin{array}{ccc} u_j^{*N}(t) & \ldots & u_j^{*N}(t+k-2) \end{array}\right] & (3.26)\\
X_k(y^*) &= \left[\begin{array}{ccc} y^{*N}(t) & \ldots & y^{*N}(t+k-1) \end{array}\right] & (3.27)\\
\Sigma_k^*(u_j^* u_j^*) &= X_k^T(u_j^*) X_k(u_j^*) & (3.28)\\
\Sigma_k^*(y^* y^*) &= X_k^T(y^*) X_k(y^*) & (3.29)\\
\Sigma_k^*(y^* u_j^*) &= X_k^T(y^*) X_k(u_j^*) = \Sigma_k^{*T}(u_j^* y^*) & (3.30)
\end{aligned}$$

where N is assumed large enough to solve the problem considered.

Let us now partition the matrix Σ_k^* as follows:

$$\Sigma_k^* = \begin{bmatrix} \Sigma_k^*(y^*y^*) & \Sigma_k^*(y^*u_1^*) & \ldots & \Sigma_k^*(y^*u_r^*) \\ \Sigma_k^*(u_1^*y^*) & \Sigma_k^*(u_1^*u_1^*) & \ldots & \Sigma_k^*(u_r^*y^*) \\ \vdots & \vdots & \ddots & \vdots \\ \Sigma_k^*(u_r^*y^*) & \Sigma_k^*(u_r^*u_1^*) & \ldots & \Sigma_k^*(u_r^*u_r^*). \end{bmatrix}. \tag{3.31}$$

To solve the realization problem it is possible to consider the sequence of increasing–dimension matrices

$$\Sigma_2^*,\ \Sigma_3^*,\ \ldots\ \Sigma_k^*,\ \ldots \tag{3.32}$$

testing their singularity. As soon as a singular matrix Σ_k^* is found then $n = k-1$ and the parameters $\alpha_0, \ldots, \alpha_{n-1}, \beta_{0j}, \ldots, \beta_{(n-1)j}$ $(j = 1, \ldots, r)$ describe the dependence relationship of the $(n+1)$-th vector of Σ_{n+1}^* on the remaining ones.

In Problem 3.3.2 it has been assumed that N is large enough to avoid unwanted linear dependence relationships due to limitations in the dimension of the involved vector spaces; this means $N \geq (r+1)n+1$.

If a lower number of samples is available then only a partial realization problem can be solved.

In the noisy case the following identification problem can be proposed.

Problem 3.3.3. Given a noisy input–output sequence $u_1(\cdot)$, …, $u_r(\cdot)$, $y(\cdot)$ uniquely determine, if possible, the order n and the parameters α_i, β_{ij} of a model as shown in Equation 3.23 of the system which has generated the noiseless sequences $u_1^*(\cdot)$, …, $u_r^*(\cdot)$, $y^*(\cdot)$.

Note that in the presence of noise, the procedure described for the solution of Problem 3.3.2 would obviously be useless since the matrices Σ_k^* would always be non–singular.

In the Frisch scheme it is normally assumed that:

$$\begin{cases} u_j(t) &= u_j^*(t) + \tilde{u}_j(t), \qquad j = 1, \ldots, r \\ y(t) &= y^*(t) + \tilde{y}(t) \end{cases} \tag{3.33}$$

where every noise term $\tilde{u}_j(t)$, $\tilde{y}(t)$ is independent of every other term and only $u_j(t)$ and $y(t)$ are known. Without loss of generality, all the variables may be assumed to be zero-mean.

Consequently the generic positive-definite matrix Σ_k associated with the input-output noise-corrupted sequences may always be expressed as the sum of two terms $\Sigma_k = \Sigma_k^* + \tilde{\Sigma}_k$ where

$$\tilde{\Sigma}_k = \mathrm{diag}[\tilde{\sigma}_y I_k, \tilde{\sigma}_{u_1} I_{k-1}, \ldots, \tilde{\sigma}_{u_r} I_{k-1}] \geq \mathbf{0} \tag{3.34}$$

since no correlation has been assumed among the noise samples at different times. This condition is verified for additive white noise with variance $\tilde{\sigma}_y$ and $\tilde{\sigma}_{u_j}$ on the input-output sequences.

Problem 3.3.4. Given a sequence of increasing–dimension $\Big(((r+1)k-r) \times ((r+1)k-r)\Big)$ symmetric positive-definite covariance matrices with:

$$\Sigma_2, \ \Sigma_3, \ \ldots \ \Sigma_k, \ \ldots \tag{3.35}$$

find, for each k, all diagonal non–negative definite matrices

$$\tilde{\Sigma}_k = \mathrm{diag}[\tilde{\sigma}_y I_k, \tilde{\sigma}_{u_1} I_{k-1}, \ldots, \tilde{\sigma}_{u_r} I_{k-1}] \tag{3.36}$$

such that

$$\Sigma_k^* = \Sigma_k - \mathrm{diag}[\tilde{\sigma}_y I_k, \tilde{\sigma}_{u_1} I_{k-1}, \ldots, \tilde{\sigma}_{u_r} I_{k-1}] \geq 0\,. \tag{3.37}$$

Note that, unlike the algebraic case, for each k the noise space is always $\Re_+^{(r+1)}$, while the parameter space is defined on $\Re^{(r+1)k-r}$.

It can be seen that for each k the solution set of relation 3.37 describes, in the first orthant of the $(\tilde{\sigma}_y, \tilde{\sigma}_{u_1}, \ldots, \tilde{\sigma}_{u_r})$ hyperplane, a hypersurface whose concavity faces the origin [Beghelli *et al.*, 1990].

In the noise space, the $(r+1)$ solutions $(\tilde{\sigma}_y, 0, \tilde{\sigma}_{u_2}, \ldots, \tilde{\sigma}_{u_r})$, $(\tilde{\sigma}_y, \tilde{\sigma}_{u_1}, 0, \ldots, \tilde{\sigma}_{u_r})$, $\cdots$, $(\tilde{\sigma}_y, \tilde{\sigma}_{u_1}, \ldots, \tilde{\sigma}_{u_{r-1}}, 0)$ correspond to the limit case of noise affecting only the output or the input sequences. This case can be considered as the natural extension to the dynamic case of the computation of least-square solutions.

Previous results hold for every value of k. Since determination of the system order requires the increasing values of k to be tested, it is relevant to analyse the behaviour of the associated curves when k varies.

This corresponds to a comparison of the admissible solution sets for different model orders. In this context the following result can be proved [Beghelli *et al.*, 1990].

Theorem 3.3.3. *The solution sets of condition 3.37 for different values of k are non–crossing curves.*

It is also important to observe that, since it is assumes that a system of the form given by Eq. 3.23 has generated the noiseless data, for $k > n$ all the hyper–surfaces of the type described by Eq. 3.37 have necessarily at least one common point, *i.e.* point $(\tilde{\sigma}_{y^*}, \tilde{\sigma}_{u_1^*}, \ldots, \tilde{\sigma}_{u_r^*})$ corresponding to the true variances $\tilde{\sigma}_{y^*}$ and $\tilde{\sigma}_{u_j^*}$ of the noise affecting the output and the inputs of the system.

The search for a solution for the identification problem can thus start by determining this point in the noise space. The following considerations can now be stated.

With reference to the diagonal non–negative definite matrices

$$\tilde{\Sigma}_k = \mathrm{diag}[\tilde{\sigma}_{y^*} I_k, \tilde{\sigma}_{u_1^*} I_{k-1}, \ldots, \tilde{\sigma}_{u_r^*} I_{k-1}] \tag{3.38}$$

the following properties hold:

- If $k \leq n$, the matrices Σ_k^* are positive-definite.

- If $k > n$, the dimension of the null space of Σ_k^* and, consequently, the multiplicity of its least eigenvalue, is equal to $(k-n)$.

- For $k = (n+1)$, matrix Σ_k^* is characterised by a linear dependence relation among its $(r+1)k-r$ vectors and the coefficients which link the k-th vector of Σ_k^* to the remaining ones are the parameters α_i, β_{ij}, with $i = 0, \ldots, n-1$ and $j = 1, \ldots, r$, of the system given by Eq. 3.23 which has generated the noiseless sequences.

– For $k > (n+1)$, all linear dependence relations among the vectors of the matrix Σ_k^* are characterised by the same $(r+1)n$ coefficients α_i, β_{ij}.

As an example, Figure 3.2 shows the above properties for a second order $(n = 2)$ SISO dynamic system.

The point marked by a circle corresponds to the input-output noise variances $\tilde{\sigma}_{y^*}$ and $\tilde{\sigma}_{u^*}$ affecting the measurements.

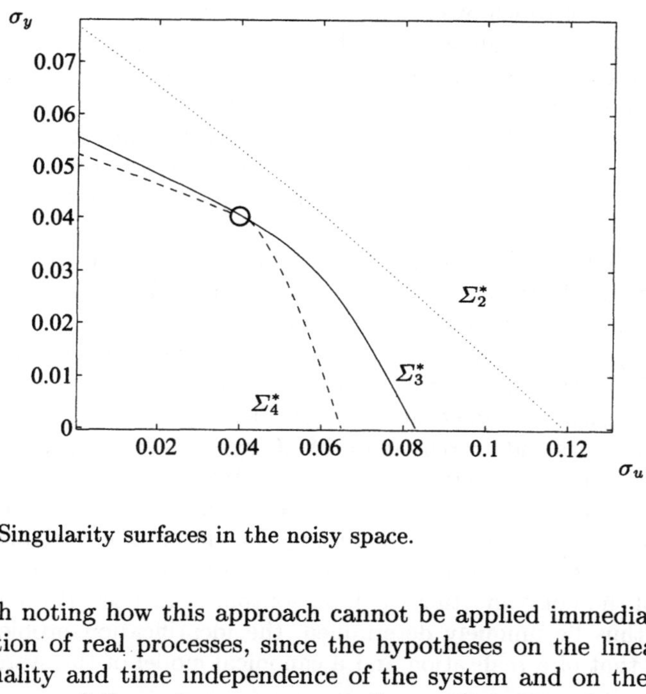

Fig. 3.2. Singularity surfaces in the noisy space.

It is worth noting how this approach cannot be applied immediately to the identification of real processes, since the hypotheses on the linearity, finite dimensionality and time independence of the system and on the additivity and whiteness of the noise are not usually verified. Therefore, the hypersurfaces defined by Equation 3.37 have no common point for $k > n$.

The definition of a suitable criterion of model selection in such cases was suggested in Beghelli *et al.* [Beghelli *et al.*, 1994a].

3.3.3 The Frisch Scheme in the MIMO Case

A multivariable system can be represented by means of canonical models of the form:

$$y_i^*(t+\nu_i) = \sum_{j=1}^{m} \sum_{k=0}^{\nu_{ij}-1} \alpha_{ijk} y^*(t+k) + \sum_{j=1}^{r} \sum_{k=0}^{\nu_i - 1} \beta_{ijk} u_j^*(t+k) \tag{3.39}$$

where $i = 1, \cdots, m$. r and m are the number of inputs and outputs of the system, respectively [Guidorzi, 1981]. The indices $\nu_{ij} - 1$ satisfy the following relations:

$$\begin{cases} \nu_{ij} = \nu_i & \text{for } i = j \\ \nu_{ij} = \min(\nu_i + 1, \nu_j) & \text{for } i > j \\ \nu_{ij} = \min(\nu_i, \nu_j) & \text{for } i < j. \end{cases} \tag{3.40}$$

The model of Eq. 3.39 decomposes the system into m interconnected subsystems the orders of which are given by the integers ν_i.

Such integers completely define the system structure and are coincident with the Kronecker observability invariants of any realization of the system.

Given the sequences $y_i(t)$ $(i = 1, \cdots, m)$ and $u_j(t)$ $(j = 1, \cdots, r)$ generated by the system of the type given by Eq. 3.39, the identification problem consists of uniquely determining both the ***structure***, *i.e.* the set of integers $\nu_1, \cdots, \nu_m$, as well as the characteristic ***parameters*** α_{ijk} and β_{ijk} of the model 3.39.

The solution of the identification problem is described by Guidorzi in [Guidorzi, 1981] with reference to canonical models, but can easily be generalised to multistructural (overlapping) models.

Advantages associated with the use of these identified models with reference to FDI concern the minimal parametrisation [Delmaire *et al.*, 1999], reduced storage, computing time and high efficiency of the related algorithms.

The techniques and properties for MISO systems identification presented in Section 3.3.2 can be generalised for the MIMO case. Because of these properties, it is possible to conclude that if, starting from a certain structure, the hypersurfaces associated to the increasing dimension of covariance matrices have only a common point in the noise space, then this point represents the variances of the noise affecting the input-output sequences. If the noise variances can thus be uniquely determined, the identification problem can be reduced to that of a realisation and a canonical model of the system can be obtained.

From a computational point of view, it can be noted that the search for the noise variances is made in an $(m + r)$-dimensional space and may, therefore, be time expensive.

The results given in Section 3.3.2 and extended to the MIMO case do not exclude the possibility that the previously considered hypersurfaces may be coincident and consequently that non–unique solution exists.

In conclusion, it can be stated that no conceptual differences exist between the application of the Frisch scheme to the identification of SISO and MIMO dynamic systems.

3.4 Models for Non-linear Dynamic Systems

In this section the problem of finding a suitable non–linear model for real processes is presented. With the aim to provide a general treatment of the problem, we focus our attention on the non–linear "black–box" parametric models.

In fact, input–output process measurements can be successfully used to infer an analytical description of the system in the framework of a parametric structure which possess approximation properties with respect to the complex, non–linear, unknown analytical functions that are amenable as candidate to describe the real behaviour of the observed process [Juditsky *et al.*, 1995].

So, the choice of the parametric structure becomes an important and difficult step towards the system identification, expecially when the behaviour of the target process is non-linear, as it is common by far in real world applications [Billings and Voon, 1983b].

The approach chosen refers to non–linear processes that operate at different regimes, as this is the usual assumption in several engineering fields, in which distinct models are associated to each admissible operating condition.

A switching function governs the transition among different models or interpolations of models. Such mathematical descriptions are referred to in current literature as *piecewise models* or *hybrid models*, as suggested by several authors [Johansen and Foss, 1993, Billings and Voon, 1986, Chen and Billings, 1989, Chen *et al.*, 1990b, Friedmann, 1991, Skeppstedt *et al.*, 1992, Hathaway and Bezdek, 1983].

In particular, great attention has been paid in the current literature to the piecewise linear systems, formed by a collection of *linear* [Bemporad and Morari, 1999] or *affine* [Pettit and Wellstead, 1995, Sontag, 1981, Banks and Kathur, 1989] *dynamic local models* and a switching function that selects the appropriate model depending on operating point.

The following concerns a structure that merges dynamic, time–invariant and affine submodels describing locally the behaviour of the observed process in its different operating regimes. This type of models has been formerly introduced by Priestly [Priestly, 1988] as state–dependent models and used in a stochastic environment for time series model fitting. In particular, we discuss the interpolation properties of such models with respect to non–linear discrete–time regression functions [Leontaritis and Billings, 1985a, Leontaritis and Billings, 1985b].

3.4.1 Piecewise Affine Model

The main idea underlying the mathematical description of non–linear dynamic systems is based on the interpretation of single input–single output, non–linear, time–invariant regression models

[Leontaritis and Billings, 1985a, Leontaritis and Billings, 1985b] such as:

$$y(t+n) = F\big(y(t+n-1), \cdots, y(t), u(t+n-1), \cdots, u(t)\big), \quad t = 0, 1, \ldots \quad (3.41)$$

where $u(\cdot)$ and $y(\cdot)$ belong to the bounded input $\mathcal{U}$ and output $\mathcal{Y}$ sets, respectively, n is the finite system memory (*i.e.* the model order) and $F(\cdot)$ is a continuous non–linear function defining a hypersurface from a $\mathcal{D}_n$ to $\mathcal{Y}$, $\mathcal{D}_n$ being the Cartesian product $\mathcal{U}^n \times \mathcal{Y}^n$.

The identification of the non–linear system can be translated to the approximation of the mathematical model given by Equation 3.41 using a parametric structure that exhibits arbitrary accuracy interpolation properties.

A piecewise model defined through the composition of simple models having local validity is the natural candidate to perform this task, as it combines function interpolation properties with mathematical tractability.

Hybrid Model Structure. The piecewise SISO model is formed by a collection of parametric submodels of the type:

$$y(t+n) = \sum_{j=0}^{n-1} \alpha_j^{(i)} y(t+j) + \sum_{j=0}^{n-1} \beta_j^{(i)} u(t+j) + b^{(i)}, \quad t = 0, 1, \ldots \quad (3.42)$$

in which the system operating point is described by the input and output samples $y(t+n-1), \cdots, y(t)$ and $u(t+n-1), \cdots, u(t)$, that can be collected with a vector $\boldsymbol{x}_n(t) = [y(t), \cdots, y(t+n-1), u(t), \cdots, u(t+n-1)]^T$. The *switching* function $\chi_i(\boldsymbol{x}_n(t)), i = 1, \ldots, M$ is:

$$\chi_i\big(\boldsymbol{x}_n(t)\big) = \begin{cases} \chi_i\big(\boldsymbol{x}_n(t)\big) = 1 \text{ if } \boldsymbol{x}_n(t) \in \mathcal{A}_n^{(i)} \\ \chi_i\big(\boldsymbol{x}_n(t)\big) = 0 \text{ otherwise} \end{cases} \quad (3.43)$$

where $\{R_n^{(1)}, \ldots, R_n^{(M)}\}$ is a partition of $\mathcal{D}_n$, whose structure will be characterised in the following.

Thus, the output $y(t+n)$ of the non–linear dynamic system described by Eq. 3.41 can be approximated by the *piecewise affine model* $f(\cdot)$ in the form:

$$y(t+n) = f\big(\boldsymbol{x}_n(t)\big) = \sum_{i=1}^{M} \chi_i\big(\boldsymbol{x}_n(t)\big)[\boldsymbol{x}_n(t), 1]^T \boldsymbol{a}^{(i)} \quad (3.44)$$

where the model parameters are collected in the vector $\boldsymbol{a}^{(i)} = [\alpha_0^{(i)}, \ldots, \alpha_{n-1}^{(i)}, \beta_0^{(i)}, \ldots, \beta_{n-1}^{(i)}, b^{(i)}]^T$. It is worthwhile noting that the model is affine in each $R_n^{(i)}$, $\boldsymbol{a}^{(i)}$ being the affine submodel parameters.

Non–linear System Approximation with Affine Models. Thus, it is now very important to understand the approximation capabilities of this piecewise affine model with respect to the target function.

In particular, our interest goes beyond the simple approximation capabilities under the usual Lebesgue norm, but is also focused on the first derivative of the target function, as this avoids any jitteriness and actually assures higher approximation quality.

We shall give a metric structure to the set of functions $F : \mathcal{D}_n \mapsto \mathcal{Y}$ defining first the usual Lebesgue norm of order w as

$$||F||_w = \begin{cases} \left[\int_{\mathcal{D}_n} |F(\boldsymbol{x}_n)|^w d\boldsymbol{x}_n\right]^{\frac{1}{w}} & \text{if } 1 \leq w < \infty \\ \sup_{\boldsymbol{x}_n \in \mathcal{D}_n} |F(\boldsymbol{x}_n)| & \text{if } w = \infty \end{cases}$$

and then a stricter Sobolev norm which also takes into account the first derivatives of the function

$$^1||F||_w = ||F||_w + \sum_{j=1}^{2n} \left\| \frac{\partial F}{\partial x_n^j} \right\|_w \tag{3.45}$$

x_n^j being the j–th component of vector $\boldsymbol{x}_n$.

If $\mathcal{C}^v$ is the set of those functions $f : \mathcal{D}_n \mapsto \mathcal{Y}$ featuring continuous v-th order derivatives, we may state the following result.

Property 3.4.1. The set of input-output relationships f given by Equation 3.44 is dense in $\mathcal{C}^1$ in the sense of $^1||\cdot||_w$ for any $w > 0$. Thus, for any $\epsilon > 0$, any $w > 0$ and any $F \in \mathcal{C}^1$ there is a piecewise affine system f such that

$$^1||F - f||_w < \epsilon$$

The proof of the previous property is based on the result on Ritz's piecewise polynomial function, which leads to the formulation of the following theorem.

Theorem 3.4.1. *(Adapted from [Strang and Fix, 1973, Theorem 3.1]) Let $F : \mathcal{D}_n \mapsto \mathcal{Y}$ be given such that $F \in \mathcal{C}^1$. Assume also that the value of F is given at certain points in $\mathcal{D}_n$ called nodes and that $\mathcal{D}_n$ itself is partitioned into convex regions between these nodes. Moreover, a polynomial function f interpolating the values of F at the nodes is given in each of these regions.*

If the functional form of f is such that any first-order polynomial could be exactly reproduced in each region, then for each $\epsilon > 0$ a length $\hat{d}$ exists such that, in any non–degenerating region γ in which any two points can be connected with a segment of length less than $\hat{d}$, the following relation holds

$$\max_{\boldsymbol{x}_n \in \gamma} |F(\boldsymbol{x}_n) - f(\boldsymbol{x}_n)| \leq \epsilon$$

and

$$\max_{\boldsymbol{x}_n \in \gamma} \left| \frac{\partial F}{\partial x_n^j}(\boldsymbol{x}_n) - \frac{\partial f}{\partial x_n^j}(\boldsymbol{x}_n) \right| \leq \epsilon,$$

$$j = 1, \ldots, 2n$$

Moreover, if F has a bounded Hessian $\nabla\nabla F$, $\hat{d}$ has a simple relationship with ϵ, since two constants C_0 and C_1 exist such that

$$\max_{\boldsymbol{x}_n \in \gamma} |F(\boldsymbol{x}_n) - f(\boldsymbol{x}_n)| \leq C_0 \hat{d}^2 \|\nabla\nabla F\|_\infty$$

and

$$\max_{\boldsymbol{x}_n \in \gamma} \left| \frac{\partial F}{\partial x_n^j}(\boldsymbol{x}_n) - \frac{\partial f}{\partial x_n^j}(\boldsymbol{x}_n) \right| \leq C_1 \hat{d} \|\nabla\nabla F\|_\infty$$

$$j = 1, \ldots, 2n,$$

where $\|\nabla\nabla F\|_\infty = \max_{\boldsymbol{x}_n \in \mathcal{D}_n} \max_{j,i} \left| \frac{\partial^2 f}{\partial x_n^j x_n^i}(\boldsymbol{x}_n) \right|, \quad j, i = 1, \ldots, 2n$ □

As $\mathcal{D}_n$ is compact, Theorem 3.4.1 is sufficient to prove density in the sense of ${}^1\|\cdot\|_w$ for any $w > 0$. In fact, for any function g, be it $(F - f)$ itself or one of its derivatives,

$$\sup_{\boldsymbol{x}_n \in \mathcal{D}_n} |g(\boldsymbol{x}_n)| = \max_{\gamma \subset \mathcal{D}_n} \max_{\boldsymbol{x}_n \in \gamma} |g(\boldsymbol{x}_n)|$$

and

$$\int_{\mathcal{D}_n} |g(\boldsymbol{x}_n)|^w d\boldsymbol{x}_n \leq \mu(\mathcal{D}_n) \max_{\boldsymbol{x}_n \in \mathcal{D}_n} |g(\boldsymbol{x}_n)|^w,$$

where the $\mu(\cdot)$ is a measure defined over $\mathcal{D}_n$.

Thus, as long as the target function has a bounded Hessian, a piecewise affine model is sufficient to approximate its values and its derivatives providing that $\hat{d}$ is small enough.

In principle, this property can also be reversed. In fact, if a bound on the Hessian of the target function were known, then an upper bound on the diameter of the affine regions (and thus an hint on their number) would be available depending on the desired accuracy.

Regrettably, in practical cases this rarely happens and we are only guaranteed that decreasing the coarseness of the partition will eventually improve the identification.

As a final remark, note that even if the derivatives are discontinuous at region boundaries, an easy corollary to Theorem 3.4.1 ensures that as regions surrounding a given point shrink, then in that point the derivative of the approximation along any direction tends to the derivative of the target.

The consistency of the piecewise-polynomial approach to derivative approximation is therefore guaranteed even along region boundaries.

3.4.2 Model Continuity and Domain Partitioning

Since the model given by Equation 3.41 is assumed continuous, $f(\cdot)$ is also continuous over the whole $\mathcal{D}_n$ [Fantuzzi *et al.*, 2002].

In such a case the parameter vectors are constrained to satisfy the following relation:

$$\lim_{\substack{\boldsymbol{x}_n(t)\to\bar{\boldsymbol{x}}_n \\ \boldsymbol{x}_n(t)\in R_n^{(i')}}} f(\boldsymbol{x}_n(t)) = \lim_{\substack{\boldsymbol{x}_n(t)\to\bar{\boldsymbol{x}}_n \\ \boldsymbol{x}_n(t)\in R_n^{(i'')}}} f(\boldsymbol{x}_n(t)) \tag{3.46}$$

$\bar{\boldsymbol{x}}_n$ being an accumulation point for both $R_n^{(i')}$ and $R_n^{(i'')}$, *i.e.* if

$$[\bar{\boldsymbol{x}}_n(t), 1]^T\, \boldsymbol{a}^{(i')} = [\bar{\boldsymbol{x}}_n(t), 1]^T\, \boldsymbol{a}^{(i'')} \tag{3.47}$$

The straightforward application of Equation 3.47 to all the accumulation points common to neighbouring regions leads to an infinite number of constraints.

Yet, the following Theorem shows that the adoption of regions with straight borders guarantees that only a finite number of them are linearly independent.

Theorem 3.4.2. *Let $R_n^{(i',i'')}$ be the set of all the accumulation points of two neighbouring regions $R_n^{(i')}$ and $R_n^{(i'')}$. If $R_n^{(i',i'')}$ is convex, and p points $\boldsymbol{x}_n^1(t), \ldots, \boldsymbol{x}_n^p(t) \in R_n^{(i',i'')}$ exist for which Equation 3.47 is satisfied, then Equation 3.47 is also satisfied by any point $\bar{\boldsymbol{x}}_n(t)$ of their convex hull.*

Proof. If $\bar{\boldsymbol{x}}_n(t)$ belongs to the convex hull of $\boldsymbol{x}_n^1(t), \ldots, \boldsymbol{x}_n^p(t)$ then p non-negative scalars $\delta_1, \ldots, \delta_p$ exist such that

$$\sum_{k=1}^{p} \delta_k = 1 \tag{3.48}$$

and

$$\bar{\boldsymbol{x}}_n(t) = \sum_{k=1}^{p} \delta_k \boldsymbol{x}_n^k(t). \tag{3.49}$$

Then the continuity constraints $[\boldsymbol{x}_n^k, 1]^T(t)\, \boldsymbol{a}^{(i')} - [\boldsymbol{x}_n^k, 1]^T(t)\boldsymbol{a}^{(i'')} = 0$ for $k = 1, 2, \ldots, p$, can be combined by means of Equations 3.49 and 3.48 to obtain the result.

Theorem 3.4.2 suggests that regions whose boundaries are convex polyhedra should be considered. In this case, in fact, continuity can be ensured simply by setting the value of the local models only on the vertices of the boundaries.

In this case, Theorem 3.4.2 guarantees that the continuity constraints (one for each polyhedral vertex) can be collected in a finite matrix C_n such that:

$$C_n A_n = \mathbf{0}. \tag{3.50}$$

being

$$A_n = \begin{bmatrix} \boldsymbol{a}^{(1)} \\ \vdots \\ \boldsymbol{a}^{(M)} \end{bmatrix}$$

In particular, it is undoubtedly convenient to "triangulate" the domain $\mathcal{D}_n$, *i.e.* to partition it into $2n$–dimensional simpleces.

Moreover, we will assume that the triangulation is such that two simpleces are either disjoint, or have in common a whole k–dimensional boundary, with $k = 0, 1, \ldots, 2n-1$.

In this way, the local affine model of Equation 3.44 can be forced to assume given values at most in $2n+1$ vertices of each simplex, which are affinely independent points.

If we adopt this point of view our approach depends on the availability of a systematic procedure to triangulate $2n$-dimensional domains.

If we assume that both inputs and outputs are confined within certain intervals such a domain is actually a $2n$-dimensional hyper–rectangle which we will triangulate by means of a *recursive* procedure.

Hence, given the triangulation of a $2(n-1)$-dimensional hyper–rectangular domain the procedure gives us the triangulation of a $2n$-dimensional hyper–rectangular domain by applying two conceptually identical steps.

This approach yields the same final triangulation as the non–recursive approach proposed in [Rovatti *et al.*, 1998a] and thus, even if it is only asymptotically efficient in terms of number of generated simpleces [Mara, 1976, Cottle, 1982, Sallee, 1982, Sallee, 1984, Haiman, 1991] it benefits from the fast simplex searching feature that is highlighted in [Rovatti *et al.*, 1998a].

Moreover, its statement in recursive terms helps us to cope with the fact that we do not know from the beginning which order n we will choose for the final model. Instead, we try with increasing n until we obtain satisfactory results. In this case, recursive triangulation allow us to recycle part of the work done for lower dimensional domains.

To begin with, we assume that $\mathcal{U} = \bigcup_i \mathcal{I}_i^{\mathcal{U}}$ and $\mathcal{Y} = \bigcup_i \mathcal{I}_i^{\mathcal{Y}}$ where the $\mathcal{I}_i^{\mathcal{U}}$ and $\mathcal{I}_i^{\mathcal{Y}}$ are intervals.

When $n = 1$ we resort to the natural partition of each rectangle which is the Cartesian product $\mathcal{I}^{\mathcal{U}}_{i'} \times \mathcal{I}^{\mathcal{Y}}_{i''}$ in two triangles separated by one of the two diagonals.

Figure 3.3 shows an example of this partition when only one interval is present both in $\mathcal{Y}$ and in $\mathcal{U}$ [Fantuzzi *et al.*, 2002].

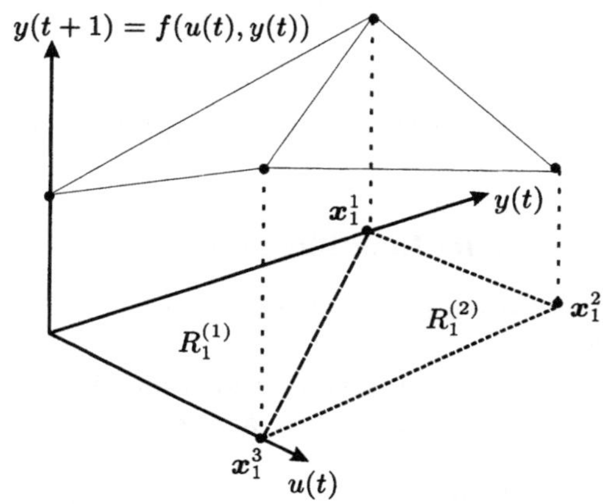

Fig. 3.3. An example of a partition of a 2–dimensional space $\mathcal{D}_1 = \mathcal{U} \times \mathcal{Y}$. Regions $R_1^{(1)}$ and $R_1^{(2)}$ have triangular shape.

Then we assume to have a triangulation of the $2n$-dimensional domain $\mathcal{U}^n \times \mathcal{Y}^n$ into simpleces. For any given simplex s of that triangulation and any given interval $\mathcal{I}^{\mathcal{Y}}_i$ we consider the Cartesian product $p = s \times \mathcal{I}^{\mathcal{Y}}_i$.

What we obtain is a $(2n+1)$-dimensional rectangular prism whose faces are the two $2n$-dimensional simpleces s and s', the latter being the translation of s along the direction of the last axis and for a length equal to the length of $\mathcal{I}^{\mathcal{Y}}_i$.

Now let $\boldsymbol{\xi}_0, \ldots, \boldsymbol{\xi}_{2n}$ be the vertices of s and $\boldsymbol{\xi}'_0, \ldots, \boldsymbol{\xi}'_{2n}$ be the vertices of s'. We triangulate the prism p by considering all the $(2n+1)$-dimensional simpleces that are the convex hulls of the $2n+2$ vertices $\boldsymbol{\xi}_0, \ldots, \boldsymbol{\xi}_i, \boldsymbol{\xi}'_i, \ldots, \boldsymbol{\xi}'_{2n}$ for $i = 0, \ldots, 2n$. With this, the prism is triangulated into $2n+1$ simpleces all having the same volume.

If this is done for every simplex s of the triangulation of $\mathcal{U}^n \times \mathcal{Y}^n$ and for every interval $\mathcal{I}^{\mathcal{Y}}_i$ we obtain a triangulation of $\mathcal{U}^n \times \mathcal{Y}^{n+1}$.

Most naturally, we may now consider all the $(2n+2)$-dimensional prisms which are the Cartesian products of a simplex of this last triangulation and an interval $\mathcal{I}^{\mathcal{U}}_i$ to obtain a triangulation of $\mathcal{U}^{n+1} \times \mathcal{Y}^{n+1}$.

Figure 3.4 shows how each of the steps for the triangulation of prisms works in the case of a 3-dimensional domain.

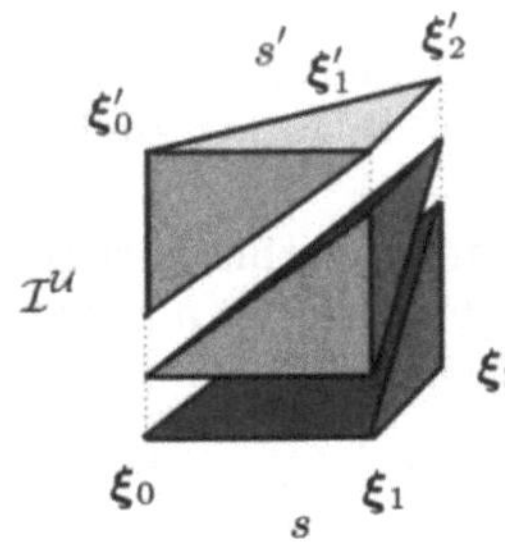

Fig. 3.4. The elementary step of triangulating a prism in the three dimensional case.

3.4.3 Local Affine Model Identification

In this section we discuss the local model identification procedure [Fantuzzi *et al.*, 2002].

The basic idea stems from the system identification with noisy measurement presented in Section 3.3.2, slightly adapted to handle affine instead of homogeneous models.

Let us assume that the input–output data $u(t)$ and $y(t)$, $(t = 0, 1, , \dots, L_i)$ generated by a SISO system of the type in Equation 3.42 are available.

Restricting our investigation to find order n and parameters $\boldsymbol{a}^{(i)}$ for local model of Equation 3.42 in region $R_n^{(i)}$, the following matrix should be defined:

$$\begin{aligned} X_k^{(i)} &= \begin{bmatrix} y(k) & \boldsymbol{x}_k^T(0) & 1 \\ y(k+1) & \boldsymbol{x}_k^T(1) & 1 \\ \vdots & \vdots & \\ y(k+N_i-1) & \boldsymbol{x}_k^T(N_i-1) & 1 \end{bmatrix} \\ \Sigma_k^{(i)} &= \left(X_k^{(i)}\right)^T X_k^{(i)} \end{aligned} \tag{3.51}$$

with $k + N_i - 1 \leq L_i$ and N_i is chosen so that $k + N_i - 1$ is large enough to avoid unwanted linear dependence relationships due to limitations in the dimension of the vector spaces involved.

To determine the model order n in region $R_n^{(i)}$, it is possible to consider the sequence of increasing–dimension positive-definite or positive-semidefinite $\big((2k+2) \times (2k+2)\big)$ symmetric matrices:

$$\Sigma_2^{(i)}, \;\; \Sigma_3^{(i)}, \;\; \dots \;\; \Sigma_k^{(i)}, \;\; \dots \tag{3.52}$$

testing their singularity as k increases. As soon as a singular matrix $\Sigma_k^{(i)}$ is found then $n = k$, and the parameters $\boldsymbol{a}^{(i)}$ describe the dependence relationship of the first vector of $\Sigma_n^{(i)}$ on the remaining ones as

$$\Sigma_n^{(i)} \begin{bmatrix} -1 \\ \boldsymbol{a}^{(i)} \end{bmatrix} = 0 \tag{3.53}$$

It's worth noting that the vectors $\boldsymbol{x}_n(0), \boldsymbol{x}_n(1), \ldots, \boldsymbol{x}_n(N_i - 1)$ in Equation 3.51 must belong to the region $R_n^{(i)}$ according to the partition defined in Equation 3.43.

Note also that in the presence of noise the above procedure for determining the order and model parameters would obviously be useless since matrices Σ_k would always be non–singular (positive-definite).

In order to solve the problem in a mathematical framework, it is necessary to characterise the noise affecting the input-output data.

According to Frisch [Frisch, 1934], Kalman [Kalman, 1982b] and Beghelli *et al.* [Beghelli *et al.*, 1990], the following assumptions are made. The noises $\tilde{u}(t)$ and $\tilde{y}(t)$ are assumed additive on the input–output data $u^*(t)$ and $y^*(t)$ and region independent, so that

$$u(t) = u^*(t) + \tilde{u}(t) \tag{3.54}$$

$$y(t) = y^*(t) + \tilde{y}(t) \tag{3.55}$$

Clearly, only $u(t)$ and $y(t)$ are available for the identification procedure, and moreover every noise term $\tilde{u}(t)$ and $\tilde{y}(t)$ is modelled with a zero–mean white process and is supposed to be independent of every other term. These structures are the well–known Errors–In-Variables models.

Under these assumptions, and furthermore that $\bar{\tilde{\sigma}}_u$ and $\bar{\tilde{\sigma}}_y$ are the input and output noise variances respectively, the generic positive-definite matrix $\Sigma_k^{(i)}$ associated with the input-output noise-corrupted sequences can always be expressed as the sum of two terms $\Sigma_k^{(i)} = \Sigma_k^{*(i)} + \bar{\tilde{\Sigma}}_k$ where

$$\bar{\tilde{\Sigma}}_k = \text{diag}[\bar{\tilde{\sigma}}_y I_{k+1}, \bar{\tilde{\sigma}}_u I_k, 0] \geq \mathbf{0} \tag{3.56}$$

Thus, it is again possible to determine the order and parameters of the model in region $R_n^{(i)}$ from the analysis of the sequence of increasing–dimension $\big((2k+2) \times (2k+2)\big)$ symmetric positive-definite matrices

$$\Sigma_2^{(i)}, \;\; \Sigma_3^{(i)}, \;\; \ldots \;\; \Sigma_k^{(i)}, \;\; \ldots \tag{3.57}$$

The solution to the above identification problem requires the computation of the unknown noise covariances $\bar{\tilde{\sigma}}_u$ and $\bar{\tilde{\sigma}}_y$, that can be achieved solving the following relation:

$$\Sigma_k^{*(i)} = \Sigma_k^{(i)} - \tilde{\Sigma}_k \geq 0. \tag{3.58}$$

in the variables $\tilde{\sigma}_u, \tilde{\sigma}_y$, where $\tilde{\Sigma}_k = \text{diag}[\tilde{\sigma}_y I_{k+1}, \tilde{\sigma}_u I_k, 0]$.

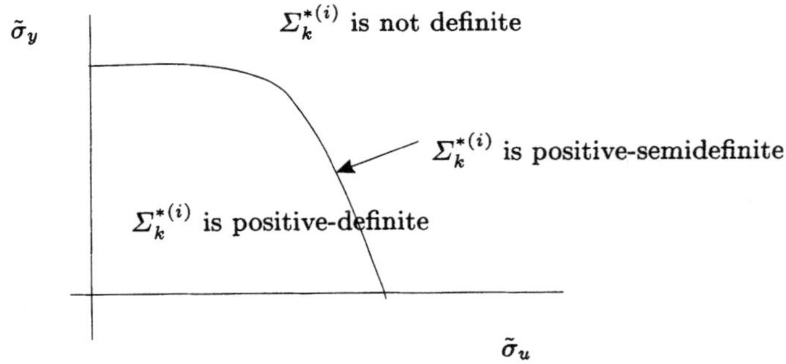

Fig. 3.5. A possible example of the singularity curve for matrix $\Sigma_k^{*(i)}$.

It is worth noting that the set of values of variables $\tilde{\sigma}_u, \tilde{\sigma}_y$ which make matrix $\Sigma_k^{*(i)}$ positive-semidefinite forms a curve, as depicted by Figure 3.5.

Unfortunately, the relation 3.58 admits for any k an infinite solution set describing a curve $\Gamma_k^{(i)}(\tilde{\sigma}_y, \tilde{\sigma}_u) = 0$ in the first orthant of the noise plane whose concavity faces the origin. In [Beghelli *et al.*, 1990] a constructive methodology to numerically compute this curve is given.

Since determination of the system order requires the increasing values of k to be tested, it is relevant to analyse the behaviour of the associated curves when k varies.

As proved by Beghelli *et al.* [Beghelli *et al.*, 1990], the solution sets of condition 3.58 for different values of k are non–crossing curves in the noise plane $(\tilde{\sigma}_y, \tilde{\sigma}_u)$.

It is also important to observe that, since we assume that a system of type given by Equation 3.42 has generated the noiseless data, for $k \geq n$ all the curves of type given by Equation 3.58 have necessarily at least one common point, *i.e.* point $(\bar{\tilde{\sigma}}_u, \bar{\tilde{\sigma}}_y)$ corresponding to the true variances of the noise affecting the input and the output data.

The search for a solution for the identification problem can thus start from the determination of this point in the noise space. This task can be achieved on the basis of the following properties:

Property 3.4.2. With reference to the diagonal non–negative definite matrices $\bar{\bar{\Sigma}}_k$, the following properties hold:

1. If $k < n$ the matrices $\Sigma_k^{*(i)}$ are positive-definite.

2. If $k > n$ the dimension of the null space of $\Sigma_k^{*(i)}$ and consequently, the number of eigenvalues equal to zero is $(k - n + 1)$.

3. For $k = n$, the matrix $\Sigma_k^{*(i)}$ is characterised by a linear dependence relation among its $2k + 2$ vectors, and the coefficients which link the first vector of $\Sigma_k^{*(i)}$ to the remaining ones are the parameters $\boldsymbol{a}^{(i)}$, of the system 3.42 which has generated the noiseless sequences.

4. For $k \geq (n + 1)$, all the $k - n + 1$ linear dependence relations among the vectors of the matrix Σ_k^* are characterised by the same $2n+2$ coefficients $\boldsymbol{a}^{(i)}$.

Figure 3.2 shows the above properties for a system such as 3.42 with $n = 3$. The point marked by a circle corresponds to the input-output noise variances $\bar{\tilde{\sigma}}_y$ and $\bar{\tilde{\sigma}}_u$ affecting the measurements.

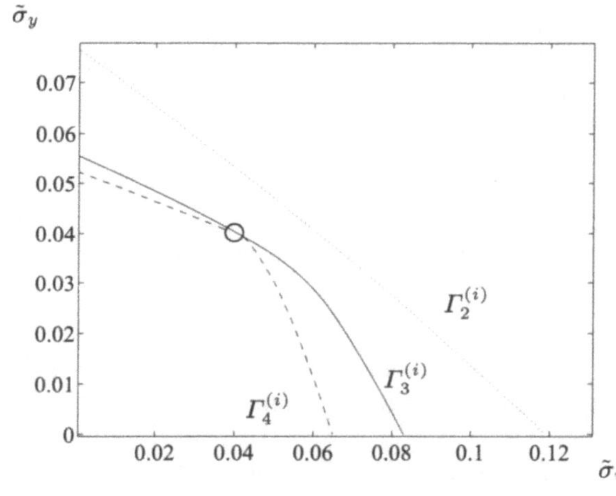

Fig. 3.6. Singularity curves in the noise space for a third order system. The example shows that for $k = 3$ and $k = 4$ the curves share the common point $\bar{\tilde{\sigma}}_u, \bar{\tilde{\sigma}}_y$ (marked by a circle).

3.4.4 Multiple-Model Identification

In Section 3.4.3 we discussed a procedure for the identification of the noise variances $\bar{\tilde{\sigma}}_u$ and $\bar{\tilde{\sigma}}_y$ and of the system order n, with respect to a particular region $\mathcal{A}_n^{(i)}$ [Fantuzzi *et al.*, 2002].

If the noise characteristics are common to all the regions $\mathcal{A}_n^{(i)}$, since the physical nature of the process generating the noise is independent of the model structure and of the partition of $\mathcal{D}_n$, and all assumptions regarding the Frisch scheme are fulfilled, a common point $(\bar{\tilde{\sigma}}_y, \bar{\tilde{\sigma}}_u)$ in the noise plane exists for the singularity surfaces.

Under these conditions, as an example, the singularity surfaces regarding two regions $\mathcal{R}_n^{(i)}$ and $\mathcal{R}_n^{(j)}$ for a model with order $n = 3$ are depicted in Figure 3.7. The curves share the common point $(\bar{\bar{\sigma}}_u, \bar{\bar{\sigma}}_y)$ representing the variances of the true noises which affect the data.

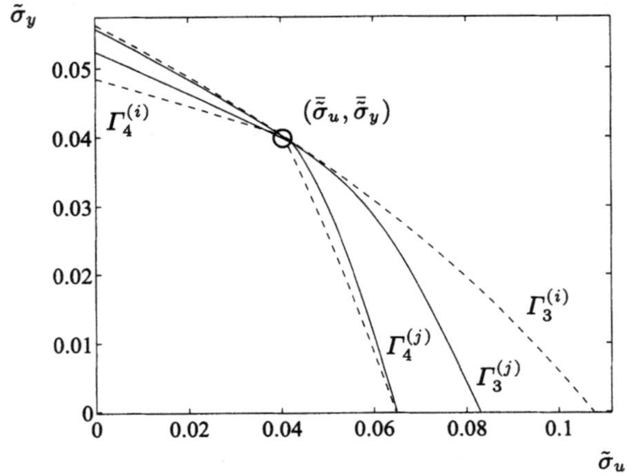

Fig. 3.7. An example of singularity curves in two regions $\mathcal{R}_3^{(i)}$ and $\mathcal{R}_4^{(j)}$.

When the order n has been determined, the parameters $\boldsymbol{a}^{(i)}, i = 1, \ldots, M$ can be identified solving the following equation

$$(\Sigma_n^{(i)} - \bar{\bar{\Sigma}}_n)\boldsymbol{a}^{(i)} = \mathbf{0} \quad \text{for } i = 1, \ldots, M. \tag{3.59}$$

The previous result can be fully applied when the assumptions behind the Frisch scheme are satisfied (independence between input–output sequences, additive noise, noise whiteness).

In real applications, we are forced to relax these assumptions, thus no common point can be determined among the surfaces $\Gamma_n^{(i)} = 0$ in the noise plane and a unique solution to the identification problem can be obtained only by introducing a criterion to select a different noisy point for each region as best approximation of the ideal case.

With reference to the identification of the system order n in the i–th region $\mathcal{A}_n^{(i)}$, it must be noted that the $\Gamma_{n+1}^{(i)} = 0$ curve has a single point in common with the $\Gamma_n^{(i)} = 0$ curve in ideal conditions, which corresponds to a double singularity of the matrix $\Sigma_{n+1}^{*(i)}$.

In real cases, the order n can be computed, finding the point $(\tilde{\sigma}_u, \tilde{\sigma}_y) \in \Gamma_{n+1}^{(i)} = 0$ that makes $\Sigma_{n+1}^{*(i)}$ closer to the double singular condition (*i.e.* minimal eigenvalue equal to zero and the second minimum eigenvalue near to zero).

As n is unknown, increasing system orders k must be tested, and the value of k associated to the minimum of the second eigenvalue of the matrix $\Sigma_{k+1}^{*(i)}$ corresponds to the order n. This criterion is consistent as it leads to the common point of the surfaces when the assumptions of the Frisch scheme are not violated.

Note that since the order n of the piecewise model 3.44 is region independent, it can be determined by choosing a k that fulfils the following inequality

$$\max_{i=1,\ldots,M_k} \lambda_k^{(i)} < \epsilon \tag{3.60}$$

when ϵ is an arbitrary positive constant and $\lambda_k^{(i)}$ is the minimal eigenvalue different from zero of matrix $\Sigma_{k+1}^{*(i)}$.

This result leads to the derivation of the following algorithm for selection of the model order [Fantuzzi *et al.*, 2002]:

1. Fix ϵ, k and M_k (k is the initial hypothesis on model order).
2. Construct partitions $\{\mathcal{A}_k^{(1)}, \ldots, \mathcal{A}_k^{(M_k)}\}$.
3. Cluster data into partitions.
4. Compute matrices $\Sigma_{k+1}^{*(i)}$ from data clustered in region $\mathcal{A}_k^{(i)}$
5. Compute test 3.60
   – *If success*: $n = k$, *exit*
   – *else* $k = k + 1$, *goto* 2

Once the model order n is selected, the parameters $\boldsymbol{a}^{(i)}, i = 1, \ldots, M$ cannot be computed from Equation 3.59, because the surfaces $\Gamma_n^{(i)} = 0$ do not share the common point $(\bar{\bar{\sigma}}_u, \bar{\bar{\sigma}}_y)$.

In this case, for each region different input–output noise variances $(\bar{\bar{\sigma}}_u^{(i)}, \bar{\bar{\sigma}}_y^{(i)})$ must be considered and the relation 3.58 should be rewritten as:

$$\Sigma_n^{*(i)} = \Sigma_n^{(i)} - \tilde{\Sigma}_n^{(i)} \geq 0 \tag{3.61}$$

where $\tilde{\Sigma}_n^{(i)} = \text{diag}[\bar{\bar{\sigma}}_u^{(i)} I_{n+1}, \bar{\bar{\sigma}}_y^{(i)} I_n, 0]$.

The values $(\bar{\bar{\sigma}}_u^{(i)}, \bar{\bar{\sigma}}_y^{(i)})$ can be computed by solving an optimisation problem that minimises both the distances between $(\bar{\bar{\sigma}}_u^{(i)}, \bar{\bar{\sigma}}_y^{(i)})$ and $(\bar{\bar{\sigma}}_u^{(j)}, \bar{\bar{\sigma}}_y^{(j)})$ with $i \neq j$ and the continuity constraints proposed in Equation 3.50

$$\begin{aligned} J\left((\bar{\bar{\sigma}}_u^{(1)}, \bar{\bar{\sigma}}_y^{(1)}), \ldots, (\bar{\bar{\sigma}}_u^{(M)}, \bar{\bar{\sigma}}_y^{(M)})\right) &= d\left((\bar{\bar{\sigma}}_u^{(1)}, \bar{\bar{\sigma}}_y^{(1)}), \ldots, (\bar{\bar{\sigma}}_u^{(M)}, \bar{\bar{\sigma}}_y^{(M)})\right) + \\ &+ (C_n A_n)^T \boldsymbol{H} C_n A_n \end{aligned} \tag{3.62}$$

$\boldsymbol{H}$ being a positive-definite weighting matrix and d a distance defined as:

$$\begin{aligned} d\left((\bar{\tilde{\sigma}}_u^{(1)}, \bar{\tilde{\sigma}}_y^{(1)}), \ldots, (\bar{\tilde{\sigma}}_u^{(M)}, \bar{\tilde{\sigma}}_y^{(M)})\right) &= \\ = \sum_{i=1}^{M} \sum_{j=i+1}^{M} \sqrt{(\bar{\tilde{\sigma}}_u^{(i)} - \bar{\tilde{\sigma}}_u^{(j)})^2 + (\bar{\tilde{\sigma}}_y^{(i)} - \bar{\tilde{\sigma}}_y^{(j)})^2}. \end{aligned} \tag{3.63}$$

It is worthwhile observing that the matrix A_n collects the parameters $\boldsymbol{a}^{(i)}, i = 1, \ldots, M$ which depend on $(\bar{\tilde{\sigma}}_u^{(i)}, \bar{\tilde{\sigma}}_y^{(i)})$.

Now, let us take into account the problem of determining the model order n. In the real case the item 2 of Property 3.4.2 is only approximately fulfilled (*i.e.* for $k > n$ null eigenvalue has algebraic multiplicity one, whereas the *second* minimum eigenvalue is very close to zero).

Minimisation of cost function 3.62 can be computationally difficult, as it depends on $2M$ independent variables.

Therefore, in order to decrease the complexity of the problem, a common parametrisation can be defined for all the surfaces $\Gamma_n^{(i)}(\bar{\tilde{\sigma}}_u^{(i)}, \bar{\tilde{\sigma}}_y^{(i)}) = 0$ by introducing polar coordinates:

$$\begin{cases} \bar{\tilde{\sigma}}_u^{(i)} = \rho^{(i)} \cos \frac{\pi}{2} q \\ \bar{\tilde{\sigma}}_y^{(i)} = \rho^{(i)} \sin \frac{\pi}{2} q \end{cases} \tag{3.64}$$

where $\rho^{(i)}$ is determined so that $\Gamma_n^{(i)}(\rho^{(i)} \cos \frac{\pi}{2} q, \rho^{(i)} \sin \frac{\pi}{2} q) = 0$ and $q \in [0, 1]$.

The cost function has the form:

$$\begin{aligned} J(q) &= d\left((\bar{\tilde{\sigma}}_u^{(1)}(q), \bar{\tilde{\sigma}}_y^{(1)}(q)), \ldots, (\bar{\tilde{\sigma}}_u^{(M)}(q), \bar{\tilde{\sigma}}_y^{(M)}(q))\right) + \\ &\quad + (C_n A_n)^T H C_n A_n. \end{aligned} \tag{3.65}$$

The parametrisation chosen to simplify the minimisation problem leads to consistent results. In fact, when the data are generated by a continuous piecewise affine dynamic system, all assumptions regarding the Frisch scheme being fulfilled and noise being region-independent, the surfaces $\Gamma_n^{(i)} = 0$ share a common point in the noise plane. In these conditions, the cost function $J(q) = 0$ and the variances $(\bar{\tilde{\sigma}}_u, \bar{\tilde{\sigma}}_y)$ are identified exactly.

Finally, one should note how once the parameter q minimising the cost function 3.65 is computed, the matrices $\tilde{\Sigma}_n^{(i)}$ can be constructed and the model parameter $\boldsymbol{a}^{(i)}, i = 1, \ldots, M$ determined by means of relation

$$(\Sigma_n^{(i)} - \tilde{\Sigma}_n^{(i)})\boldsymbol{a}^{(i)} = \mathbf{0} \quad \text{for } i = 1, \ldots, M. \tag{3.66}$$

This completes the multiple-model identification procedure.

In Chapter 5 application examples concerning the identification of a real non–linear process using piecewise affine models are presented and exploited for the generation of residual diagnostic signals [Fantuzzi *et al.*, 2002].

3.5 Fuzzy Modelling and Identification

Since its introduction in 1965, fuzzy set theory has found applications in a wide variety of disciplines. Modelling and control of dynamic systems belong to the fields in which fuzzy set techniques have received considerable attention, not only from the scientific community but also from industry. Many systems are not amenable to conventional modelling approaches due to the lack of precise, formal knowledge about the system, due to strongly non–linear behaviour, due to the high degree of uncertainty, or due to the time varying characteristics [Babuška, 1998].

Fuzzy modelling along with other related techniques such as neural networks have been recognised as powerful tools which can facilitate the effective development of models. One of the reasons for this is the capability of fuzzy systems to integrate information from different sources, such as physical laws, empirical models, or measurements and heuristics.

Fuzzy models can be seen as logical models which use "IF-THEN" rules to establish qualitative relationships among the variables in the model. Fuzzy sets serve as a smooth interface between the qualitative variables involved in the rules and the numerical data at the inputs and outputs of the model. The rule-based nature of fuzzy models allows the use of information expressed in the form of natural language statements and consequently makes the models transparent to interpretation and analysis. At the computational level, fuzzy models can be regarded as flexible mathematical structures, similar to neural networks, that can approximate a large class of complex non–linear systems to a desired degree of accuracy.

Recently, a great deal of research activity has focused on the development of methods to build or update fuzzy models from numerical data. As discussed in Chapter 2, most approaches are based on neuro–fuzzy systems, which exploit the functional similarity between fuzzy reasoning systems and neural networks. This "marriage" of fuzzy systems and neural networks enables a more effective use of optimisation techniques for building fuzzy systems, especially with regard to their approximation accuracy. However, the aspects related to the transparency and interpretation tend to receive considerably less attention. Consequently, most neuro-fuzzy models can be regarded as "grey–box" models which provide little insight to help understand the underlying process.

The approach exploited in this Section focuses on the identification of transparent rule–based fuzzy models which can accurately predict the quantities of interest, and at the same time provide insight into the system that generated the data. Attention is paid to the selection of appropriate model structures in terms of the dynamic properties, as well as the internal structure of the fuzzy rules (in particular, Takagi–Sugeno type) [Takagi and Sugeno, 1985]. From the system identification point of view, a fuzzy model is regarded as a composition of local affine sub models. Fuzzy sets naturally provide smooth transitions between the submodels, and enable the integration of various types of knowledge within a common framework.

In order to generate fuzzy models automatically from measurements, a comprehensive methodology is developed. This employs fuzzy clustering techniques to partition the available data into subsets characterised by a linear behaviour. The relationships between the presented identification method and linear regression are exploited, allowing for the combination of fuzzy logic techniques with system identification tools.

Using the concepts of model–based fault detection, the design of a residual generator based on a fuzzy model of a non–linear dynamic process is addressed. The orientation of the section is towards methodologies that in the author's experience proved to be practically useful. The presentation reflects theoretical and practical issues in a balanced way, aiming at readership from the academic world and also from industrial practice. Simulation examples are given in Section 5 where three selected real-world applications are presented in detail.

In addition, an implementation in a *MATLAB Toolbox* of the Fuzzy Modelling and Identification techniques presented in the following is available [Babuška, 2000]. This toolbox can be obtained from Robert Babuška [Babuška, 1998].

3.5.1 Fuzzy Multiple Inference Identification

The term fuzzy identification usually refers to techniques and algorithms for constructing fuzzy models from data. Two main approaches to the integration of knowledge and data in a fuzzy model can be distinguished [Babuška, 1998]:

1. The expert knowledge expressed in a verbal form is translated into a collection of **IF-THEN** rules. In this way, a certain model structure is created. Parameters in this structure (membership functions, weights of the rules, etc.) can be fine–tuned using input–output data. The particular tuning algorithms exploit the fact that at the computational level, a fuzzy model can be seen as a layered structure (network), similar to artificial neural networks, to which standard learning algorithms can be applied. This approach is usually called neuro-fuzzy modelling.

2. No prior knowledge about the system under study is initially used to formulate the rules, and a fuzzy model is constructed using numerical data only. It is expected that the extracted rules and membership functions

can provide an *a posteriori* interpretation of the system's behaviour. An expert can confront this information with his own knowledge, can modify the rules, or supply new ones, and can design additional experiments in order to obtain more informative data.

These two techniques, of course, can be combined, depending on the particular application. This section focuses mainly on the presentation of methods and algorithms for the second approach, *i.e.*, for the automated acquisition of fuzzy models from data.

It is believed that this technique is more useful in practice, as it can obviate the process of knowledge acquisition which is a well–known bottleneck for the practical applications of knowledge–based systems [McGraw and Harbisson-Briggs, 1989]. Instead, the expert is invited to assume a more active role of model analysis and validation, which may lead to revealing new pieces of information, and may result in a kind of emergent knowledge acquisition.

To date, relatively little attention has been devoted to the identification of transparent fuzzy models from data. Most of the techniques reported in the literature aim at obtaining numerical models that simply fit the data with the best possible accuracy, without paying attention to the interpretation of the results [Takagi and Sugeno, 1985, Sugeno and Kang, 1988, Johansen, 1996, Wang, 1995]. Many other identification techniques can be used for completely "grey–box" modelling, such as standard non–linear regression [Seber and Wild, 1989], spline techniques [de Boor, 1978, Brown and Harris, 1994a], or neural networks [Hunt *et al.*, 1992b]. In many cases, a natural requirement is that a model not only predicts accurately the system's outputs but also provides some insights into the working of the system. Such a model can be used not only for the given situation, but can also be more easily adapted to changing design parameters and operating conditions.

In this Section, fuzzy models are viewed as a class of local modelling approaches, which attempt to solve a complex modelling problem by decomposing it into a number of simpler subproblems. The theory of fuzzy sets offers an excellent tool for representing the uncertainty associated with the decomposition task, for providing smooth transitions between the individual local sub models, and for integrating various types of knowledge within one common framework.

In particular, fuzzy logic is exploited to define a Takagi–Sugeno (TS) fuzzy model [Takagi and Sugeno, 1985]. The TS fuzzy model for non–linear dynamic systems is described by a collection of local linear or affine submodels, each one approximating the system behaviour around a single working point. The scheduling of the submodels is achieved through a smooth function of the system state, the behaviour of which is defined using fuzzy set theory [Klir and Yuan, 1995].

Recalling comments at the beginning of Section 3.4, it can be recognised that such a structure fits the definition of the ***multiple-model*** as stated by Billings and its co–workers [Leontaritis and Billings, 1985a, Leontaritis and Billings, 1985b]. In fact, the basic approach to fuzzy modelling is similar to that presented in Section 3.4.1, in which a number of local models are designed and the estimate of outputs is given by a smooth (fuzzy) fusion of local outputs.

A large part of fuzzy modelling and identification algorithms (see [Babuška and Verbruggen, 1995, Babuška *et al.*, 1997, Babuška, 1998] and references therein) share a common two–step procedure, in which at first, the operating regions are determined using heuristics or data clusterings techniques. Then, in the second stage, the identification of the parameters of each submodel is achieved using Least–Squares algorithm or Frisch scheme.

From this perspective, fuzzy identification can be regarded as a search for a decomposition of a non–linear system, which gives a desired balance between the complexity and the accuracy of the model, effectively exploring the fact that the complexity of systems is usually not uniform. Since it cannot be expected that sufficient prior knowledge is available concerning this decomposition, methods for automated generation of the decomposition, primarily from system data, are developed. A suitable class of fuzzy clustering algorithms is used for this purpose.

3.5.2 Takagi-Sugeno Multiple-Model Paradigm

A fuzzy rule–based model suitable for the approximation of a large class of non–linear systems was introduced by Takagi and Sugeno [Takagi and Sugeno, 1985].
In the TS fuzzy model Figure 3.8, the rule consequents are crisp functions of the model inputs:

$$R_i : \ \ \textbf{IF } \boldsymbol{x}(t) \text{ is } A_i \text{ THEN } y_i = f_i\big(\boldsymbol{x}(t)\big), \quad i = 1, 2, \cdots, K, \tag{3.67}$$

where $\boldsymbol{x}(t) \in \Re^p$ is the input (antecedent) variable and $y_i \in \Re$ is the output (consequent) variable. R_i denotes the i–th rule, and K is the number of rules in the rule base. A_i is the antecedent fuzzy set of the i–th rule, defined by a (multivariate) membership function:

$$\mu_{A_i}(\boldsymbol{x}) : \ \Re^p \to [0, 1]\,. \tag{3.68}$$

As in the linguistic model, the antecedent proposition "$\boldsymbol{x}(t)$ is A_i" is usually expressed as a logical combination of simple propositions with univariate fuzzy sets defined for the individual components of $\boldsymbol{x}(t)$, often in the conjunctive form:

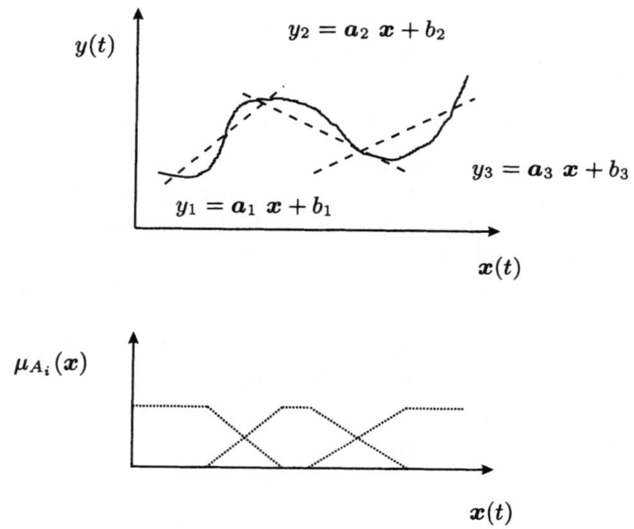

Fig. 3.8. Fuzzy model diagram.

$$R_i: \quad \textbf{IF } x_1 \text{ is } A_{il} \textbf{ and } x_2 \text{ is } A_{i2} \textbf{ and } \cdots \textbf{ and } x_p \text{ is } A_{ip}$$
$$\textbf{THEN } y_i = f_i(\boldsymbol{x}), i = 1, 2, \cdots, K. \tag{3.69}$$

The consequent functions f_i are typically chosen as instances of a suitable parameterised function, whose structure remains equal in all the rules and only the parameters vary. A simple and practically useful parametrisation is the affine linear form:

$$y_i = \boldsymbol{a}_i \, \boldsymbol{x} + b_i, \tag{3.70}$$

where a_i is a parameter vector and b_i is a scalar offset. We refer to this model as an *affine TS model.* The consequents of the affine TS model are hyperplanes (p-dimensional linear subspaces) in $\Re^{p+1}$. The antecedent of each rule defines a (fuzzy) validity region for the corresponding affine consequent model. The global model is composed of a concatenation of the local models, and can be seen as a smoothed piecewise approximation of a non–linear surface. Approximation properties of the affine TS model were investigated by Rovatti [Rovatti, 1996].

A special case of the consequent function occurs when $b_i = 0, i = 1, \cdots, K$. Then the model is called a homogeneous TS model:

$$\textbf{IF } \boldsymbol{x} \text{ is } A_i \text{ THEN } y_i = \boldsymbol{a}_i \boldsymbol{x}, \; i = 1, 2, \cdots, K. \tag{3.71}$$

This model has more limited approximation capabilities than the affine TS model [Fantuzzi and Rovatti, 1996].

When $a_i = 0$, $i = 1, \cdots, K$, the consequents in model 3.70 are constant functions, and the ***singleton model***, is obtained:

$$\textbf{IF } \boldsymbol{x} \text{ is } A_i \textbf{ THEN } \; y_i = b_i, \; i = 1, 2, \cdots, K. \tag{3.72}$$

Before the output can be inferred, the degree of fulfilment of the antecedent denoted by $\beta_i(\boldsymbol{x})$ must be computed. For rules with multivariate antecedent fuzzy sets given by Equations 3.67 and 3.68, the degree of fulfilment is simply equal to the membership degree of the given input $\boldsymbol{x}$, *i.e.*, $\beta_i = \mu_{A_i}(\boldsymbol{x})$. When logical connectives are used, the degree of fulfilment of the antecedent is computed as a combination of the membership degrees of the individual propositions using the fuzzy logic operators [Jager, 1995, Babuška, 1998].

In the Takagi–Sugeno model, the inference is reduced to a simple algebraic expression, similar to the fuzzy–mean defuzzification formula [Takagi and Sugeno, 1985]:

$$y = \frac{\sum_{i=1}^{K} \beta_i(\boldsymbol{x}) y_i}{\sum_{i=1}^{K} \beta_i(\boldsymbol{x})} \tag{3.73}$$

By denoting the normalised degree of fulfilment

$$\lambda_i(\boldsymbol{x}) = \frac{\beta_i(\boldsymbol{x})}{\sum_{j=1}^{K} \beta_j(\boldsymbol{x})} \tag{3.74}$$

the affine TS model with a common consequent structure can be expressed as a pseudo–linear model with input–dependent parameters:

$$y = \Big(\sum_{i=1}^{K} \lambda_i(\boldsymbol{x}) \boldsymbol{a}_i^T \Big) \boldsymbol{x} + \sum_{i=1}^{K} \lambda_i(\boldsymbol{x}) b_i = \boldsymbol{a}^T(\boldsymbol{x}) \boldsymbol{x} + b(\boldsymbol{x}). \tag{3.75}$$

The parameters $\boldsymbol{a}(\boldsymbol{x})$, $b(\boldsymbol{x})$ are convex linear combinations of the consequent parameters $\boldsymbol{a}_i$ and b_i, *i.e.*:

$$\boldsymbol{a}(\boldsymbol{x}) = \sum_{i=1}^{K} \lambda_i(\boldsymbol{x}) \boldsymbol{a}_i^T, \; b(\boldsymbol{x}) = \sum_{i=1}^{K} \lambda_i(\boldsymbol{x}) b_i. \tag{3.76}$$

Because of this property, a TS model can be regarded as a mapping from the antecedent (input) space to a convex region (polytope) in the space of the parameters of a quasi–linear system 3.75.

Consider, for instance, a dynamic system described by the following TS rules:

$$
\begin{aligned}
R_i: \quad & \textbf{IF } y(k) \text{ is } A_{il} \textbf{ and } y(k-1) \text{ is } A_{i2}\textbf{and} \quad \cdots \; y(k-n_y+1) \text{ is } A_{in_y} \\
& \textbf{and } u(k) \text{ is } B_{il}\textbf{and } u(k-1) \text{ is } B_{i2}\textbf{and} \quad \cdots \; u(k-n_u+1) \text{ is } B_{in_u} \\
& \textbf{THEN } y(k+1) = \textstyle\sum_{j=1}^{n_y} y(k-j+1) + \sum_{j=1}^{n_u} u(k-j+1),
\end{aligned}
\tag{3.77}
$$

where the consequents are linear ARX models (n_u and n_y are integers related to the order of the system).

3.5.3 Fuzzy Clustering for Fuzzy Identification

An effective approach to the identification of complex non–linear systems is to partition the available data into subsets and approximate each subset by a simple model [Babuška and Verbruggen, 1995]. Fuzzy clustering can be used as a tool to obtain a partitioning of data where the transitions between the subsets are gradual rather than abrupt. This section gives an introduction to the basic concepts of fuzzy clustering [Babuška, 1998].

The aim of this section is to explain clustering at a level necessary to understand the subsequent applications. For a more detailed treatment of the subject, the reader may refer to the classical monographs by Duda and Hart [Duda and Hart, 1973], Bezdek [Bezdek, 1981] and Jain and Dubes [Jain and Dubes, 1988]. A more recent overview can be found in a collection of Bezdek and Pal [Bezdek and Pal, 1992], and the monograph by Backer [Backer, 1995]. The notation and terminology in this chapter closely follows [Bezdek, 1981].

Cluster Analysis and Methods. The aim of cluster analysis is the classification of objects according to similarities among them, and the organising of data into groups. Clustering techniques are among the unsupervised (learning) methods, since they do not use prior class identifiers. Most clustering algorithms also do not rely on assumptions common to conventional statistical methods, such as the underlying statistical distribution of data, and therefore they are useful in situations where little prior knowledge exists. The potential of clustering algorithms to reveal the underlying structures in data can be exploited, not only for classification and pattern recognition, but also for the reduction of complexity in modelling and optimisation.

Clustering techniques can be applied to data which are typically observations of some physical process. Each observation consists of n measured variables, grouped into an n-dimensional column vector $\boldsymbol{z}_k = [z_{lk}, \cdots, z_{nk}]^T$, $\boldsymbol{z}_k \in \Re^n$. A set of N observations is denoted by $\boldsymbol{Z} = \{\boldsymbol{z}_k | k = 1, 2, \cdots, N\}$, and is represented as an $n \times N$ matrix:

$$
\boldsymbol{Z} = \begin{bmatrix} z_{ll} & z_{l2} & \cdots & z_{lN} \\ z_{2l} & z_{22} & \cdots & z_{2N} \\ \vdots & \vdots & \vdots & \vdots \\ z_{nl} & z_{n2} & \cdots & z_{nN} \end{bmatrix}
\tag{3.78}
$$

In the pattern recognition terminology, the columns of this matrix are called patterns or objects, the rows are called the features or attributes, and $\boldsymbol{Z}$ is called the pattern or data matrix. The meaning of the columns and rows of $\boldsymbol{Z}$ depends on the context.

When clustering is applied to the modelling and identification of dynamic systems, the columns of $\boldsymbol{Z}$ contain samples of time signals, and the rows are, for instance, physical variables observed in the system (position, velocity, temperature, etc.). In order to represent the system's dynamics, past values of the variables are typically included in $\boldsymbol{Z}$ as well.

Generally, a cluster is a group of objects that are more similar to one another than to members of other clusters [Bezdek, 1981, Jain and Dubes, 1988]. In metric spaces, similarity is often defined by means of a distance norm. Distance can be measured among the data vectors themselves, or as a distance from a data vector to some prototypical object of the cluster. The prototypes are usually not known beforehand, and are sought by the clustering algorithms simultaneously with the partitioning of the data. The prototypes may be vectors of the same dimension as the data objects, but they can also be defined as geometrical objects, such as linear or non–linear subspaces or functions. Data can reveal clusters of different geometrical shapes, sizes and densities

Algorithms that can detect subspaces of the data space are of particular interest for identification and will be summarised in the following.

Many clustering algorithms have been introduced in the literature. Since clusters can formally be seen as subsets of the data set, one possible classification of clustering methods can be according to whether the subsets are fuzzy or crisp (hard). Hard clustering methods are based on classical set theory, and require that an object either does or does not belong to a cluster. *Fuzzy clustering* methods, however, allow the objects to belong to several clusters simultaneously, with different degrees of membership. In many situations, fuzzy clustering is more natural than hard clustering, as objects on the boundaries between several classes are not forced to fully belong to one of the classes, but rather are assigned membership degrees between O and 1 indicating their partial memberships.

Another classification of clustering techniques can be related to the algorithmic approach of the different techniques [Bezdek, 1981]. In particular, the class of clustering algorithms presented here exploits an objective function to measure the desirability of partitions. Non–linear optimisation algorithms are used to search for local extrema of the objective function.

Therefore, in the following, fuzzy clustering algorithms with objective function will be presented. These methods lead to least–squares optimisation, and hence there are close relationships between clustering with fuzzy objective function and statistical regression and systems identification methods [Babuška, 1998]. In more detail, the clustering algorithm presented in this section is based on optimisation of the basic *c–means* objective function and it is known as *fuzzy c–means clustering* algorithm [Dunn, 1974].

Fuzzy c–Means Clustering Algorithms. This fuzzy clustering algorithm is based on minimisation of the fuzzy functional formulated as [Bezdek, 1981]:

$$J(\boldsymbol{Z};\boldsymbol{U},\boldsymbol{V}) = \sum_{i=1}^{c}\sum_{k=1}^{N}(\mu_{ik})^m \| \boldsymbol{z}_k - \boldsymbol{v}_i \|^2 \qquad (3.79)$$

where

$$\boldsymbol{U} = [\mu_{ik}] \in M_{f_c} \qquad (3.80)$$

is a fuzzy partition matrix of $\boldsymbol{Z}$,

$$\boldsymbol{V} = [\boldsymbol{v}_l,\ \boldsymbol{v}_2,\ \ldots,\ \boldsymbol{v}_c]\ ,\ \boldsymbol{v}_i \in \Re^n \qquad (3.81)$$

is a vector of cluster prototypes (centres), which have to be determined,

$$D^2_{ik\boldsymbol{A}} = \| \boldsymbol{z}_k - \boldsymbol{v}_i \|_{\boldsymbol{A}} = (\boldsymbol{z}_k - \boldsymbol{v}_i)^T \boldsymbol{A}\,(\boldsymbol{z}_k - \boldsymbol{v}_i) \qquad (3.82)$$

is a squared distance norm, and

$$m \in [1, \infty) \qquad (3.83)$$

is a weighting exponent which determines the fuzziness of the resulting clusters. The measure of dissimilarity in Equation 3.79 is the squared distance between each data point $\boldsymbol{z}_k$ and the cluster prototype $\boldsymbol{v}_i$. This distance is weighted by the power of the membership degree of that point $(\mu_{ik})^m$. The value of the cost function 3.79 can be seen as a measure of the total variance of $\boldsymbol{z}_k$ from $\boldsymbol{v}_i$.

The *fuzzy c–means (FCM) algorithm* consists of the minimisation of the c–means functional 3.79. It represents a non–linear optimisation problem that can be solved by using a variety of available methods [Bezdek *et al.*, 1987, DeSarbo, 1982, Babu and Murty, 1994].

The most popular method, however, exploits the first–order conditions for stationary points of Equation 3.79. They can be found by adjoining the constraint of the fuzzy partition [Babuška, 1998]

$$\sum_{i=1}^{c}\mu_{ik} = 1\ ,\ 1 \le k \le N \qquad (3.84)$$

to J by means of Lagrange multipliers:

$$\bar{J}(\boldsymbol{Z};\boldsymbol{U},\boldsymbol{V},\boldsymbol{\lambda}) = \sum_{i=1}^{c}\sum_{k=1}^{N}(\mu_{ik})^m D^2_{ik\boldsymbol{A}} + \sum_{k=1}^{N}\left[\sum_{i=1}^{c}\mu_{ik} - 1\right], \qquad (3.85)$$

and by setting the gradients of $\bar{J}$ with respect to $\boldsymbol{U}$, $\boldsymbol{V}$ and $\boldsymbol{\lambda}$ to zero. If $D^2_{ik\boldsymbol{A}} > 0$, $\forall i, k$ and $m > 1$, then $(\boldsymbol{U}, \boldsymbol{V}) \in M_{f_c} \times \Re^{n \times c}$ may minimise Function 3.79 only if

$$\mu_{ik} = \frac{1}{\sum_{j=1}^{c} \left(D^2_{ik\boldsymbol{A}} / D^2_{ik\boldsymbol{A}} \right)^{2/(m-1)}} \ , 1 \le i \le c \ , 1 \le k \le N \tag{3.86}$$

and

$$\boldsymbol{v}_i = \frac{\sum_{k=1}^{N} (\mu_{ik})^m \boldsymbol{z}_k}{\sum_{k=1}^{N} (\mu_{ik})^m} \tag{3.87}$$

This solution also satisfies the fuzzy partition constraints defined by Equations 3.88 and 3.89

$$\mu_{ik} \in [0, \ 1] \ , 1 \le i \le c \ , 1 \le k \le N \tag{3.88}$$

and

$$0 < \sum_{k=1}^{N} \mu_{ik} < N \ , 1 \le i \le c. \tag{3.89}$$

It is worthwhile noting that Equation 3.87 gives $\boldsymbol{v}_i$ as the weighted mean of the data items that belong to a cluster, where the weights are the membership degrees. That is why the algorithm is called "c–means".

The alternating optimisation scheme used by the FCM algorithm loops through the estimates of $\boldsymbol{U}^{(l)}$ and $\boldsymbol{V}^{(l)}$ at the step l. The termination error is usually given by choosing a value of ϵ so that $\max_{ik}(|\mu_{ik}^{(l)} - \mu_{ik}^{(l-1)}|) < \epsilon$. Convergence of the algorithm has been proven by Bezdek [Bezdek, 1980].

The number of clusters c is the most important parameter, in the sense that the remaining parameters have secondary effects on $\boldsymbol{U}$, compared with the effects of the number of clusters. The choice of the number of clusters is discussed in the following.

Gustafson–Kessel Clustering Algorithm. A family of algorithms can be derived from the basic FCM scheme and the most important will be presented in this section: the Gustafson–Kessel algorithm.

Gustafson and Kessel [Gustafson and Kessel, 1979] (GK) extended the standard fuzzy c–means algorithm by employing an adaptive distance norm, in order to detect clusters of different geometrical shapes in one data set.

Each cluster has its own norm-inducing matrix $\boldsymbol{A}_i$ which yields the following norm:

$$D^2_{ik\boldsymbol{A}_i} = (\boldsymbol{z}_k - \boldsymbol{v}_i)^T \boldsymbol{A}_i (\boldsymbol{z}_k - \boldsymbol{v}_i). \tag{3.90}$$

The matrices $\boldsymbol{A}_i$ are used as optimisation variables in the c–means functional, thus allowing each cluster to adapt the distance norm to the local topological structure of the data. Let $\boldsymbol{A}$ denote a c-tuple of matrices: $\boldsymbol{A} = (\boldsymbol{A}_l, \boldsymbol{A}_2, \cdots, \boldsymbol{A}_c)$. The objective functional of the GK algorithm is defined by:

$$J(\boldsymbol{Z};\boldsymbol{U},\boldsymbol{V},\boldsymbol{A}) = \sum_{i=1}^{c}\sum_{k=1}^{N}(\mu_{ik})^m D_{ik\boldsymbol{A}_i}^2 \tag{3.91}$$

where $\boldsymbol{U} \in M_{f_c}$, $V \in \Re^{n\times c}$ and $m > 1$. The solutions,

$$(\boldsymbol{U},\boldsymbol{V},\boldsymbol{A}) = \arg\min_{M_{f_c}\times\Re^{n\times c}\times \mathrm{PD}^n} J(\boldsymbol{Z};\boldsymbol{U},\boldsymbol{V},\boldsymbol{A}), \tag{3.92}$$

are stationary points of J, where PD^n denotes a space of $n \times n$ positive-definite matrices. For a fixed $\boldsymbol{A}$, conditions 3.86 and 3.87 can be directly applied. However, the objective function 3.91 cannot be directly minimised with respect to $\boldsymbol{Ai}$, since it is linear in $\boldsymbol{Ai}$. J could be made as small as desired by making $\boldsymbol{Ai}$ less positive-definite. To obtain a feasible solution, $\boldsymbol{Ai}$ must be constrained in some way. The usual way of accomplishing this is to constrain the determinant of $\boldsymbol{Ai}$. Allowing the matrix $\boldsymbol{Ai}$ to vary with its determinant fixed corresponds to optimising the cluster's shape while its volume remains constant:

$$|\boldsymbol{A}_i| = \rho_i \;,\; \rho_i > 0 \;,\; \forall\, i \tag{3.93}$$

Using the Lagrange multiplier method, the following expression for $\boldsymbol{Ai}$ is obtained:

$$\boldsymbol{A}_i = [\rho_i \det(\boldsymbol{F}_i)]^{1/n}\boldsymbol{F}_i^{-1}, \tag{3.94}$$

where $\boldsymbol{F}_i$ is the fuzzy covariance matrix of the ith cluster defined by:

$$\boldsymbol{F}_i = \frac{\sum_{k=1}^{N}(\mu_{ik})^m(\boldsymbol{z}_k - \boldsymbol{v}_i)(\boldsymbol{z}_k - \boldsymbol{v}_i)}{\sum_{k=1}^{N}(\mu_{ik})^m} \tag{3.95}$$

It is worth noting that without any prior knowledge, the cluster volumes ρ_i are simply fixed at 1 for each cluster. A drawback of the GK algorithm is that due to the constraint defined by Equation 3.93, it only can find clusters of approximately equal volumes.

The eigenstructure of the cluster covariance matrix provides information about the shape and orientation of the cluster. The ratio of the lengths of the cluster's hyperellipsoid axes is given by the ratio of the square roots of the eigenvalues of $\boldsymbol{F}_i$. The directions of the axes are given by the eigenvectors of $\boldsymbol{F}_i$ [Babuška, 1998]. Linear subspaces of the data space are represented by flat hyperellipsoids, that can be seen as hyperplanes. The eigenvector corresponding to the smallest eigenvalue determines the normal to the hyperplane, and can be used to compute optimal local linear models from the covariance matrix [Babuška, 1998].

Optimal Number of Clusters. When clustering real data without any *a priori* information about the data structure, one usually has to make assumptions about the number of underlying subgroups (clusters) c in the data. The chosen clustering algorithm then searches for c clusters, regardless of whether they are really present in the data or not. Two main approaches to determining the appropriate number of clusters in data can be distinguished [Babuška, 1998]:

1. Clustering data for different values of c, and using validity measures to assess the goodness of the obtained partitions. Different scalar validity measures have been proposed in the literature.

2. Starting with a sufficiently large number of clusters, and successively reducing this number by merging clusters that are similar (compatible) with respect to some predefined criteria.

3.5.4 Product Space Clustering and Fuzzy Model Identification

This section addresses the decomposition of a non–linear identification problem into a set of locally linear models by means of product space fuzzy clustering. The identification procedure is first outlined and the structure selection and the choice of regressors in the modelling of dynamic systems are therefore discussed. Moreover, the principle of identification of non–linear systems by product space clustering is of particular interest.

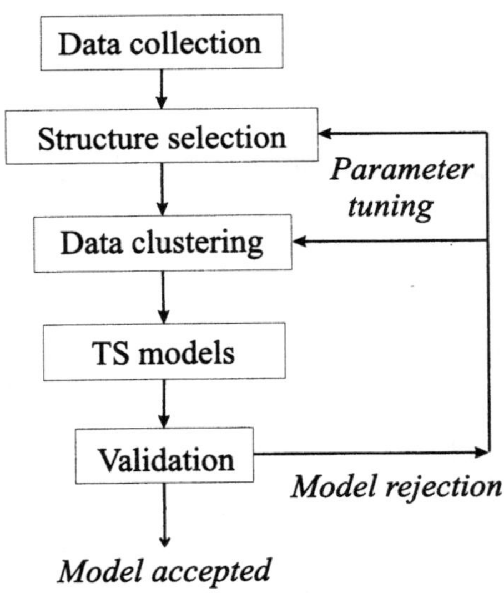

Fig. 3.9. Steps of the fuzzy clustering identification approach.

Figure 3.9 outlines the individual steps of the iterative identification procedure. As in a typical identification session, some of the steps may be repeated for different choices of the various parameters. The different steps and the meaning of the blocks in Figure 3.9 are summarised below.

1. **Data collection**. This is an important initial step for any identification method, since it determines the information content of the identification data set. Although the choice of the excitation signal may be problem dependent, the input data should preferably excite the system in the entire range of the considered variables both in amplitude and in frequency. Moreover, the choice of a suitable sampling period, the design of (anti–aliasing) filters, the duration of the experiments, etc., are other important issues of the experiment design [Norton, 1986, Ljung, 1999].

2. **Structure selection**. The purpose of this step is to determine the relevant input and output variables with respect to the aim of the modelling experiment. When identifying dynamic systems, the structure and the order of the model dynamics must be chosen. Structure selection allows us to translate the identification of a dynamic system into a regression problem that can be solved in a quasi–static manner. The structure can be selected in an automated way by comparing different candidate structures in terms of some performance measures. In most cases, a reasonable choice can be made by the user.

3. **Data Clustering**. Structure selection leads to a non–linear static regression problem, which is then approximated by a collection of local linear submodels. The location and the parameters of the sub models are found by partitioning the available data into clusters. Each of the clusters defines a fuzzy region in which the system can be approximated locally by an affine submodel. Moreover, an appropriate number of clusters can be found. This stage can require several repetitions of Step 3 for a different number of clusters.

4. **TS model**. Fuzzy clustering divides the available data into groups in which local affine relations exists between the inputs and the output. In order to obtain a model suitable for prediction, the TS fuzzy model parameters are identified from the available fuzzy partition matrix and from the cluster prototypes. Therefore, the membership functions and other parameters that constitute the fuzzy model are extracted in an automated way.

5. **Validation**. By means of validation, the final model is either accepted as appropriate for the given purpose, or it is rejected. In the latter case, some steps of the identification loop shown in Figure 3.9 may be repeated with a different setting, as it is usual also in other approaches to linear and non–

linear system identification [Ljung, 1999, Norton, 1986]. In addition to the usual numerical validation by means of simulation, interpretation of fuzzy models plays an important role in the validation step. The coverage of the input space by the rules can be analysed, and, for an incomplete rule base, additional rules can be provided based on prior knowledge, local linearisation, or first-principle models.

In the fuzzy modelling and identification procedure, the problem of structure estimation can consists of several subproblems:

1. *Input and output variable selection.* Although most identification methods assume that the input and output variables of the process are known [Ljung, 1999, Norton, 1986], in reality, especially for multivariable and closed-loop systems, it is often not clear which variables should be considered as the model inputs. The selection of the input and output variables is based on the aim of the modelling and identification experiment. This stage can also be partially automated. Several candidate models with different input variables can be compared in terms of some performance measure, and the best one is then selected [Guidorzi, 1975, Guidorzi, 1981].
2. *System's order representation.* A common approach is to transform the identification of a dynamic system into a static regression problem [Billings and Voon, 1983a, Leontaritis and Billings, 1985a, Leontaritis and Billings, 1985b, Chen and Billings, 1989, Sjöberg *et al.*, 1995]. The choice of this particular transformation is usually based on a combination of *a priori* knowledge with intuition, insight, and understanding of the process behaviour. Physical modelling of the relationships and physical laws can guide the selection of the relevant variables, and of the model's order. This transformation can be regarded as a mapping from the domain of time signals into a space of variables that fully determine the state of the system. These variables are called the regressors. The system's behaviour can be predicted by means of a static mapping from the space of regressors to the space of the model output (regressand). The choice of the regressors is a crucial step, as an inappropriate choice may lead to useless models.
3. *Fuzzy model complexity.* This choice is related to the number of rules in the model. In the fuzzy modelling literature, the term "structure identification" is often applied to this step [Sugeno and Kang, 1988]. When fuzzy clustering is applied to identify fuzzy models from data, the number of clusters is the primary parameter that must be chosen. It should be noted that the complexity of the final model is also related to the particular type of the model used. The ways in which these models approximate functions differ, which means that the model complexity needed to achieve a required level of accuracy may differ. In practice, a trade–off is usually chosen between the accuracy and the complexity of the model.

3.5.5 Non-linear Regression Problem and Black-Box Models

Fuzzy systems are general function approximators and therefore can be applied to general non–linear regression problems [Wang, 1992, Kosko, 1994, Zeng and Singh, 1996]. Non–linear regression is the modelling of the static dependence of a response variable, called the regressand, $y \in Y \subset \Re$ on the regression vector $\boldsymbol{x} = [x_l, \cdots, x_p]^T$ over some domain $X \subset \Re^p$.

The elements of the regression vector will be called the regressors and the domain X the regressor space. The system that generated the data is presumed to be described by:

$$y \approx f(\boldsymbol{x}) \tag{3.96}$$

The deterministic function $f(\cdot)$ captures the dependence of y on $\boldsymbol{x}$, and $\approx$ reflects the fact that y will not be an exact function of $\boldsymbol{x}$.

As an example, if the dynamics of the system under observation can be described by an Equation Error (EE) model [Kalman, 1982b, Kalman, 1990],

$$y = f(\boldsymbol{x}) + \varepsilon \tag{3.97}$$

where ε can represent the error term.

The aim of regression is to use the data to construct a function $F(\boldsymbol{x})$ that can serve as a reasonable approximation of $f(\boldsymbol{x})$ not only for the given data, but over the entire domain X. The definition of "reasonable approximation" depends on the purpose for which the model is constructed. If the aim of modelling is to obtain predictions of y, accuracy is the most relevant criterion. Lack of accuracy is usually defined by the integral error

$$I = \int_X \| f(\boldsymbol{x}) - F(\boldsymbol{x}) \| \, \mathrm{d}\boldsymbol{x}, \tag{3.98}$$

over the entire domain X. In general, this error cannot be computed, since the value of f is known only at the available data points. Therefore, the average prediction error over the available data is often used

$$J = \frac{1}{N} \sum_{i=1}^{N} \| f(\boldsymbol{x}_i) - F(\boldsymbol{x}_i) \| , \tag{3.99}$$

where N denotes the number of data samples. The attainment of the minimum of I in Equation 3.98 implies the best model possible with the selected structure. This is, however, not the case with the criterion J, which only guarantees that the model fits the available data with the least–squares error. A separate validation step is hence necessary, in order to assess the goodness of the model over the entire region of interest X.

Apart from accurate predictions, the goal may be to obtain a model that can be used to analyse and understand the properties of the real system that generated the data. A strong potential of fuzzy models is that they describe systems as a collection of simple local submodels that are expressed as rules. Rules can also be combined with analytical models, such as the local linear models in the TS structure described in Section 3.5.2.

There are a number of possibilities for the choice of regressors in non–linear black–box identification. Since extensive literature is available on this topic [Leontaritis and Billings, 1985a, Leontaritis and Billings, 1985b, Chen and Billings, 1989, Sjöberg *et al.*, 1995], only a brief review is given here with respect to the use of the particular structures for clustering-based identification.

The NARX (Non–linear AutoRegressive with eXogenous input) model is frequently used with many non–linear identification methods, such as neural networks [Hunt *et al.*, 1992a], radial basis functions [Chen *et al.*, 1991], CMAC [Brown and Harris, 1994a], fuzzy [Lee *et al.*, 1994] and neuro–fuzzy models (Section 2.8.3). The NARX model establishes a relation between the past input-output data and the predicted output:

$$\hat{y}(k+1) = F\big(y(k),\ \cdots,\ y(k-n_y+1),\ u(k),\ \cdots,\ u(k-n_u+1)\big), \quad (3.100)$$

where k denotes discrete time samples, n_u and n_y are integers related to the system's order, and F denotes a fuzzy model. In the NARX model, the regression vector is a collection of a finite number of past inputs and outputs, $x(k) = [y(k),\ \cdots,\ y(k-n_y+1),\ u(k),\ \cdots,\ u(k-n_u+1)]^T$. The regressand is the predicted output $\hat{y}(k+1)$. Hence, from a set of observed inputs and outputs of an unknown dynamic system, the function $F(\cdot)$ in Equation 3.100 can be approximated by using static non–linear regression.

Identification of the input–output model structure corresponds to the choice of the model type and of the related structural parameters n_y, n_u. *A priori* knowledge is typically used to make a first guess of the range of these parameters and a structure is then sought within this range that minimises a certain criterion. The most straightforward approach is to use the mean–square prediction error directly:

$$J = \frac{1}{N}\sum^{N} i = 1\big(y(i) - \hat{y}(i)\big)^2, \quad (3.101)$$

evaluated on a different data set from that used to identify the system, in order to avoid fitting the noise.

Since linear identification techniques are much simpler and numerically more robust than non–linear methods, it is usually worthwhile to start with a linear model to determine the structure. A variety of tools can be used, such as the singular value and the coherence tests [Verhaegen and Dewilde, 1992] or the information–theoretic criteria [Akaike, 1974, Rissanen, 1978]. The structure of the best linear model is then used as a starting point for non–linear modelling. Also cluster validity measures can guide the selection of the model's order and the number of clusters within the given structure.

So far, it has been assumed that y is a scalar, *i.e.*, the system under study is a MISO system. With input–output models, MIMO systems can be represented in two ways: the function $\boldsymbol{F}$ is a vector–valued function, or the MIMO system is decomposed into a set of coupled MISO systems. While the former approach is typically used with neural networks, in fuzzy modelling the decomposition approach is mostly adopted. The reason is that the it is more flexible if each output is associated with a different sort of non–linearity. One output may contain a complex non–linearity in some region, while another output may be linear in the same region. By decomposing the MIMO mapping into several MISO mappings, the number of membership functions and rules can be reduced.

Identification by Product Space Clustering. The principle of identification by product space clustering is to approximate a non–linear regression problem by decomposing it into several local linear subproblems. This approach has a number of advantages in comparison with global non–linear models, such as neural networks. The model structure is easy to understand and interpret, both qualitatively and quantitatively. Various types of knowledge can be integrated in the model, including empirical knowledge, measured data and available mathematical models. In addition, the approach has computational advantages and lends itself to straightforward adaptive and learning algorithms [Murray-Smith and Johansen, 1997].

Fuzzy clustering is applied in the product space of the regressors and the regressand: $X \times Y$. Let $\boldsymbol{X}$ denote the matrix in $\Re^{N \times p}$, having the regression vectors $\boldsymbol{x}_k^T$ in its rows, and let $\boldsymbol{y}$ denote the column vector in $\Re^N$, containing the regressands y_k :

$$\boldsymbol{X} = \begin{bmatrix} \boldsymbol{x}_1^T \\ \vdots \\ \boldsymbol{x}_N^T \end{bmatrix}, \boldsymbol{y} = \begin{bmatrix} y_1 \\ \vdots \\ y_N \end{bmatrix} \tag{3.102}$$

N denotes the number of data samples, p is the dimension of the regression vector. For an input–output model of a dynamic system, the matrix $\boldsymbol{X}$ contains shifted versions of the input and output data.

The decomposition of a global non–linear mapping into a set of locally linear models is based on a geometrical interpretation of the regression problem. The unknown non–linear function $y = f(\boldsymbol{x})$ represents a non–linear surface or hypersurface in the product space: $\left(X \times Y\right) \subset \Re^{p+1}$. This surface is called the regression surface.

The available data represent a sample from the regression surface. By clustering the data, local linear models can be found that approximate the regression surface in an optimal way. The set of data to be clustered, denoted $\boldsymbol{Z}$, is constructed by concatenating the regressor data matrix X and the regressand vector $\boldsymbol{y}$:

$$\boldsymbol{Z}^T = [\boldsymbol{X}, \boldsymbol{y}] \tag{3.103}$$

This data set is a subset of the Cartesian product space $X \times Y$ defined by the non–linear functional relationship 3.96:

$$\boldsymbol{Z} \subset X \times y \quad \text{such that} \quad \boldsymbol{y} \approx f(\boldsymbol{X}) \tag{3.104}$$

The data set $\boldsymbol{Z}$ is partitioned into fuzzy subsets by applying fuzzy clustering algorithms capable of detecting linear substructures in data, according to Sections 3.5.3.

The membership of the data samples in the clusters is described by the fuzzy partition matrix. Each cluster is characterised by its centre and covariance matrix which represents the variance of the data in the cluster. The fuzzy clustering algorithm $\mathcal{C}$ can be regarded as a mapping $\mathcal{C} : (Z \times \mathbb{N}) \rightarrow (M_{f_c} \times \Re^{n \times c} \times \mathrm{PD}^n)$:

$$(\boldsymbol{U}, \boldsymbol{V}, \boldsymbol{F}) = \mathcal{C}(\boldsymbol{Z}, c; \boldsymbol{U}^o, m, \epsilon), \tag{3.105}$$

where c is the number of clusters, $\boldsymbol{U}^o$ is the initial partition matrix, whilst m and ϵ are the parameters of the clustering algorithm, according to Section 3.5.3. The partition matrix $\boldsymbol{U}$ contains the membership degrees of the data points in the clusters with prototypes $\boldsymbol{V}$.

It is worth noting that the cluster covariance matrix $\boldsymbol{F}_i$ contains information about the shape and orientation of the i–th cluster. The j–th eigenvalue and the j–th unit eigenvector of $\boldsymbol{F}_i$ are denoted λ_{ij} and ϕ_{ij}, respectively [Babuška, 1998].

The eigenvalues of the cluster covariance matrix $\boldsymbol{F}_i$ are arranged such that:

$$\lambda_{i1} \geq \lambda_{i2} \geq \cdots \geq \lambda_{in} \tag{3.106}$$

and the eigenvectors are labelled accordingly. The eigenvectors ϕ_{i1} to $\phi_{i(n-1)}$ span the i–th cluster's linear subspace and the nth eigenvector ϕ_{in} is the normal to this linear subspace. Since λ_{in} is the smallest eigenvalue, ϕ_{in} is therefore called the smallest eigenvector.

When the intrinsic dimension of the data is p, λ_{in} is in orders of magnitude smaller than the remaining eigenvalues. This means that the n-dimensional data can be locally represented by a linear combination of $n-1$ variables. Hence, the proportions between the eigenvalues can be used to check whether an appropriate structure has been chosen. If all the eigenvalues are in the same order of magnitude, then no functional relationship between the regressors and the regressand has been detected (the chosen structure may not be rich enough). However, when several relatively small eigenvalues are found, the regression problem may be (locally) of a lower dimension than was assumed, or the data might not be rich enough [Babuška, 1998].

3.5.6 Fuzzy Model Identification From Clusters

As shown in the previous sections, fuzzy clustering algorithms can be used to approximate a set of data by local linear models. Each of these models is represented by a fuzzy subset in the data set available for identification. In order to obtain a model useful for prediction, an additional step must be applied to generate a model independent of the identification data. Such a model can be represented either as a rule base or as a fuzzy relation [Babuška, 1998]. This section reviews algorithms for constructing fuzzy TS models from the fuzzy partitions obtained by product space clustering. In particular, the construction of Takagi–Sugeno models is addressed and methods for generating the antecedent membership functions and estimating the consequent parameters are presented below.

Each cluster obtained by product space clustering of the identification data set can be regarded as a local linear approximation of the regression hypersurface. The global model can be conveniently represented as a set of affine Takagi–Sugeno rules:

$$R_i : \textbf{IF } \boldsymbol{x} \text{ is } A_i \textbf{ THEN } y_i = \boldsymbol{a}_i^T \boldsymbol{x} + b_i \ , \ i = 1, 2, \cdots, K. \qquad (3.107)$$

The *antecedent* fuzzy sets A_i can be computed analytically in the antecedent product space, or can be extracted from the fuzzy partition matrix. The *consequent parameters* $\boldsymbol{a}_i$ and b_i are estimated from the data using the methods presented in Section 3.3 (*i.e.*, Frisch scheme procedure), or they can be extracted from the eigenstructure of the cluster covariance matrices.

Antecedent Membership Function Generation. The antecedent membership functions can be obtained by projecting the fuzzy partition onto the antecedent variables, or by computing the membership degrees directly in the product space of the antecedent variables. These two methods are described in the following but the second one will be exploited for the identification of the TS models by means of the Fuzzy Modelling and Identification Toolbox (*FMID*), for *Matlab* [Babuška, 2000] developed by Robert Babuška [Babuška, 1998].

- *Antecedent Membership Function Estimation by Projection.* The principle of this method consists of projecting the multi–dimensional fuzzy sets defined pointwise in the rows of the partition matrix $\boldsymbol{U}$ onto the individual antecedent variables of the rules. These variables can be the original regression variables, in which case the projection is an orthogonal projection of the data. New, transformed antecedent variables can be obtained by means of eigenvector projection, using the p largest eigenvectors of the cluster covariance matrices. The eigenvector projection is useful for clusters which are opaque to the axis of the regression space, and cannot be represented by axis-orthogonal projection with a sufficient accuracy.

By projecting the i–th row $\boldsymbol{\mu}_i$ of the fuzzy partition matrix $\boldsymbol{U}$ onto the antecedent variable x_j, a point–wise definition of the fuzzy set A_{ij} is obtained. In order to obtain a prediction model, the antecedent membership functions must be expressed in a form that allows computation of the membership degrees, also for input data not contained in the data set $\boldsymbol{Z}$. This is achieved by approximating the point–wise defined membership function by some suitable parametric function.

The piece-wise exponential membership functions proved to be suitable for the accurate representation of the actual cluster shape. This function is fitted to the envelope of the projected data by numerically optimising its parameters. An advantage of this method over the *multi–dimensional membership functions*, summarised below, is that the projected membership functions can always be approximated such that convex fuzzy sets are obtained. Moreover, asymmetric membership functions can be used to reflect the actual partition of the considered non–linear regression problem.

– *Multi–dimensional Antecedent Membership Functions.* In this second method, the antecedent membership functions are represented analytically by computing an inverse of the distance from the cluster prototype. The membership degree is computed directly for the entire input vector (without the decomposition). The antecedents of the TS rules are simple propositions with multi–dimensional fuzzy sets given by Equation 3.107, and $\beta_i(\boldsymbol{x}) = \mu_{A_i}(\boldsymbol{x})$.

Recall that $\boldsymbol{F}^x = [f_{ij}]$, $1 \leq i,\ j \leq p$ includes all but the last column of the cluster covariance matrix. The corresponding norm–inducing matrix is given by:

$$\boldsymbol{A}_i^x = [\rho_i \det(\boldsymbol{F}_i^x)]^{1/p} (\boldsymbol{F}_i^x)^{-1}. \tag{3.108}$$

Let $\boldsymbol{v}_i^x = [v_{1i}, \cdots, v_{pi}]^T$ denote the projection of the cluster centre onto X. The norm

$$D_{\boldsymbol{A}_i^x}(\boldsymbol{x}, \boldsymbol{v}_i^x) = (\boldsymbol{x} - \boldsymbol{v}_i^x)^T \boldsymbol{A}_i^x (\boldsymbol{x} - \boldsymbol{v}_i^x). \tag{3.109}$$

measures the distance of the antecedent vector $\boldsymbol{x}$ from the projection of the cluster centre $\boldsymbol{v}_i^x$. To transform the distance into the membership degree, two methods can be applied which are described below.

One possibility is to use the formula employed by the clustering algorithms:

$$\beta_i(\boldsymbol{x}) = \frac{1}{\sum_{j=1}^{c} \left(D_{\boldsymbol{A}_i^x}(\boldsymbol{x}, \boldsymbol{v}_i^x) / D_{\boldsymbol{A}_i^x}(\boldsymbol{x}, \boldsymbol{v}_i^x) \right)^{2/(m-1)}} \tag{3.110}$$

This expression computes the degree of fulfilment of one rule relative to the other rules and the sum of the membership degrees of all the rules equals one, as in fuzzy clustering. Because of this constraint we class this method as probabilistic.

On the other hand, the degree of fulfilment of a rule can also be computed independently of the remaining rules, by applying a second transformation of the distance into the membership degree:

$$\beta_i(\boldsymbol{x}) = \frac{1}{1 + D_{\boldsymbol{A}_i^x}(\boldsymbol{x}, \boldsymbol{v}_i^x)} \tag{3.111}$$

In both cases, the distance is computed in relation to the cluster centre, which implies that the obtained multi-dimensional membership functions 3.110 and 3.111 have the same position (centre) in the antecedent space. However, the shape of the membership functions may differ significantly [Babuška, 1998].

Estimating Consequent Parameters. There are several methods of obtaining the consequent parameters. Based on the geometrical interpretation of the TS model, the consequent parameters can be directly computed from the cluster prototypical points and the smallest eigenvectors of the cluster covariance matrices. This method assumes that errors are present in both the regressors and the regressand, and corresponds to the total least–squares solution of the local linearisation around the cluster centre [van Huffel and Vandewalle, 1991].

A set of optimal parameters with respect to the model output can also be estimated from the identification data set by ordinary least–squares methods or by using the approach presented in Section 3.3.2. This approach can be formulated as minimisation of the total prediction error using the TS defuzzification formula 3.73, or as minimisation of the prediction errors of the individual local models, solved as a set of c independent, weighted least-squares problems.

In the following, an example of identification of the consequent TS parameters by exploiting total least–squares algorithm. This approach is usually preferred when the TS model should serve as predictor [Babuška, 1998].

Computing Consequent Parameters by Total Least–squares Method. The consequent parameters $\boldsymbol{a}_i$ and b_i of the affine TS model 3.107 can be derived from the geometrical structure of the clusters.

Assume that the collection of c clusters approximates the regression surface. These clusters can be approximately regarded as p–dimensional linear subspaces of the regression space. The eigenvector ϕ_{in} corresponding to the smallest eigenvalue λ_{in}, determines the normal vector to the hyperplane spanned by the remaining eigenvectors of that cluster.

The smallest eigenvector of the i–th cluster will be denoted ϕ_i omitting the subscript n. Recall that $\boldsymbol{z}^T = [\boldsymbol{x}^T, y]$ is the data vector and $\boldsymbol{v}_i$ is the cluster's prototype. The implicit normal form of the consequent's hyperplane is given by:

$$\phi_i(\boldsymbol{z} - \boldsymbol{v}_i) = 0. \tag{3.112}$$

This expression states that the product of the normal vector ϕ_i with any vector belonging to the hyperplane equals zero. For the following discussion, it is convenient to partition the prototype $\boldsymbol{v}_i$ into a vector $\boldsymbol{v}_i^x$ corresponding to the regressor $\boldsymbol{x}$, and a scalar v_i^y corresponding to the regressand y: $\boldsymbol{v}_i^T = \left[\left(\boldsymbol{v}_i^x\right)^T, v_i^y\right]$. The smallest eigenvector is partitioned in the same way as the cluster centre, *i.e.* : $\phi_i^T = \left[\left(\phi_i^x\right)^T, \phi_i^y\right]$. Equation 3.112 can now be written as:

$$\left[\left(\phi_i^x\right)^T, \phi_i^y\right] \left(\left[\boldsymbol{x}^T, y\right]^T - \left[\left(\boldsymbol{v}_i^x\right)^T, v_i^y\right]^T \right) = 0. \tag{3.113}$$

Carrying out the inner product leads to the following equality:

$$\left(\phi_i^x\right)^T \left(\boldsymbol{x} - \boldsymbol{v}_i^x\right) + \phi_i^y \left(y - v_i^y\right) \tag{3.114}$$

from which, by a simple algebraic manipulation, an explicit equation for the hyperplane is obtained:

$$y = \frac{-1}{\phi_i^y} \left(\phi_i^x\right)^T \boldsymbol{x} + \frac{1}{\phi_i^y} \phi_i^T \boldsymbol{v}_i. \tag{3.115}$$

By comparing the above expression with the affine consequent of the TS rule 3.107, equations for $\boldsymbol{a}_i$ and b_i directly follow:

$$\boldsymbol{a}_i = \frac{-1}{\phi_i^y} \phi_i^x, \tag{3.116}$$

$$b_i = \frac{1}{\phi_i^y} \phi_i^T \boldsymbol{v}_i. \tag{3.117}$$

Although these equations have been derived from the geometrical interpretation of the clusters, it can be shown that $\boldsymbol{a}_i$ given by Equation 3.116 is obtained as a solution of a weighted total least-squares (TLS) problem defined locally around the cluster centre $\boldsymbol{v}_i$. The weights are the membership degrees contained in the i–th row of the fuzzy partition matrix. Hence $\boldsymbol{v}_i$ is seen as a local operating point for the model. To obtain the affine linear form used in the TS rules 3.107, the offset parameters b_i are calculated using the estimates $\boldsymbol{a}_i$ and the cluster centre $\boldsymbol{v}_i$.

Computing Consequent Parameters by Frisch Scheme Procedure. After the clustering of the data has been obtained, data subsets can be processed according the Frisch scheme identification procedure

[Frisch, 1934, Beghelli *et al.*, 1990], in order to estimate the TS parameters for each affine submodels, according to the rules presented in Section 3.4.3 [Simani *et al.*, 1998a, Fantuzzi *et al.*, 1998, Rovatti *et al.*, 2000, Simani *et al.*, 1999c, Simani, 2000b].

The system identification with noisy measurements presented in Section 3.3.2 can be applied slightly adapted to handle affine models.

In order to identify the structure of the TS SISO model of Equation 3.73 in the i–th cluster with $i = 1, \cdots, c$ and c clusters, the following matrices can be defined:

$$X_k^{(i)} = \begin{bmatrix} y(k) & \boldsymbol{x}_k^T(0) & 1 \\ y(k+1) & \boldsymbol{x}_k^T(1) & 1 \\ \vdots & \vdots & \\ y(k+N_i-1) & \boldsymbol{x}_k^T(N_i-1) & 1 \end{bmatrix} \quad (3.118)$$

and therefore

$$\Sigma_k^{(i)} = \left(X_k^{(i)}\right)^T X_k^{(i)}. \quad (3.119)$$

In order to solve the noise–rejection problem in a mathematical framework, it is necessary to follow the assumptions [Frisch, 1934] [Kalman, 1982b] [Beghelli *et al.*, 1990] that the noises $\tilde{u}(t)$ and $\tilde{y}(t)$ are additive on the input–output data $u^*(t)$ and $y^*(t)$ and region independent.

Under these assumptions, a positive-definite matrix $\Sigma_k^{(i)}$ associated to the sequences belonging to the i–th cluster can be expressed as the sum of two terms $\Sigma_k^{(i)} = \Sigma_k^{*(i)} + \bar{\tilde{\Sigma}}_k$ where

$$\bar{\tilde{\Sigma}}_k = \text{diag}[\bar{\tilde{\sigma}}_y I_{k+1}, \bar{\tilde{\sigma}}_u I_k, 0] \geq \mathbf{0}. \quad (3.120)$$

The solution of the above identification problem requires the computation of the unknown noise covariances $\bar{\tilde{\sigma}}_u$ and $\bar{\tilde{\sigma}}_y$, that can be achieved solving the following relation:

$$\Sigma_k^{*(i)} = \Sigma_k^{(i)} - \tilde{\Sigma}_k \geq 0. \quad (3.121)$$

in the variables $\tilde{\sigma}_u, \tilde{\sigma}_y$, where $\tilde{\Sigma}_k = \text{diag}[\tilde{\sigma}_y I_{k+1}, \tilde{\sigma}_u I_k, 0]$.
It is worth noting that all the surfaces of type as defined by Equation 3.121 have necessarily at least one common point, *i.e.* point $(\bar{\tilde{\sigma}}_u, \bar{\tilde{\sigma}}_y)$ corresponding to the true variances of the noise affecting the input and the output data.

The search for a solution for the identification problem can therefore start from the determination of this point in the noise space, if the noise characteristics are common to all the clusters and all assumptions regarding the Frisch scheme are satisfied (independence between input–output sequences, additive noise, noise whiteness).

In real cases, these assumptions have to be relaxed, thus no common point can be determined among surfaces $\Gamma_n^{(i)} = 0$ in the noise plane and a unique solution to the identification problem can not be obtained.

In this situation, the local fuzzy model identification can be performed by finding the point $(\tilde{\sigma}_u, \tilde{\sigma}_y) \in \Gamma_{n+1}^{(i)} = 0$ that makes $\Sigma_{n+1}^{*(i)}$ closer to the double singular condition. It leads to determine the common point of the surfaces when the assumptions of the Frisch scheme are not violated.

Moreover, for each cluster, different noises $(\bar{\tilde{\sigma}}_u^{(i)}, \bar{\tilde{\sigma}}_y^{(i)})$ and the following relation should be rewritten as

$$\Sigma_n^{*(i)} = \Sigma_n^{(i)} - \tilde{\Sigma}_n^{(i)} \geq 0 \tag{3.122}$$

where $\tilde{\Sigma}_n^{(i)} = \text{diag}[\bar{\tilde{\sigma}}_u^{(i)} I_{n+1}, \bar{\tilde{\sigma}}_y^{(i)} I_n, 0]$ whilst $(\bar{\tilde{\sigma}}_u^{(i)}, \bar{\tilde{\sigma}}_y^{(i)})$ represent the variances of input and output additive noises in the i-th cluster.

Finally, the matrices $\tilde{\Sigma}_n^{(i)}$ can therefore be built and the parameter of the model in each cluster determined by means of relation

$$(\Sigma_n^{(i)} - \tilde{\Sigma}_n^{(i)})\boldsymbol{a}^{(i)} = \mathbf{0} \quad \text{for } i = 1, \ldots, c. \tag{3.123}$$

for a number of c clusters. This completes the multiple-model identification procedure in the fuzzy environment.

In Chapter 5 examples concerning the fuzzy modelling and identification of real processes by means of TS models will be presented.

3.6 Conclusion

Several off–line identification methods have been presented in this chapter for the estimation of both linear and non–linear static and dynamic system from data affected by noise. Linear, piecewise affine and fuzzy models were discussed.

For the case of piecewise affine and fuzzy models, the multiple-model approach consists of using several local affine submodels each describing a different operating condition of the process.

The identification algorithm exploited to estimate parameters and orders of the local affine submodels is based on the well–established Frisch Scheme method for linear systems.

For the non–linear case, in order to obtain a continuous piecewise affine prototype describing the input-output behaviour of the process, continuity constraints between local linear dynamic models have to be forced.

For non–fuzzy models, such a continuity constrained problem was solved by using an optimisation technique. The properties of the solutions obtained by the Frisch Scheme enhance the fulfilment of the constraints.

On the other hand, in order to identify a non–linear process, neural networks, fuzzy and neuro–fuzzy modelling and identification methods can be exploited. By means of product space fuzzy clustering, a data set generated by the non–linear system can be partitioned into fuzzy subsets of data that are locally described by linear sub models. Prior to clustering, the regression structure of the model must be selected, in order to properly represent the system 's dynamics.

After the structure is determined, clustering in the product space of the regressors and of the regressand can be applied to partition the data. Since each cluster serves as a local linear model of the system, the clustering algorithm must be capable of detecting clusters which lie in linear subspaces.

Therefore, for the final construction of the Takagi–Sugeno fuzzy models, two main tasks can be distinguished: the generation of the antecedent membership functions, and the estimation of the consequent parameters.

The consequent parameters can be computed from the cluster centres and covariance matrices, or can be estimated by least–squares methods as well as by local application of the Frisch scheme identification procedure for linear models.

Finally, fuzzy identification algorithms exploited both to generate fuzzy partitions and to estimate TS structure were achieved by using the Fuzzy Modelling and Identification Toolbox (*FMID*) for *Matlab* developed by Robert Babuška.

CHAPTER 4

RESIDUAL GENERATION, FAULT DIAGNOSIS AND IDENTIFICATION

4.1 Introduction

The most important task in model-based FDI is the generation of residuals which are independent of disturbances. The method is based on disturbance de–coupling principle. In this approach, uncertain factors in system modelling or identification are considered to act by means of an unknown input, the disturbance, on a linear system model.

The disturbance vector is unknown but its distribution matrix is usually assumed known. However, in the following, we will outline methods of estimating the disturbance distribution matrix, under the assumption that the system can be identified with an equation error model.

Based on the disturbance distribution matrix obtained by modelling or identification procedure, the unknown input can be de–coupled form the residual.

The principle of the Unknown Input Observer (UIO) is to make the state (or output) estimation error de–coupled from the unknown inputs or disturbances. Since the residual is a weighted output estimation error, it may be de–coupled from each disturbance.

This approach was originally propose by Watanabe and Himmelblau [Watanabe and Himmelblau, 1982], who considered the sensor FDI problem for systems with modelling uncertainties. Later, the approach was generalised by Frank [Patton *et al.*, 1989, Frank, 1990] in order to perform the FDI of both sensors and actuators. Very important contributions to this subject can be found in [Chen and Patton, 1999, Liu and Patton, 1998, Isermann and Füssel, 1999, Drag and Patton, 2001].

The first step in the disturbance de–coupled residual generation consists of designing an UIO. This chapter shows how to obtain the structure of a full order UIO for FDI purpose. The design of an UIO will be presented from a mathematical point of view as well as the necessary and sufficient existence conditions.

Unlike some other works, in which the reduced order structure is exploited, this chapter is based exclusively on the use of the full order UIO. In fact, for a full order UIO, there is more design freedom available to achieve other required performance, after the disturbance de–coupling conditions have been satisfied. As an example, the remaining design of freedom can be exploited to obtain directional residuals [Chen *et al.*, 1996b, Chen and Patton, 1999].

UIO or other disturbance de–coupling based residual generation approaches require that the unknown input distribution matrix must be known *a priori*. The actual unknown input itself does not need to be known.

When uncertainties are caused by modelling errors, linearisation errors, parameter variations, etc, such a disturbance de–coupling approach cannot be directly applied because the distribution matrix $\boldsymbol{E}$ is normally unknown. To solve this problem, which is of importance in real industrial system applications, some investigators have suggested an approach exploiting *estimated* [Chen and Patton, 1999] distribution matrices. In Section 4.7, the author of this monograph will suggest a method using *identified* distribution matrices.

The last approximate strategy has extended the application of disturbance de–coupling–based residual generation to actual process FDI.

Some simulation results applied to a real industrial power plant and using identified disturbance distribution matrix technique will be shown in Chapter 5.

Finally, techniques exploiting fuzzy models and NNs are presented in order to perform residual generation and fault identification, respectively.

4.2 Output Observers for Robust Residual Generation

This subsection addresses the problem of the detection of faults on the basis of the knowledge of the measured input $\boldsymbol{u}(t)$ and output $\boldsymbol{y}(t)$ sequences by using dynamic output observers.

As previously stated in Section 2.3, the fault model comprising controller (actuator), $\boldsymbol{f}_a(t)$ and process $\boldsymbol{f}_c(t)$ failures can be described as:

$$\begin{aligned} \boldsymbol{x}(t+1) &= \boldsymbol{A}\boldsymbol{x}(t) + \boldsymbol{B}(\boldsymbol{u}(t) + \boldsymbol{f}_c(t)) + \boldsymbol{f}_s(t) \\ \boldsymbol{y}(t) &= \boldsymbol{C}\boldsymbol{x}(t), \end{aligned} \tag{4.1}$$

$\boldsymbol{A}, \boldsymbol{B}$ and $\boldsymbol{C}$ being system's matrices, $\boldsymbol{f}_s(t)$ and $\boldsymbol{f}_c(t)$ being failure affecting the process and the controller (more precisely the process' actuator), respectively.

Moreover, input and output sensor malfunctions are modelled as:

$$\left.\begin{aligned} \boldsymbol{u}(t) &= \boldsymbol{f}_u(t) + \boldsymbol{u}^*(t) \\ \boldsymbol{y}(t) &= \boldsymbol{f}_y(t) + \boldsymbol{y}^*(t) \end{aligned}\right\} \tag{4.2}$$

$\boldsymbol{u}(t)$ and $\boldsymbol{y}(t)$ being the input and output sensor measurements, respectively, $\boldsymbol{f}_u(t)$ and $\boldsymbol{f}_y(t)$ being failure on input and output sensors.

According to Figure 4.1, the fault detection and isolation of faults is therefore achieved through the processing of residual signals $\boldsymbol{r}(t)$, which are obtained comparing system measurements with Luenberger dynamic observers [Luenberger, 1971].

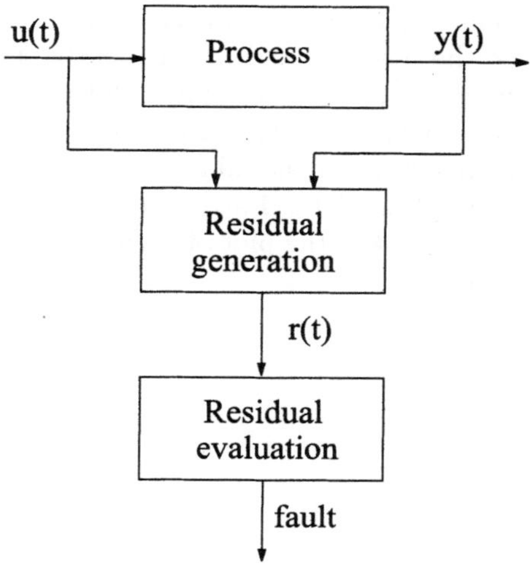

Fig. 4.1. Logic scheme of the fault detection system.

In particular, MISO dynamic observers have the following structure:

$$\boldsymbol{x}^i(t+1) = \boldsymbol{A}_i\boldsymbol{x}^i(t) + \boldsymbol{B}_i\boldsymbol{u}^*(t) + \boldsymbol{K}_i\left(\boldsymbol{y}^*(t) - \boldsymbol{C}_i\boldsymbol{x}^i(t)\right) \tag{4.3}$$

$\boldsymbol{x}^i(t)$ being the i–th observer state vector, triple $(\boldsymbol{A}_i,\boldsymbol{B}_i,\boldsymbol{C}_i)$ is a minimal state space representation (completely observable) of the link among the inputs of the process and its i–th output $y_i^*(t)$.

With reference to the FDI scheme presented in Section 4.4, the observer can be designed so that each observer output is sensitive to a particular fault. Different faults are linked to different outputs from sensitivity analysis of the residuals.

The observer eigenvalues, $\boldsymbol{p} = [p_1, \ldots, p_n]$, are chosen solving the minimisation problem

$$\min_{\boldsymbol{p}} V(\boldsymbol{p}) \tag{4.4}$$

$V(\boldsymbol{p})$ being the cost function

$$V(\boldsymbol{p}) = \frac{||r(t,\boldsymbol{p})|_h||^2}{||r(t,\boldsymbol{p})|_f||^2} \tag{4.5}$$

where $||\cdot||$ represents the 2–norm or the mean square value of the vector $r(\cdot)$, *i.e.*, the square root of the sum of the squared entries of the vector $r(\cdot)$.

Hence, according to Equation 4.5, the eigenvalues $\boldsymbol{p}$ are chosen in order to maximise the mean square error of the fault residual sensitivity $r(t,\boldsymbol{p})|_f$ (to maximise fault detection promptness) and minimise the mean square error of the residual in fault–free condition, $r(t,\boldsymbol{p})|_h$ (to minimise the occurrence of false alarms).

This pole placement procedure also provides robustness property versus measurement noise to the observer and, consequently, false alarm rejection.

Eigenvalue assignment is an indirect approach to the design robust residual of the observer. The most important direct approach to design robust (in the disturbance de–couple sense) residual generation is the use of eigenstructure assignment [Patton *et al.*, 2000, Patton and Chen, 2000].

As an example, in Figure 4.2, the plot of $V(p)$ with $p = p_n$ and $p_1 = p_2 = \cdots = p_{n-1} = 0.5$ is shown.

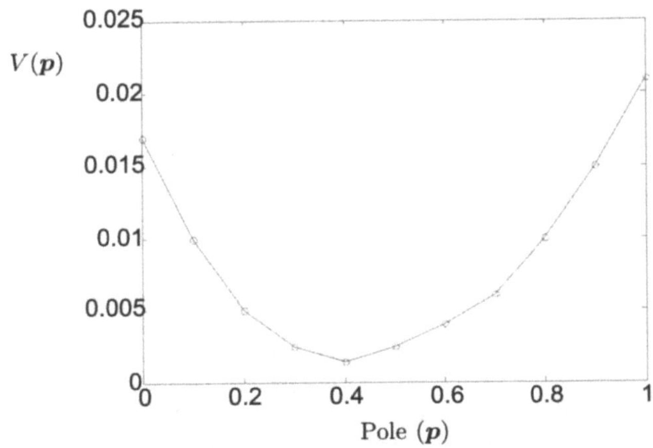

Fig. 4.2. Pole assignment cost function.

In particular, in the presence of faults, the dynamic observer for the i-th output turns in

$$\boldsymbol{x}^i(t+1) = \boldsymbol{A}_i \boldsymbol{x}^i(t) + \boldsymbol{B}_i \boldsymbol{u}(t) + \boldsymbol{K}_i \left(y_i(t) - \boldsymbol{C}_i \boldsymbol{x}^i(t) \right) \tag{4.6}$$

where $\boldsymbol{x}^i(t)$ is the i–th observer state vector and the triple $(\boldsymbol{A}_i,\boldsymbol{B}_i,\boldsymbol{C}_i)$ is a minimal state space representation (completely observable) of the link among the inputs of the process and its i–th output $y_i^*(t)$.

In the absence of faults, it can be verified that, for the i–th output, the residual $r_i(t) = y_i^*(t) - \hat{y}_i(t) = \boldsymbol{C}_i \left(\boldsymbol{x}_i(t) - \boldsymbol{x}^i(t) \right)$ is equal to zero.

These previous results were obtained using the linear model of the monitored system. Because of the residual dynamics, a simple geometrical analysis, such a fixed threshold logic can be exploited in order to detect actuator faults. Clearly, suitable threshold values have to be set under fault–free conditions.

Therefore, the inequalities of 4.7 are obtained, when the symptom evaluation in the noise–free case is performed by comparing residual signals $\boldsymbol{r}(t)$ with the fixed threshold ϵ,

$$\left.\begin{array}{ll} \boldsymbol{r}(t) \leq \epsilon & \text{for} \quad \boldsymbol{f}(t) = \mathbf{0} \\ \boldsymbol{r}(t) > \epsilon & \text{for} \quad \boldsymbol{f}(t) \neq \mathbf{0} \end{array}\right\} \tag{4.7}$$

$\boldsymbol{f}(t)$ being a generic failure vector.

4.3 Unknown Input Observer

This section deals with the design of UIO observers for discrete–time, time–invariant, linear dynamic systems with an additive unknown disturbance term. From a mathematical point of view, these systems are described by the following model

$$\left.\begin{array}{lll} \boldsymbol{x}(t+1) & = & \boldsymbol{A}\boldsymbol{x}(t) + \boldsymbol{B}\boldsymbol{u}(t) + \boldsymbol{E}\boldsymbol{d}(t) \\ \boldsymbol{y}(t) & = & \boldsymbol{C}\boldsymbol{x}(t) \end{array}\right\} \tag{4.8}$$

where, $\boldsymbol{x}(t) \in \Re^n$ is the state vector, $\boldsymbol{y}(t) \in \Re^m$ is the output vector, $\boldsymbol{u}(t) \in \Re^r$ the known input vector and $\boldsymbol{d}(t) \in \Re^q$ the unknown input vector. $\boldsymbol{A}$, $\boldsymbol{B}$, $\boldsymbol{C}$, $\boldsymbol{E}$ are known matrices with appropriate dimensions.

It is worth noting how the unknown term $\boldsymbol{E}\boldsymbol{d}(t)$ can be used to describe an additive disturbance, different kinds of modelling uncertainties (noise, unmodelled non–linear terms, time–varying dynamics, etc.) as well as fault terms.

The unknown input term may also appear in the output equation, *i.e.*,

$$\boldsymbol{y}(t) = \boldsymbol{C}\boldsymbol{x}(t) + \boldsymbol{E}_y\boldsymbol{d}(t) \tag{4.9}$$

but this case is not considered because the term $\boldsymbol{E}_y\ \boldsymbol{d}(t)$ can be nulled by using a transformation of the output signal $\boldsymbol{y}(t)$ [Chen and Patton, 1999].

For systems described by Equation 4.8, there is a term relating the control input $\boldsymbol{u}(t)$ in the output equation, *i.e.*,

$$\boldsymbol{y}(t) = \boldsymbol{C}\boldsymbol{x}(t) + \boldsymbol{D}\boldsymbol{u}(t). \tag{4.10}$$

However, the term $\boldsymbol{D}\boldsymbol{u}(t)$ is omitted in this monograph since this does not affect the generality of the discussion on the observer design.

Definition 4.3.1. *An observer is defined as an* **Unknown Input Observer** *for the system described by Equations 4.8, if its state estimation error vector* $\mathbf{e}_x(t)$ *approaches zero asymptotically, regardless of the presence of the unknown input term in the system.*

The problem of designing an observer for unknown inputs has been studied for nearly two decades and after the paper of Wang [Wang *et al.*, 1975], many approaches for the design of both full–order and reduced–order UIO have been proposed (geometric and algebraic methods, singular value decomposition and matrix inversion techniques, linear transformation algorithms) [Chen and Patton, 1999].

In this chapter, a full–order UIO structure is used and a mathematical method for designing UIO is presented. The necessary and sufficient conditions for this observer to exist are also recalled. These conditions are easy to verify and the design procedure is easy to implement.

4.3.1 UIO Mathematical Description

The full–order UIO has the following mathematical form

$$\left.\begin{aligned} \boldsymbol{z}(t+1) &= \boldsymbol{F}\boldsymbol{z}(t) + \boldsymbol{T}\boldsymbol{B}\boldsymbol{u}(t) + \boldsymbol{K}\boldsymbol{y}(t) \\ \hat{\boldsymbol{x}}(t) &= \boldsymbol{z}(t) + \boldsymbol{H}\boldsymbol{y}(t) \end{aligned}\right\} \quad (4.11)$$

where $\boldsymbol{z}(t) \in \Re^n$ is the state of the UIO, $\hat{\boldsymbol{x}}(t)$ the estimated state vector $\boldsymbol{x}(t)$, whilst $\boldsymbol{F}$, $\boldsymbol{T}$, $\boldsymbol{H}$ and $\boldsymbol{K}$ are matrices to be designed to achieve the unknown input de–coupling .

The observer described by Equations 4.11 is depicted in Figure 4.3.

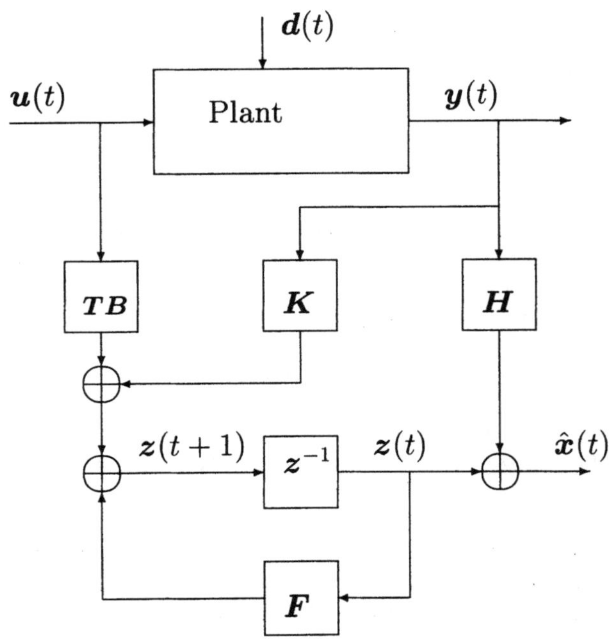

Fig. 4.3. The UIO structure.

The state estimation error obtained by the UIO 4.11 applied to the system of Equation 4.8 is described by the equation:

$$\begin{aligned} e_x(t+1) &= [A - HCA - K_1C]e_x(t) + \big[F - (A - HCA - K_1C)\big]z(t) \\ &+ \big[K_2 - (A - HCA - K_1C)\big]y(t) \\ &+ \big[T - (I - HC)\big]Bu(t) + (HC - I)Ed(t) \end{aligned} \tag{4.12}$$

where $K = K_1 + K_2$.

If the following relations hold:

$$\begin{aligned} (HC - I)E &= 0 \\ I - HC &= T \\ A - HCA - K_1C &= F \\ FH &= K_2 \end{aligned} \tag{4.13}$$

the state estimation error will then be:

$$e_x(t+1) = Fe_x(t). \tag{4.14}$$

This means that, if all the eigenvalues of F are stable, $e_x(t)$ will approach zero asymptotically, *i.e.*, $\hat{x} \to x$. Hence, according to the Definition 4.3.1, the observer described by Equations 4.11 is an UIO for the system 4.8.

The design of this UIO consists of solving Equation 4.13) and making all eigenvalues of the system matrix F be stable.

The following theorem states the existence conditions for the UIO.

Theorem 4.3.1. *Necessary and sufficient conditions for the existence of an UIO 4.11 for the system defined by Equation 4.8 are [Chen and Patton, 1999]:*

1. *rank*$(CE) =$ *rank*(E),
2. (A_1, C) *is a detectable pair,*

where $A_1 = A - E(CE)^+CA$.

A special solution for the matrix H in conditions 4.13 is given by [Chen and Patton, 1999]:

$$H^* = E(CE)^+ \tag{4.15}$$

where $(\cdot)^+$ is the pseudoinverse of the matrix CE.

It is worth noting that the number of independent row of the matrix C must not be less than the number of the independent columns of the matrix E to satisfy condition 1 in Theorem 4.3.1. It means that the maximum number of disturbances which can be de–coupled cannot be larger than the number of the independent measurements. Moreover, without unknown inputs in the system, by setting $T = I$, $H = 0$ and $E = 0$, the observer 4.11 will be a simple Luenberger observer. In this situation, condition 1 in Theorem 4.3.1 clearly holds true and condition 2 is the detectability of pair (A, C).

4.3.2 UIO Design Procedure

It can be seen how $\boldsymbol{K}_1$ is a free matrix of parameters in the design of an UIO. After $\boldsymbol{K}_1$ is computed, in order to stabilise the dynamic system matrix $\boldsymbol{F}$, other parameter matrices in the UIO can be computed by the relation $\boldsymbol{K} = \boldsymbol{K}_1 + \boldsymbol{K}_2$ and conditions 4.13. Some design freedom left in the choice of $\boldsymbol{K}_1$ may be exploited to make the diagnostic residual has directional characteristics. In this monograph, because the input–output link of the Multiple-Input Multiple-Output (MIMO) system under investigation is obtained by means of the identification of a collection of Multiple-Input Single-Output (MISO) models, this further degree of freedom will not be used in the residual design.

Under these assumptions, if the pair $(\boldsymbol{A}_1, \boldsymbol{C})$ is observable, in order to stabilise the system matrix $\boldsymbol{F} = \boldsymbol{A}_1 - \boldsymbol{K}_1\boldsymbol{C}$, the pole placement routine available in the Control System Toolbox for MATLAB [Mat, 1990] can be used.

If $(\boldsymbol{A}_1, \boldsymbol{C})$ is not observable, an observable canonical decomposition should be applied to the pair [Chen and Patton, 1999]. If $(\boldsymbol{A}_1, \boldsymbol{C})$ is detectable, the matrix $\boldsymbol{F}$ can be stabilised.

4.4 FDI Schemes Based on UIO and Output Observers

The main task of FDI is to generate residual signals which have to be sensitive to faults themselves. According to Chapter 2, a system with faults concerning system inputs and outputs can be represented as

$$\left.\begin{array}{rcl} \boldsymbol{x}(t+1) & = & \boldsymbol{A}\boldsymbol{x}(t) + \boldsymbol{B}\boldsymbol{u}(t) + \boldsymbol{B}\boldsymbol{f}_u(t) \\ \boldsymbol{y}(t) & = & \boldsymbol{C}\boldsymbol{x}(t) + \boldsymbol{f}_y(t) \end{array}\right\} \tag{4.16}$$

where $\boldsymbol{A}$, $\boldsymbol{B}$ and $\boldsymbol{C}$ are constant matrices of appropriate dimensions obtained by means of the identification procedures recalled in Chapter 3.

The vectors $\boldsymbol{f}_u(t) = [f_{u_1}(t) \ldots f_{u_r}(t)]^T$ and $\boldsymbol{f}_y(t) = [f_{y_1}(t) \ldots f_{y_m}(t)]^T$ assume values different from zero only in the presence of faults.

Usually these signals are described by step and ramp functions representing abrupt and incipient faults (bias or drift), respectively.

With reference to Chapter 2, the actual measured signals $\boldsymbol{u}(t)$ and $\boldsymbol{y}(t)$ are modelled as:

$$\left.\begin{array}{rcl} \boldsymbol{u}(t) & = & \boldsymbol{u}^*(t) + \tilde{\boldsymbol{u}}(t) \\ \boldsymbol{y}(t) & = & \boldsymbol{y}^*(t) + \tilde{\boldsymbol{y}}(t) \end{array}\right\} \tag{4.17}$$

in which, the sequences $\tilde{\boldsymbol{u}}(t)$ and $\tilde{\boldsymbol{y}}(t)$ are usually described as white, zero–mean, uncorrelated Gaussian noises.

To uniquely isolate a fault concerning one of the *system outputs*, $\boldsymbol{f}_y(t)$, under the hypothesis that inputs are fault–free, ($\boldsymbol{f}_u(t) = \boldsymbol{0}$), a bank of classical dynamic observers or Kalman filter (KF) is used, according to Figure 4.4.

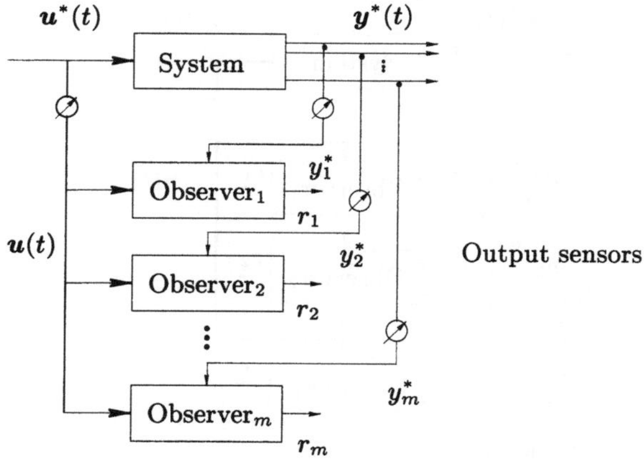

Fig. 4.4. Bank of estimators for output residual generation.

This observer configuration represents the Dedicated Observer Scheme (DOS) [Clark, 1989].

The number of these observers (estimators) is equal to the number m of system outputs, and each device is driven by a single output and all the inputs of the system.

In this case a fault on the i–th output affects only the residual function of the output observer or filter driven by the i–th output.

To uniquely isolate a fault concerning one of the *system inputs*, $\boldsymbol{f}_u(t)$, under the assumption that outputs are fault-free, ($\boldsymbol{f}_y(t) = \mathbf{0}$), a bank of UIO or UIKF is used (Figure 4.5).
Such a solution is known as the Generalised Observer Scheme (GOS) [Patton *et al.*, 1989].

The number of these observers is equal to the number r of control inputs. The i–th observer is driven by all but the i–th input and all outputs of the system and generates a residual function which is sensitive to all but the i–th input fault.

In this way the detection of single input measurement faults is possible, since a fault on the i–th input affects all the residual functions except that of the device which is insensitive to the i-th input.

In order to summarise the isolation capabilities of the schemes presented, Table 4.1 shows the "fault signatures" for the case of a single fault in each input–output signal.

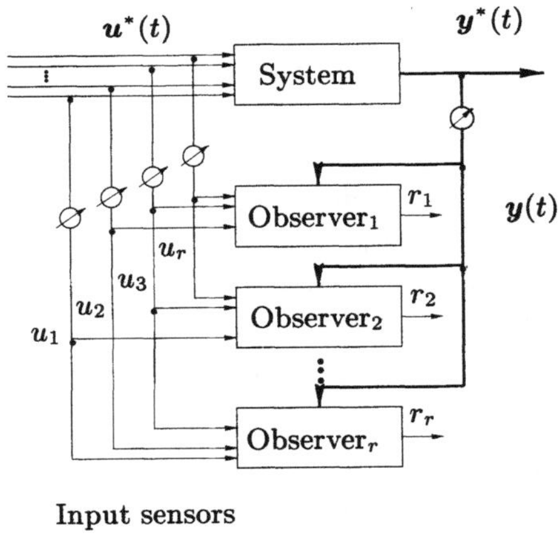

Fig. 4.5. Scheme for FDI of system inputs.

Table 4.1. Fault signatures.

	u_1	u_2	...	u_r	y_1	y_2	...	y_m
r_{UIO_1}	0	1	...	1	1	1	...	1
r_{UIO_2}	1	0	...	1	1	1	...	1
⋮	⋮	⋮	⋮	⋮	⋮	⋮	⋮	⋮
r_{UIO_r}	1	1	...	0	1	1	...	1
r_{O_1}	1	1	...	1	1	0	...	0
r_{O_2}	1	1	...	1	0	1	...	0
⋮	⋮	⋮	⋮	⋮	⋮	⋮	⋮	⋮
r_{O_m}	1	1	...	1	0	0	...	1

The residuals which are affected by the input and output faults are described by an entry '1' in the corresponding table entry, while an entry '0' means that the input or output fault does not affect the corresponding residual.

Note how multiple faults in the system outputs can be isolated since a fault on the i–th output signal affects only the residual function r_{O_i} of the output observer driven by the i–th output, but all the UIO or UIKF residual functions r_{UIO_i}. On the other hand, multiple faults on the inputs cannot be isolated by means of this technique since all the residual functions are sensitive to faults regarding different inputs.

With reference to Figure 4.4, in order to diagnose a fault on the i–th *system output* when the measurement noises are negligible ($\tilde{\boldsymbol{u}}(t) \cong \mathbf{0}$, $\tilde{\boldsymbol{y}}(t) \cong \mathbf{0}$) and $\boldsymbol{f}_u(t) = \mathbf{0}$ the model of the i–th observer ($i = 1, 2, \ldots, m$) has the form

$$\boldsymbol{x}^i(t+1) = \boldsymbol{A}^i\boldsymbol{x}^i(t) + \boldsymbol{B}^i\boldsymbol{u}(t) + \boldsymbol{K}^i\big(y_i(t) - \boldsymbol{C}^i\boldsymbol{x}^i(t)\big) \tag{4.18}$$

where $\boldsymbol{x}^i(t)$ is the observer state vector and the triple (A^i, B^i, C^i) is a minimal state space representation (completely observable) of the link among the inputs of the process and its i–th output $y_i(t)$. Such a triple can be obtained by means of the realization procedure, summarised in Chapter 3, starting from a MISO identified model.

The entries of K^i must be designed in order to assign stable eigenvalues to the matrix $(A^i - K^iC^i)$, suitably chosen within the unit circle.

In this situation and in the absence of faults, *i.e.*, $\boldsymbol{f}_y(t) = \boldsymbol{0}$, it can be verified that for the i-th output residual $r_i(t)$ the following relation holds

$$\lim_{t\to\infty} r_i(t) = \lim_{t\to\infty} \big(y_i(t) - C^i\boldsymbol{x}^i(t)\big) = 0 \tag{4.19}$$

and the rate of convergence depends on the position of the eigenvalues of the $(A^i - K^iC^i)$ matrix inside the positive real sector of the unit circle.

In the presence of a fault (step or ramp signal) on the i–th process output only the i–th output residual reaches a value different from zero and this situation leads to a complete failure diagnosis.

With reference to the devices for the FDI of the inputs, depicted in Figure 4.5, the structure of the i-th UIO ($i = 1, 2, \ldots, r$) for residual generation [Chen and Patton, 1999], under the assumptions $\tilde{\boldsymbol{u}}(t) \cong \boldsymbol{0}$, $\tilde{\boldsymbol{y}}(t) \cong \boldsymbol{0}$ and $\boldsymbol{f}_y(t) = \boldsymbol{0}$, is the following

$$\left.\begin{aligned} \boldsymbol{z}^i(t+1) &= \big(\boldsymbol{T}^i\boldsymbol{A} - \boldsymbol{K}^i\boldsymbol{C}\big)\boldsymbol{z}^i(t) + \boldsymbol{J}^i\boldsymbol{u}(t) + \boldsymbol{S}^i\boldsymbol{y}(t) \\ \boldsymbol{r}^i(t) &= \boldsymbol{L}_1^i\boldsymbol{z}^i(t) + \boldsymbol{L}_2^i\boldsymbol{y}(t) \end{aligned}\right\} \tag{4.20}$$

where $\boldsymbol{z}^i(t) \in \Re^n$ denotes the observer state vector, $\boldsymbol{r}^i(t) \in \Re^m$ is the residual vector and $\boldsymbol{F}^i$, $\boldsymbol{J}^i$, $\boldsymbol{S}^i$, $\boldsymbol{L}_1^i$ and $\boldsymbol{L}_2^i$ are matrices to be designed with appropriate dimensions. Let $\boldsymbol{T}^i$ be a linear transformation of the state $\boldsymbol{x}(t)$ of the system and define the state estimation error as $\boldsymbol{e}_x^i(t) = \boldsymbol{z}^i(t) - \boldsymbol{T}^i\boldsymbol{x}(t)$. On the suppositions $\tilde{\boldsymbol{u}}(t) = \boldsymbol{0}$, $\tilde{\boldsymbol{y}}(t) = \boldsymbol{0}$, and $\boldsymbol{f}_y(t) = \boldsymbol{0}$, it can be shown that the dynamics of the state estimation error become

$$\boldsymbol{e}_x^i(t+1) = \boldsymbol{F}^i\boldsymbol{e}_x^i(t) + \big(\boldsymbol{F}^i\boldsymbol{T}^i - \boldsymbol{T}^i\boldsymbol{A} + \boldsymbol{S}^i\boldsymbol{C}\big)\boldsymbol{x}(t) + (\boldsymbol{J}^i - \boldsymbol{T}^i\boldsymbol{B})\boldsymbol{u}(t) - \boldsymbol{T}^i\boldsymbol{B}\boldsymbol{f}_u(t), \tag{4.21}$$

whilst the residual vector is given by

$$\boldsymbol{r}^i(t) = \boldsymbol{L}_1^i\boldsymbol{e}_x^i(t) + (\boldsymbol{L}_1^i\boldsymbol{T}^i + \boldsymbol{L}_2^i\boldsymbol{C})\boldsymbol{x}(t). \tag{4.22}$$

It can be seen that if

$$\left.\begin{aligned} \boldsymbol{F}^i\boldsymbol{T}^i - \boldsymbol{T}^i\boldsymbol{A} + \boldsymbol{S}^i\boldsymbol{C} &= \boldsymbol{0}, \\ \boldsymbol{J}^i &= \boldsymbol{T}^i\boldsymbol{B}, \\ \boldsymbol{L}_1^i\boldsymbol{T}^i + \boldsymbol{L}_2^i\boldsymbol{C} &= \boldsymbol{0}, \end{aligned}\right\} \tag{4.23}$$

Equations 4.21 and 4.22 become

$$\left.\begin{aligned} \boldsymbol{e}_x^i(t+1) &= \boldsymbol{F}^i \boldsymbol{e}_x^i(t) + \boldsymbol{T}^i \boldsymbol{B} \boldsymbol{f}_u(t), \\ \boldsymbol{r}^i(t) &= \boldsymbol{L}_1^i \boldsymbol{e}_i(t). \end{aligned}\right\} \tag{4.24}$$

The matrices $\boldsymbol{T}^i$, $\boldsymbol{K}^i$, $\boldsymbol{J}^i$, $\boldsymbol{S}^i$, $\boldsymbol{L}_1^i$ and $\boldsymbol{L}_2^i$ can be constructed satisfying the following equations.

Under the hypothesis of observability of the system and in the absence of input faults, it can be seen that the i-th residual vector reaches zero as t approaches infinity and the rate of convergence depends on the position of the eigenvalues of $\boldsymbol{F}^i$ matrix inside the unit circle.

The hypothesis of system observability always holds because the transformation of the Auto Regressive eXogenous (ARX) input–output model into state space representation leads to completely observable systems [Söderström and Stoica, 1987].

If the linear transformation $\boldsymbol{T}^i$ is chosen as [Chang and Hsu, 1995]

$$\boldsymbol{T}^i = \boldsymbol{I}_n - \boldsymbol{B}_i(\boldsymbol{C}\boldsymbol{B}_i)^+\boldsymbol{C} \tag{4.25}$$

where $\boldsymbol{B}_i$ is the i-th column of B matrix and $\boldsymbol{K}^i$ is selected such that $\boldsymbol{F}^i = \boldsymbol{T}^i\boldsymbol{A} - \boldsymbol{K}^i\boldsymbol{C}$ is asymptotically stable, then, the solutions of Equation 4.23 are obtained as

$$\left.\begin{aligned} \boldsymbol{F}^i &= \boldsymbol{T}^i\boldsymbol{A} - \boldsymbol{K}^i\boldsymbol{C}, \\ \boldsymbol{S}^i &= \boldsymbol{K}^i + \boldsymbol{F}^i\boldsymbol{B}_i(\boldsymbol{C}\boldsymbol{B}_i)^+, \\ \boldsymbol{J}^i &= \boldsymbol{T}^i\boldsymbol{B}, \\ \boldsymbol{L}_1^i &= -\boldsymbol{C}, \\ \boldsymbol{L}_2^i &= [\boldsymbol{I}_m - (\boldsymbol{C}\boldsymbol{B}_i)(\boldsymbol{C}\boldsymbol{B}_i)^+]. \end{aligned}\right\} \tag{4.26}$$

The selection of the $\boldsymbol{B}_i$ matrix in Equations 4.25 and 4.26 sets to zero the i-th column of the $\boldsymbol{J}^i$ matrix. That is, the estimation error and then the residual of the i-th UIO become independent of the i-th system input.

Under the hypothesis of observability of the system 4.16 and in the absence of input faults ($\boldsymbol{f}_u(t) = \boldsymbol{0}$), it can be seen that the i–th residual vector reaches zero as t approaches infinity and the rate of convergence depends on the position of the eigenvalues of $\boldsymbol{T}^i\boldsymbol{A} - \boldsymbol{K}^i\boldsymbol{C}$ matrix inside the unit circle.

In the presence of a fault on the i–th input, the i–th residual reaches zero asymptotically while the residuals of the $r-1$ remaining observers are sensitive to the fault signal. This situation leads to the possibility of unique detection and isolation of all process input faults.

The design of this UIO requires the knowledge of a minimal form model $(\boldsymbol{A}, \boldsymbol{B}, \boldsymbol{C})$ for the system 4.16. Such a triple can be computed by using a realization procedure from a MIMO identified model. On the other hand, if the process is described mathematically by m MISO models, the triple $(\boldsymbol{A}, \boldsymbol{B}, \boldsymbol{C})$ can be directly obtained by grouping the $(\boldsymbol{A}^i, \boldsymbol{B}^i, \boldsymbol{C}^i)$ representations ($i = 1, 2, \ldots, m$).

4.5 Sliding Mode Observers for FDI

It is well known that the core element of model–based fault detection in control systems is the generation of residual signals which act as indicators of faults. The residual signals are generated using measurement estimates and a comparison with real measured quantities. For the design of residual generators, various approaches have been discussed in the literature.

In particular, the basic idea behind the use of the observer for fault detection is to estimate the outputs of the system from the measurements by using some type of observer, and then construct the residual by a properly weighted output estimate error. The residual is then examined for the likelihood of faults by using a fixed or adaptive threshold.

When, in the observer–based approaches, a full order observer is used in residual generator design, the main design procedure in fault detection becomes an equivalent state feedback control problem because of the dual relation between the state feedback control and the full order observer design. Based on this idea, some well–established approaches for state feedback control can be readily applied to robust fault detection using full order unknown input observers.

This subsection considers the use of a particular class of non–linear observer, a so–called *sliding mode observer*, which is also able to reconstruct the fault rather than detect the presence of a fault through a residual signal [Edwards *et al.*, 2000].

This problem of fault estimation is a powerful alternative to the detection of a fault via the use of a residual signal as long as the location of the fault effect on the system is known. The residual approach is more suited to the combined problem of fault detection and fault isolation, when the structure of the fault influence on the system is not perfectly known. A bank of dissimilar (but redundant) residual signals can then be used to infer the location of the fault in the system. On the other hand, the fault estimation approach is a direct way of providing fault information which, when compared with other fault estimation signals (from the same system), can be used to isolate all faults. The fault estimation method also provides a direct estimate of the size and severity of the fault, which can be important in many applications.

There has been a substantial body of new work in the field of fault estimation using a number of deterministic approaches, based upon observers with input signal reconstruction or de–convolution [Patton *et al.*, 1992] and H_∞ estimation [Stoustrup *et al.*, 1997]. More recently, Patton and Hou [Patton and Hou, 1998] have provided necessary and sufficient conditions for fault observability and reconstruction, using a matrix pencil approach and based on a study of input signal reconstruction for the unknown input observer problem. They use a numerically stable orthogonal transformation method to generate the required estimator. There is no loss of generality in their approach in the sense that it can be extended for application to certain non–linear system problems. A disadvantage, however, is the requirement for derivatives of measurement signals (in continuous time), although the discrete–time equivalent is quite realistic. An alternative strategy is to make direct use of non–linear observer structures [Patton *et al.*, 2000, Chap. 6]. However, non–linear observer approaches are limited in that the structure and parameters of the model must be known accurately.

This subsection describes an alternative philosophy of using variable structure and sliding mode theory to obviate some of the restrictions that must apply to most of the methods of fault estimation found in the literature. This subsection outlines some developments in the use of sliding mode observer theory for de–coupling the effects of fault signals from the response of the system estimated outputs. The work is based upon the sliding mode observer theory proposed by Edwards and Spurgeon [Edwards and Spurgeon, 1994, Tan and Edwards, 2001].

Sliding modes have been previously used for fault detection: Sreedhar, Fernandez and Masada [Sreedhar *et al.*, 1993] consider a model-based sliding mode observer approach although in their design procedure it is assumed that the states of the system are available. A different approach is adopted by Hermans and Zarrop [Hermans and Zarrop, 1996], who attempt to design an observer in such a way that in the presence of a fault the sliding motion is destroyed.

This subsection considers the practical situation when the system states are not available. The observer is designed to maintain a sliding motion even in the presence of faults which are detected by analysing the so–called *equivalent output injection*. The novelty lies in the manipulation of the equivalent output injection signal to explicitly reconstruct fault signals. (This may be allied to the equivalent control signal which appears in the analysis of sliding mode based feedback control systems).

4.5.1 Sliding Mode Observers

The concept of a sliding mode emerged from the Soviet Union in the late sixties where the effects of introducing discontinuous control action into dynamical systems were explored. By the use of a judicious switched control law it was found that the system states could be forced to reach and subsequently remain on a pre-defined surface in the state space. Whilst constrained to this surface, the resulting reduced–order motion, referred to as the *sliding motion*, was shown to be insensitive to any uncertainty or external disturbance signals which were implicit in the input channels. This inherent robustness property has resulted in world wide interest and research in the area of sliding mode control.

These ideas have subsequently been employed in other situations including the problem of state estimation via an observer. The earliest work by Utkin utilising a discontinuous switched component within an observer is described in [Utkin, 1977, Utkin, 1992].

A similar approach, which includes a linear output feedback term, appears in the work of Slotine and co–workers [Slotine *et al.*, 1987].

Walcott and Zak use a Lyapunov–based approach to formulate an observer which, under appropriate assumptions, exhibits asymptotic state error decay in the presence of bounded non–linearities/uncertainties in the input channel [Walcott and Żak, 1988]. The strategy of Walcott and Zak, although intuitively appealing, necessitates the use of algebraic manipulation tools to effectively solve an associated constrained Lyapunov problem for systems of reasonable order. Edwards and Spurgeon [Edwards and Spurgeon, 1994] propose an observer strategy, similar in style to that of Walcott and Zak, which circumvents the use of symbolic manipulation and offers an explicit design algorithm. This approach will be outlined here for fault detection purposes [Edwards *et al.*, 2000].

Consider the nominal linear discrete–time MISO system subject to certain faults described by:

$$\left.\begin{array}{lcl} \boldsymbol{x}(t+1) & = & \boldsymbol{A}\boldsymbol{x}(t) + \boldsymbol{B}\boldsymbol{u}(t) + \boldsymbol{D}f_i(t), \\ y(t) & = & \boldsymbol{C}\boldsymbol{x}(t) + f_o(t) \end{array}\right\} \tag{4.27}$$

where $\boldsymbol{A} \in \Re^{n\times n}$, $\boldsymbol{B} \in \Re^{n\times m}$, $\boldsymbol{C} \in \Re^{p\times n}$, $\boldsymbol{D} \in \Re^{n\times q}$ with $q \le p < n$ and the matrices $\boldsymbol{C}$ and $\boldsymbol{D}$ are of full rank.

The functions $f_i(t)$ and $f_o(t)$ are deemed to represent actuator and sensor faults, respectively, and are assumed to be bounded.

It is further assumed that the states of the system are unknown and only the signals $\boldsymbol{u}(t)$ and $y(t)$ are available.

The objective is to synthesise an observer to generate a state estimate $\hat{\boldsymbol{x}}(t)$ and output estimate $\hat{y}(t) = \boldsymbol{C}\hat{\boldsymbol{x}}$ such that a sliding mode is attained in which the output error:

$$e_y(t) = \hat{y}(t) - y(t) \tag{4.28}$$

is forced to zero in finite time.

The particular observer structure that will be considered can be written in the form:

$$\hat{\boldsymbol{x}}(t+1) = \boldsymbol{A}\hat{\boldsymbol{x}}(t) + \boldsymbol{B}\boldsymbol{u}(t) - \boldsymbol{G}_l e_y(t) + \boldsymbol{G}_n \nu, \tag{4.29}$$

where G_l and $G_n \in \Re n \times p$ are appropriate gain matrices and ν represents a discontinuous switched component to induce a sliding motion. It is shown that, provided a sliding motion can be attained, estimates of $f_i(t)$ and $f_o(t)$ can be computed from approximating the so–called equivalent output injection signal required to maintain sliding motion [Edwards *et al.*, 2000, Tan and Edwards, 2001].

4.6 Kalman Filtering and FDI from Noisy Measurements

With reference to Equations 4.17, when the signal to noise ratios $\|\boldsymbol{u}^*(t)\|_2^2/\|\tilde{\boldsymbol{u}}(t)\|_2^2$ and $\|\boldsymbol{y}^*(t)\|_2^2/\|\tilde{\boldsymbol{y}}(t)\|_2^2$ are low, a bank of KF must be employed to improve the performance of the FDI system. Even in this situation, the mathematical formulation of the classical Kalman Filter (KF) and of the Unknown Input Kalman Filter (UIKF) is similar to the one described by Equations 4.18 and 4.20 [Chen and Patton, 1999].

The essential difference concerns the feedback matrix $\boldsymbol{K}^i$ which becomes time–dependent and is computed by solving a Riccati equation. The solution of this equation requires the knowledge of the covariance matrices of the input and the output noises which can be identified by means of the dynamic Frisch scheme [Diversi and Guidorzi, 1998].

With reference to the time–invariant, discrete–time, linear dynamic system described by Equation 2.1 the i-th KF for the i-th output has the structure [Jazwinski, 1970]:

$$\boldsymbol{x}_F^i(t+1|t) = \boldsymbol{A}\boldsymbol{x}_F^i(t|t) + \boldsymbol{B}\boldsymbol{u}(t) \tag{4.30}$$

$$y_F^i(t+1|t) = \boldsymbol{C}_i\boldsymbol{x}_F^i(t+1|t) \tag{4.31}$$

$$\boldsymbol{P}(t+1|t) = \boldsymbol{A}\boldsymbol{P}(t|t)\boldsymbol{A}^T + \boldsymbol{Q} \tag{4.32}$$

$$\boldsymbol{K}_i(t+1) = \boldsymbol{P}(t+1|t)\boldsymbol{C}_i^T\left[\boldsymbol{C}_i\boldsymbol{P}(t+1|t)\boldsymbol{C}_i^T + \boldsymbol{R}\right]^{-1} \tag{4.33}$$

$$\boldsymbol{x}_F^i(t+1|t+1) = \boldsymbol{x}_F^i(t+1|t) + \boldsymbol{K}_i(t+1)\left[y_i(t+1) - \hat{y}_F^i(t+1|t)\right] \tag{4.34}$$

$$\begin{aligned}\boldsymbol{P}(t+1|t+1) = {} & \left[\boldsymbol{I} - \boldsymbol{K}_i(t+1)\boldsymbol{C}_i\right]\boldsymbol{P}(t+1|t)\left[\boldsymbol{I} - \boldsymbol{K}_i(t+1)\boldsymbol{C}_i\right]^T + \\ & +\boldsymbol{K}_i(t+1)\boldsymbol{R}\boldsymbol{K}_i^T(t+1).\end{aligned} \tag{4.35}$$

The variables $\boldsymbol{x}_F^i(t+1|t)$ and $y_F^i(t+1|t)$ are the one step prediction of the state and of the output of the process, respectively. $\boldsymbol{x}_F^i(t|t)$ is the state estimation given by the filter, $\boldsymbol{C}_i$ the i-th row of the output distribution matrix $\boldsymbol{C}$, $\boldsymbol{P}(t+1|t)$ is the covariance matrix of the one step prediction error $\boldsymbol{x}(t+1) - \boldsymbol{x}_F^i(t+1|t)$ whilst $P(t|t)$ is the covariance matrix of the filtered state error $\boldsymbol{x}(t) - \boldsymbol{x}_F^i(t|t)$. $\boldsymbol{Q}$ is the covariance matrix of the input vector noise $\tilde{\boldsymbol{u}}(t)$ and $\boldsymbol{R}$ is the variance of the i-th component of the output noise $\tilde{\boldsymbol{y}}(t)$. $\boldsymbol{K}_i(t+1)$ is the time-variant gain of the filter and $y_i(t)$ is the i-th component of the measured output $\boldsymbol{y}(t)$.

It can be proved that the innovation $e_i(t+1) = y_i(t+1) - y_F^i(t+1|t) = y_i(t+1) - \boldsymbol{C}_i\boldsymbol{x}_F^i(t+1|t)$ is a zero-mean white process when all the assumptions regarding the system 2.1 and the statistical characteristics of the noises described by Equation 2.4 are completely fulfilled. A Riccati equation is obtained by substituting Equation 4.32 into Equation 4.35. The solution of this equation converges to a steady state solution when the pair (A, C_i) is completely observable and the pair $(\boldsymbol{A}, \boldsymbol{D})$ is completely reachable, where $\boldsymbol{D}$ is a matrix such that $\boldsymbol{Q} = \boldsymbol{D}\boldsymbol{D}^T$.

In the presence of a fault on the i-th output ($f_{y_i}(t) \neq 0$), the stochastic properties (mean–value, variance and whiteness, etc) of the innovation process $e_i(t)$ change abruptly so that the fault detection can be based on these variations [Basseville, 1988].

Finally, note how multiple faults in outputs can be isolated since a fault on the i-th output affects only the innovation of the KF driven by the i-th output and all the innovation of the filters with unknown input.

On the other hand, with reference to the UIKF [Xie *et al.*, 1994, Xie and Soh, 1994], a single fault on the i-th input affects all the filter innovations except that of the filter with unknown input which is insensitive to the i-th input. A UIKF design procedure similar to that of Equation 4.26 can be found in [Xie *et al.*, 1994, Xie and Soh, 1994].

4.7 Residual Robustness to Disturbances

As discussed in earlier sections, the main and most challenging task of model–based FDI is the generation of residuals in which outputs and inputs of the system are processed by an appropriate algorithm (a processor) to generate a fault indicator signal (residual). Ideally, this signal should be near zero for the fault-free case, and should increase significantly when a fault appears in the system. This means that the residual should be de–coupled from the system inputs and modelling uncertainty. The residual that has this property can then be used to detect and isolate faults reliably.

In general, both faults and uncertainty affect the residual, and discrimination between these two effects is difficult. The effects of disturbances act as a source of false alarms which must be minimised. The ideal case is to make the residual itself become de–coupled from disturbances (robust residual generation). This is the principle of a ***robust residual generator*** which can be achieved by minimising the effect of disturbances on residuals.

In particular, we have to consider that the basis of our fault diagnosis technique is the use of mathematical models. Hence, the model should have a certain accuracy. In order to make a diagnosis algorithm robust against modelling uncertainty, we should also have some knowledge about modelling uncertainty. Otherwise, if an FDI algorithm can be made robust without ***a priori*** knowledge of the modelling, a model would clearly not be required in the first place.

The information of modelling uncertainty is normally represented by assumptions on uncertainty. These assumptions should be easy to handle by the robust design in a systematic manner, otherwise it does not provide any assistance for robust design. The disturbance representation of uncertainty can be handled by the unknown input observer or the eigenstructure assignment. However, this assumption is not realistic, *i.e.*, the distribution matrix cannot be obtained directly in practice.

In real situation, we can obtain some descriptions about uncertainty, for example, parameters of the system are within a certain bound. However, these descriptions are not easy to handle in designing robust FDI algorithms. The aim of the following sections is to present some techniques to bridge the gap between theoretical assumption and practical reality. This aim is fulfilled by uncertainty modelling, in which a disturbance description with an approximate distribution matrix is used. A number of situations covering a wide range of possibilities of uncertainty are considered [Patton *et al.*, 2000].

4.7.1 Determination, Optimisation and Estimation of Disturbance Distribution Matrix

From Eq. 4.8, it can be seen that the distribution matrix $\boldsymbol{E}$ must be known *a priori* for achieving disturbance de–coupling robust FDI. Furthermore, this matrix must be a row rank deficient matrix [Patton *et al.*, 2000, chap. 7].

In most practical systems the modelling uncertainty is unstructured, *i.e.*, the disturbance distribution matrix $\boldsymbol{E}$ is unknown. To apply the disturbance de–coupling robust FDI technique to the system with unstructured uncertainty, an approximate disturbance distribution matrix $\boldsymbol{E}$ is needed to represent effects of modelling uncertainty on the system.

In the following part of this section, we outline how to estimate this matrix for real uncertain systems. Note that, in the unknown input observer approach for robust FDI, the determination of the optimal disturbance distribution matrix $\boldsymbol{E}$ is a common problem for all disturbance de–coupling approaches, including the orthogonal parity equation approach of Gertler [Gertler, 1998].

We have to consider that the basis of our fault diagnosis technique is the use of mathematical models. Hence, the model should have a certain accuracy. In order to make a diagnosis algorithm robust against modelling uncertainty, we should also have some knowledge about modelling uncertainty. Otherwise, if an FDI algorithm can be made robust without *a priori* knowledge of the modelling, a model would clearly not be required in the first place. The information of modelling uncertainty is normally represented by assumptions on uncertainty. These assumptions should be easy to handle by the robust design in a systematic manner, otherwise it does not provide any assistance for robust design.

The disturbance representation of uncertainty can be handled by the unknown input observer or the eigenstructure assignment.

Sections 4.7.2 to 4.7.8 summarise the work by Patton and Chen [Patton and Chen, 1993]. On the other hand, a novel approach based on dynamic system identification is presented in Section 4.7.9 [Simani *et al.*, 1999a, Fantuzzi *et al.*, 2001a, Fantuzzi *et al.*, 2001b, Fantuzzi and Simani, 2002].

4.7.2 Additive Non-linear Disturbance and Noise

In certain situations, some *a priori* knowledge about uncertainty is available for determining the distribution matrix $\boldsymbol{E}$.

Consider the dynamic equation of the monitored system:

$$\boldsymbol{x}(t+1) = \boldsymbol{A}\boldsymbol{x}(t) + \boldsymbol{B}\boldsymbol{u}(t) + \boldsymbol{N}\mu(t) + \boldsymbol{M}f\left(\boldsymbol{x}(t), \boldsymbol{u}(t), t\right) \tag{4.36}$$

where $\mu(t)$ is a noise or external disturbance vector.

In this equation, the non–linearity is considered as an additive term $f\left(\boldsymbol{x}(t), \boldsymbol{u}(t), t\right)$, *i.e.*, the system dynamics can be separated into two parts as linear and non–linear. This kind of non–linear dynamic structure exists in some non–linear chemical processes [Patton *et al.*, 2001a, Patton *et al.*, 2001b, Simani and Patton, 2002a].

For the system described above, the uncertainty can be modelled as an additive term $\boldsymbol{E}\,\boldsymbol{d}(t)$ with $\boldsymbol{d}(t)$ an unknown input signal and where:

$$\boldsymbol{E}\boldsymbol{d}(t) = \begin{bmatrix} \boldsymbol{N} & \boldsymbol{M} \end{bmatrix} \begin{bmatrix} \mu(t) \\ f\left(\boldsymbol{x}(t), \boldsymbol{u}(t), t\right) \end{bmatrix}. \tag{4.37}$$

4.7.3 Model Complexity Reduction

Most systems can have significantly higher order dynamics than their useful models. Consider, for example, the discrete–time linear dynamic system described by a higher order model as:

$$\begin{bmatrix} \boldsymbol{x}(t+1) \\ \boldsymbol{x}_h(t+1) \end{bmatrix} = \begin{bmatrix} \boldsymbol{A}_{11} & \boldsymbol{A}_{12} \\ \boldsymbol{A}_{21} & \boldsymbol{A}_{22} \end{bmatrix} + \begin{bmatrix} \boldsymbol{B}_1 \\ \boldsymbol{B}_2 \end{bmatrix} \boldsymbol{u}(t) \tag{4.38}$$

where $\boldsymbol{x}(t) \in \Re^n$ is a partial state vector corresponding to dominate dynamic part of the system, $\boldsymbol{x}_h(t)$ represents the higher order dynamics in the system and frequently neglected in practice.

For ease of design and implementation in control and fault diagnosis, a low order model is used to approximate this system:

$$\begin{aligned} \boldsymbol{x}(t+1) &= \boldsymbol{A}\boldsymbol{x}(t) + \boldsymbol{B}\boldsymbol{u}(t) + \left(\boldsymbol{A}_{11} - \boldsymbol{A}\right)\boldsymbol{x}(t) + \left(\boldsymbol{B}_1 - \boldsymbol{B}\right)\boldsymbol{u}(t) + \\ &\quad + \boldsymbol{A}_{12}\boldsymbol{x}_h(t) = \boldsymbol{A}\boldsymbol{x}(t) + \boldsymbol{B}\boldsymbol{u}(t) + \boldsymbol{E}\boldsymbol{d}(t) \end{aligned} \tag{4.39}$$

where:

$$\boldsymbol{E}\boldsymbol{d}(t) = \begin{bmatrix} \left(\boldsymbol{A}_{11} - \boldsymbol{A}\right) & \left(\boldsymbol{B}_1 - \boldsymbol{B}\right) & \boldsymbol{A}_{12} \end{bmatrix} \begin{bmatrix} \boldsymbol{x}(t) \\ \boldsymbol{u}(t) \\ \boldsymbol{x}_h(t) \end{bmatrix} \tag{4.40}$$

A typical application of this partitioned state space structure arises when comparing a reduced order model with the full–order system, for example, in an observer used for FDI. For this case, the nominal model $(\boldsymbol{A}, \boldsymbol{E})$ is the reduced order model and the remaining modelling errors are considered as an additive term $\boldsymbol{Ed}(t)$.

4.7.4 Parameter Uncertainty

A system model with parameter uncertainty can be described as:

$$\boldsymbol{x}(t+1) = (\boldsymbol{A} + \Delta\boldsymbol{A})\,\boldsymbol{x}(t) + (\boldsymbol{B} + \Delta\boldsymbol{B})\,\boldsymbol{u}(t) \tag{4.41}$$

The parameter perturbations considered in robust control are sometimes approximated as:

$$\Delta\boldsymbol{A} \approx \sum_{i=1}^{N} a_i \boldsymbol{A}_i \quad \text{and} \quad \Delta\boldsymbol{B} \approx \sum_{i=1}^{N} b_i \boldsymbol{B}_i \tag{4.42}$$

where $\boldsymbol{A}_i$ and $\boldsymbol{B}_i$ are known matrices with proper dimensions, a_i and b_i are scalar factors.

In this case, the ***unstructured ncertainty*** can be approximated by the structured uncertainty as:

$$\boldsymbol{Ed}(t) = \Delta\boldsymbol{A}\boldsymbol{x}(t) + \Delta\boldsymbol{B}\boldsymbol{u}(t) = \begin{bmatrix} \boldsymbol{A}_1 & \ldots & \boldsymbol{A}_N & \boldsymbol{B}_1 & \ldots & \boldsymbol{B}_N \end{bmatrix} \begin{bmatrix} \boldsymbol{a}_1\boldsymbol{x}(t) \\ \vdots \\ \boldsymbol{a}_N\boldsymbol{x}(t) \\ \boldsymbol{b}_1\boldsymbol{u}(t) \\ \vdots \\ \boldsymbol{b}_N\boldsymbol{u}(t) \end{bmatrix} \tag{4.43}$$

Now, consider the situation where the system matrices are functions of the parameter vector $\alpha \in Re^g$:

$$\boldsymbol{x}(t+1) = \boldsymbol{A}(\alpha)\boldsymbol{x}(t) + \boldsymbol{B}(\alpha)\boldsymbol{u}(t) \tag{4.44}$$

If the parameter vector varies around the nominal condition $\alpha = \alpha_0$, Equation 4.44 can be rewritten as:

$$\boldsymbol{x}(t+1) = \boldsymbol{A}(\alpha_0)\boldsymbol{x}(t) + \boldsymbol{B}(\alpha_0)\boldsymbol{u}(t) + \sum_{i=1}^{g}\left(\frac{\partial\boldsymbol{A}}{\partial\alpha_i}\delta\alpha_i\boldsymbol{x} + \frac{\partial\boldsymbol{B}}{\partial\alpha_i}\delta\alpha_i\boldsymbol{u}\right). \tag{4.45}$$

In this case, the distribution matrix and unknown input vector are:

$$\boldsymbol{E} = \left[\; \frac{\partial\boldsymbol{A}}{\partial\alpha_1} \Big| \;\; \frac{\partial\boldsymbol{B}}{\partial\alpha_1} \Big| \;\; \ldots \Big| \;\; \frac{\partial\boldsymbol{A}}{\partial\alpha_g} \Big| \;\; \frac{\partial\boldsymbol{B}}{\partial\alpha_g} \;\right], \tag{4.46}$$

$$\boldsymbol{d}(t) = \left[\; \delta\alpha_1\boldsymbol{x}^T | \;\; \delta\alpha_1\boldsymbol{u}^T | \;\; \ldots | \;\; \delta\alpha_g\boldsymbol{x}^T | \;\; \delta\alpha_g\boldsymbol{u}^T \;\right]^T. \tag{4.47}$$

4.7.5 Distribution Matrix Low Rank Approximation

It has been show that the distribution matrix can be derived directly from the available uncertainty information. For most situations, this $n \times n_1$ matrix has full row rank, *i.e.*, $\text{rank}\,(\boldsymbol{E}) = n$. A necessary condition for disturbance de–coupling is to find a matrix $\boldsymbol{H}$ to satisfy the first condition in Eqs. 4.13 [Patton *et al.*, 2000, chap. 7]. If $\text{rank}\,(\boldsymbol{E}) < n$, the first relation in 4.13 has solutions and exact de–coupling is possible. If, however, $\text{rank}\,(\boldsymbol{E}) = n$, the first condition in Eqs. 4.13 has no solutions and exact de–coupling is impossible.

An approximate de–coupling must be accepted as a compromise. The procedure will be to compute a matrix $\boldsymbol{E}^*$ that is as close as possible to $\boldsymbol{E}$, and $\text{rank}\,(\boldsymbol{E}^*) = q \leq n-1$, *i.e.*, to find the solution of following optimisation problem:

$$\min ||\boldsymbol{E} - \boldsymbol{E}^*||_F^2 \quad \text{subject to:}\ \ \text{rank}\,(\boldsymbol{E}^*) = q \leq n-1 \tag{4.48}$$

Here $||\cdot||_F^2$ denotes the Frobenius norm, defined as the root of the sum of squares of the entries of the associated matrix. The matrix $\boldsymbol{E}^*$ is thus chosen so that the sum of the squared distances between the columns of $\boldsymbol{E}$ and $\boldsymbol{E}^*$ is minimised, subject to the constraint that $\text{rank}\,(\boldsymbol{E}^*) < n$. This optimisation problem can be solved by the *Singular Value Decomposition* (SVD) of $\boldsymbol{E}$ [Patton *et al.*, 2000, chap. 7].

4.7.6 Model Estimation with Bounded Uncertainty

Now, consider the case when the full-order system model is not available. An identification procedure is used to obtain the nominal model $(\boldsymbol{A}, \boldsymbol{B}, \boldsymbol{C})$ with the estimation error $\{\Delta\boldsymbol{A}, \Delta\boldsymbol{B}\}$. Normally, $\Delta\boldsymbol{A}$ and $\Delta\boldsymbol{B}$ are unknown but bounded:

$$\begin{aligned} \boldsymbol{A}_m \leq \Delta\boldsymbol{A} \leq \boldsymbol{A}_M \\ \boldsymbol{B}_m \leq \Delta\boldsymbol{B} \leq \boldsymbol{B}_M \end{aligned} \tag{4.49}$$

where $\boldsymbol{A}_m$, $\boldsymbol{A}_M$, $\boldsymbol{B}_m$, $\boldsymbol{B}_M$ are known and, for example, $\Delta\boldsymbol{A} \leq \boldsymbol{A}_M$ denotes that each element of $\Delta\boldsymbol{A}$ is not larger than the corresponding element of $\boldsymbol{A}_M$. This represents an unstructured and bounded uncertainty.

Consider $\Delta\boldsymbol{A}$ and $\Delta\boldsymbol{B}$ in a finite set of possibilities, say $\{\Delta\boldsymbol{A}_i, \Delta\boldsymbol{B}_i\}$ $(i = 1, 2, \ldots, M)$ within the interval $\boldsymbol{A}_m \leq \Delta\boldsymbol{A} \leq \boldsymbol{A}_M$ and $\boldsymbol{B}_m \leq \Delta\boldsymbol{B} \leq \boldsymbol{B}_M$. This might involve choosing representative points, reflecting desired weighting on the likelihood or importance of particular sets of parameters. In this situation, a set of unknown input distribution matrices is obtained:

$$\boldsymbol{E}_i = [\Delta\boldsymbol{A}_i, \Delta\boldsymbol{B}_i] \quad \text{with} \quad i = 1, 2, \ldots, M \tag{4.50}$$

In order to make the disturbance de–coupling valid for a wide range of model parameter variations, an optimal matrix $\boldsymbol{E}^*$ should be made to be near all $\boldsymbol{E}_i$ $(i = 1, 2, \ldots, M)$ as closely as possible. The optimisation problem is thus defined as:

$$\min \|\boldsymbol{E}^* - [\boldsymbol{E}_1 \boldsymbol{E}_2 \dots \boldsymbol{E}_M]\|_F^2 \quad \text{subject to: } \operatorname{rank}(\boldsymbol{E}^*) = q \leq n-1 \tag{4.51}$$

$\boldsymbol{E}^*$ is then used to design disturbance de–coupling robust residual generators. As $\boldsymbol{E}^*$ is close to all $\boldsymbol{E}_i$, the approximate de–coupling is achieved over whole range of parameter variations.

4.7.7 Disturbance Vector and Disturbance Matrix Estimation

In most cases, there is insufficient accurate information available about the system state space model and all that can be obtained is a linearised low order model with matrices $(\boldsymbol{A}, \boldsymbol{B}, \boldsymbol{C})$.

In order to account unavoidable modelling errors, we assume the system is described as Eqs. 4.8 where $\boldsymbol{E}\boldsymbol{d}(t)$ is used to represent modelling errors. If the term $\boldsymbol{E}\boldsymbol{d}(t)$ can be derived, we may be also able to estimate the structured matrix $\boldsymbol{E}$. It seems reasonable to add $\boldsymbol{E}\boldsymbol{d}(t)$ to account for all uncertainties in the model.

Firstly, assume that the vector $\boldsymbol{d}_c(t) = \boldsymbol{E}\boldsymbol{d}(t)$ is a slowly time–varying vector, so that the system model can be rewritten in augmented form as:

$$\begin{bmatrix} \boldsymbol{x}(t+1) \\ \boldsymbol{d}_c(t+1) \end{bmatrix} = \begin{bmatrix} \boldsymbol{A} & \boldsymbol{I} \\ \boldsymbol{0} & \boldsymbol{0} \end{bmatrix} \begin{bmatrix} \boldsymbol{x}(t) \\ \boldsymbol{d}_c(t) \end{bmatrix} + \begin{bmatrix} \boldsymbol{B} \\ \boldsymbol{0} \end{bmatrix} \boldsymbol{u}(t), \tag{4.52}$$

$$\boldsymbol{y}(t) = \begin{bmatrix} \boldsymbol{C} & \boldsymbol{0} \end{bmatrix} \begin{bmatrix} \boldsymbol{x}(t) \\ \boldsymbol{d}_c(t) \end{bmatrix} \boldsymbol{u}(t) \tag{4.53}$$

If we have the true system input and output data $\{\boldsymbol{u}(t), \boldsymbol{y}(t)\}$, an observer based on Eqs. 4.52 and 4.53 can be used to estimate the disturbance vector $\boldsymbol{d}_c(t)$. Once we have $\hat{\boldsymbol{d}}_c(t)$, it is possible to obtain some information about the distribution matrix $\boldsymbol{E}$.

It is worth noting how the system described by Eqs. 4.52 and 4.53 is observable if and only if rank $(\boldsymbol{C}) = n$ and the matrix pair $(\boldsymbol{A}, \boldsymbol{C})$ is observable. This constraint can limit the use of this technique for estimating the disturbance vector, as it requires that the system has more than n (state dimension $\boldsymbol{x}(t) \in \Re^n$) independent measurements. It is clear that if we want to estimate the modelling uncertainty without any *a priori* information about it, additional measurements are required.

Generally speaking, when the vector $\boldsymbol{d}_c(t)$ is estimated, there are many combinations of $\boldsymbol{E}$ and $\boldsymbol{d}(t)$, but for the current robust FDI methods, we only need to know the structure of $\boldsymbol{E}$, and $\boldsymbol{d}(t)$ can be chosen arbitrarily.

There are two possibilities: one is that $\boldsymbol{E}$ is a vector and $d(t)$ is an arbitrary scalar function; another is that $\boldsymbol{E}$ is a matrix and $\boldsymbol{d}(t)$ is an arbitrary vector function. Using the augmented observer, we can get the estimation of the disturbance vector $\boldsymbol{d}(t)$ as $\left\{\hat{d}_c(1), \dots, \hat{d}_c(N)\right\}$. If the direction of the vector $\hat{\boldsymbol{d}}_c(t)$ changes slightly for all $t = 1, 2, \dots, N$, it is feasible for $\boldsymbol{E}$ to be a vector and $d(t)$ an arbitrary scalar function. In this case, the matrix $\boldsymbol{E}$ can be approximated as:

$$\boldsymbol{E} = \frac{1}{N}\sum_{t=1}^{N} \hat{d}_c(t) \tag{4.54}$$

It is very likely to be the case that $d_c(t)$ cannot be assumed as a constant direction vector, *i.e.* , the directions of $\hat{d}_c(t)$ are then very much different for all $t = 1, 2, \ldots, M$. In this case, it is still possible to express the vector $\boldsymbol{d}_c(t)$ as: $\boldsymbol{d}_c(t) = \boldsymbol{E}\boldsymbol{d}(t)$, where $\boldsymbol{E} \in \Re^{n\times q}$ is a constant matrix, with $\boldsymbol{d}(t) \in \Re^q$, $\boldsymbol{d}_c(t) \in \Re^n$ and $q \leq n$. In the robust FDI problem, $\boldsymbol{E}$ must be row rank deficient in order to have an annihilating matrix $\boldsymbol{H}$ such that the equation $\boldsymbol{H}\boldsymbol{E} = \boldsymbol{0}$ holds as one of the conditions for achieving robust FDI (see relations 4.13) [Patton *et al.*, 2000, chap. 7].

It is worthwhile noting that the FDI algorithm design and the determination of the disturbance distribution matrix in the discrete–time domain can be carried out by using de–convolution technique [Patton *et al.*, 2000, chap. 7], since some special properties exist in discrete–time design.

Here, we consider the discrete–time model of the system as described by Eqs. 4.8, the vector $\boldsymbol{E}\boldsymbol{d}(t)$ is used to account for all uncertainties in the model. Let us suppose that the matrices $(\boldsymbol{A}, \boldsymbol{B}, \boldsymbol{C})$ are known nominal model parameters. $\boldsymbol{u}^*(t)$ is the model input which equals to the system input $\boldsymbol{u}(t)$. $\boldsymbol{y}(t)$ is the model output which is normally not equal to true system output $\boldsymbol{y}^*(t)$ due to modelling uncertainty.

The task here is to determine the additional term $\boldsymbol{E}\boldsymbol{d}(t)$ using the nominal model parameters $(\boldsymbol{A}, \boldsymbol{B}, \boldsymbol{C})$ and real system inputs and outputs $\{\boldsymbol{u}^*(t), \boldsymbol{y}^*(t)\}$. After a good estimate of the vector $\boldsymbol{d}_c(t)$ is obtained, it is possible to decompose into $\boldsymbol{E}\boldsymbol{d}(t)$ with $\boldsymbol{E}$ as a structured matrix.

It can be shown that the modelling output error (*i.e.*, the difference between true system output and model output) is:

$$\boldsymbol{e}(t) = \boldsymbol{y}^*(t) - \boldsymbol{y}(t) = \boldsymbol{y}^*(t) - \boldsymbol{C}\boldsymbol{A}^t\boldsymbol{x}(0) - \sum_{i=1}^{t} \boldsymbol{C}\boldsymbol{A}^{i-1}\boldsymbol{B}\boldsymbol{u}^*(t) - \sum_{i=1}^{t} \boldsymbol{C}\boldsymbol{A}^{i-1}\boldsymbol{d}_c(t). \tag{4.55}$$

with $t = 1, \ldots, N$. If the model is "good", it should represent the system behaviour accurately so that the output modelling error will be zero, *i.e.*,

$$\boldsymbol{e}(t) \to \boldsymbol{0} \quad , t = 1, \ldots, N. \tag{4.56}$$

This is the point for computing the disturbance vector $\boldsymbol{d}_c$ [Patton *et al.*, 2000, chap. 7]. In particular, when the number of independent measurements m is not smaller than the state number n, and the disturbance is a constant bias vector, *i.e.* $\boldsymbol{d}_c(t) = \boldsymbol{d}_c \; \forall \; t$, it can be shown that

$$\begin{bmatrix} C \\ C+CA \\ \vdots \\ C+CA+\ldots+CA^{N-1} \end{bmatrix} d_c(t) = \begin{bmatrix} y_e(1) \\ y_e(2) \\ \vdots \\ y_e(N) \end{bmatrix} \tag{4.57}$$

or,

$$Gd_c(t) = Y \tag{4.58}$$

where

$$y_e(t) = y^*(t) - CA^t x(0) - \sum_{i=1}^{t} CA^{i-1}Bu^*(t) \tag{4.59}$$

and $G \in \Re^{m\ N \times n}$, $d_c(t) \in \Re^n$, $Y \in \Re^{Nm}$.

It is easily to see that *rank* $(G) = n$ if and only if $N \geq n$ and the matrix pair (A, C) is observable. Hence, for an observable system, we can estimate the constant disturbance vector using the following expression when the number of observations N is not a very large:

$$d_c = \left(G^T G\right)^{-1} G^T Y. \tag{4.60}$$

In this section, in Eq. 4.55, we have assumed that the initial state vector $x(0)$ is known *a priori*. However, this is not always true and some approximation must be made in practice. For a large μ and for $t > \mu$, we have

$$\sum_{i=1}^{\mu} CA^{i-1} d_c(t-i) = y_e(t) \tag{4.61}$$

and

$$y_e(t) \approx y(t) - \sum_{i=1}^{\mu} CA^{i-1}Bu^*(t-i) = y_e(t). \tag{4.62}$$

On the other hand, if we assume that the disturbance vector $d_c(t)$ is piece–wise constant vector, *i.e.*,

$$d_c(t-1) = d_c(t-2) = \ldots = d_c(t-\mu) \tag{4.63}$$

Eq. 4.61 now can be rewritten as:

$$\left[C \sum_{i=1}^{\mu} A^{i-1} \right] d_c(t-1) = y_e(t). \tag{4.64}$$

Once again, a unique solution for the disturbance vector $d_c(t)$ exists *iff* rank$(C) = n$. This requires that the independent output dimension m is not smaller than the state dimension n.

This section has presented some method for estimating the disturbance vector. However, there are certain limitations in these methods and some further research is still needed. As an example, in the following Section 4.7.9, a novel approach exploiting identification technique is developed for the estimation of the E matrix.

4.7.8 Distribution Matrix Optimisation for Varying Operating Point Cases

The operating point of the system varies according to the various plant conditions, and different operating points correspond to different unknown input matrices, $\boldsymbol{E}_i$ $(i = 1, 2, \ldots, M)$.

It is attractive to be able design a single FDI scheme for a whole range (or a set) of operating points. The success of the single FDI design depends on its robustness properties.

In order to make the disturbance de-coupling hold for all operating points, we must make:

$$\boldsymbol{H}\ \boldsymbol{E}_i = \boldsymbol{0},\ \text{for } i = 1, 2, \ldots, M \tag{4.65}$$

or:

$$\boldsymbol{H}\,[\boldsymbol{E}_1,\ \boldsymbol{E}_2,\ \ldots,\ \boldsymbol{E}_M] = \boldsymbol{H}\ \boldsymbol{P} = \boldsymbol{0}. \tag{4.66}$$

If $rank(\boldsymbol{P}) \leq n-1$, Eq. 4.66 has solutions and the exact de–coupling at all operating points is achievable. Otherwise, approximate de–coupling must be used. Hence, it can be solved by defining an optimisation problem:

$$\min\|\boldsymbol{P} - \boldsymbol{P}^*\| \ \text{ subject to: } \ \text{rank}\,(\boldsymbol{P}^*) \leq n-1. \tag{4.67}$$

This problem can be solved using singular value decomposition (SVD). Matrices $\boldsymbol{H}$ and $\boldsymbol{P}^*$ should ensure that a fixed FDI scheme is effective for different operating points .

4.7.9 Disturbance Distribution Matrix Identification

All model–based FDI methods use a model of the monitored system to produce the symptom generator. If the system is not complex and can be described accurately by the mathematical model, FDI is directly performed by using a simple geometrical analysis of residuals.

In real industrial systems however, the modelling uncertainty is unavoidable. The design of a reliable and robust FDI scheme should take into account of the modelling uncertainty with respect to the sensitivity of the faults.

The model–based FDI technique requires a high accuracy mathematical description of the monitored system. The better the model represents the dynamic behaviour of the system, the better will be the FDI precision. If an FDI method can be developed which is insensitive to modelling uncertainty, a very accurate model is not necessarily needed.

All uncertainties can be summarised as disturbances acting on the system. Although the disturbance vector is unknown, its distribution matrix can be obtained by an identification procedure. Under this assumption, the disturbance de–coupling principle can be exploited to design a fault detection scheme using UIOs.

Under the hypothesis that the system can be described as an equation error model, this section has studied the method of obtaining the disturbance distribution matrix from the fault-free system data, by taking into account the equation error term.

The UIO performing the disturbance de–coupling can be designed from the equation error model.

In the following, in fact, it is assumed that the monitored system, depicted in Figure 4.6, can be described by a linear, discrete-time equation error model of the type

$$y_i^*(t) = \sum_{k=1}^{n} \alpha_{ik} y_i^*(t-k) + \sum_{j=1}^{r} \sum_{k=1}^{n} \beta_{ikj} u_j^*(t-k) + \varepsilon_i(t). \tag{4.68}$$

where $y_i^*(t)$ $(i = 1, \ldots, m)$ is the i-component of the system output vector $\boldsymbol{y}^*(t)$, whilst u_j^* the j-component of the control input vector $\boldsymbol{u}^* \in \Re^r$. n, α_{ik} and β_{ikj} are the parameters to be determined by an identification approach. The term $\varepsilon_i(t)$ takes into account the modelling error, which is due to process noises, parameter variations, etc.

As depicted in Figure 4.6, in real applications the input and output sensor signals $\boldsymbol{u}(t)$ and $\boldsymbol{y}(t)$ are affected by faults.

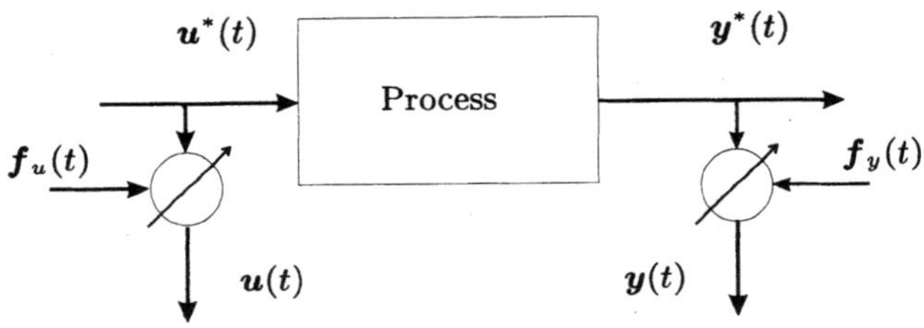

Fig. 4.6. The monitored system.

By using the transfer function description, system 4.68 can be rewritten in the form

$$y_i^*(t) = \boldsymbol{F}_i(z)\boldsymbol{u}^*(t) + \boldsymbol{G}_i(z)\varepsilon_i(t) \tag{4.69}$$

and its structure is depicted in Figure 4.7, in which z is the unitary advance operator.
The symptom generation is implemented by means of dynamic observers with unknown inputs, in order to produce a set of signals from which it will be possible to diagnose faults associated to outputs. This choice should minimise the effects of disturbances, which act as a source of false alarms.

The design of the UIO requires the knowledge of a state space model of the system under investigation. In particular, in this section, in order to design the UIO, the identification of a number of MISO models ($m = 1$), of the type of Equation 4.69 equal to the number of the output variables has been chosen.

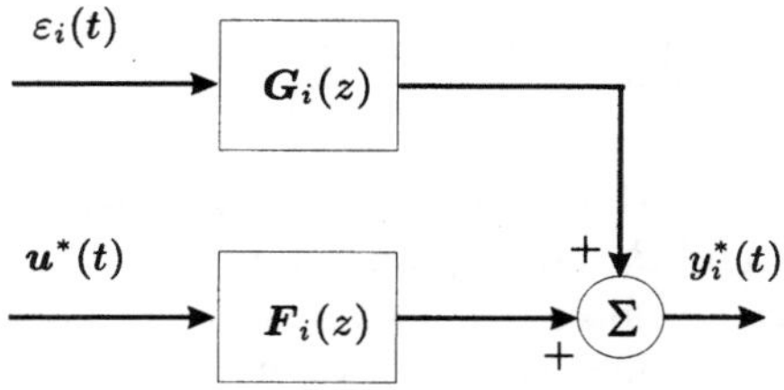

Fig. 4.7. The structure of the equation error model.

Under no–fault conditions, it can be proved that a state space formulation of the input-output equation error model 4.68 for the i-th output becomes

$$\left.\begin{array}{lcl} \boldsymbol{x}_i(t+1) & = & \boldsymbol{A}_i\boldsymbol{x}_i(t) + \boldsymbol{B}_i\boldsymbol{u}(t) + \boldsymbol{E}_i\varepsilon_i(t) \\ y_i(t) & = & \boldsymbol{C}_i\boldsymbol{x}_i(t) + \boldsymbol{F}_i\varepsilon_i(t), \quad t = 1, 2, \ldots \end{array}\right\} \tag{4.70}$$

where the matrices $\boldsymbol{A}_i(n \times n)$, $\boldsymbol{B}_i(n \times r)$, $\boldsymbol{C}_i(1 \times n)$, $\boldsymbol{E}_i(n \times 1)$ and $\boldsymbol{F}_i$ are functions of the α_{ik} and β_{ikj} parameters [Söderström and Stoica, 1987].

If the vector $\varepsilon_i(t)$ is considered as a disturbance and $\boldsymbol{E}_i$, $\boldsymbol{F}_i$ its distribution matrices, terms $\boldsymbol{E}_i\varepsilon_i(t)$ and $\boldsymbol{F}_i\varepsilon_i(t)$ represent uncertainties acting upon the system.

The i-th residual (symptom) generator using an UIO is thus described as

$$\left.\begin{array}{lcl} \boldsymbol{z}_i(t+1) & = & \boldsymbol{N}_i\boldsymbol{z}_i(t) + \boldsymbol{L}_i y_i(t) + \boldsymbol{G}_i\boldsymbol{u}(t) \\ \boldsymbol{r}_i(t) & = & y_i(t) - \boldsymbol{C}_i\big(\boldsymbol{z}_i(t) - \boldsymbol{D}_i y_i(t)\big) \end{array}\right\} \tag{4.71}$$

where $\boldsymbol{z}_i(t) \in \Re^n$ denotes the i-th observer state vector, $\boldsymbol{C}_i\big(\boldsymbol{z}_i(t) - \boldsymbol{D}_i y_i(t)\big)$ represents the estimate of $y_i(t)$ whilst $\boldsymbol{r}_i(t)$ is the residual vector. A design procedure is used for finding suitable matrices $\boldsymbol{N}_i$, $\boldsymbol{L}_i$, $\boldsymbol{G}_i$ and $\boldsymbol{D}_i$ with appropriate dimension.

With the choices:

$$\left.\begin{array}{lcl} \boldsymbol{D}_i & = & -\boldsymbol{E}_i(\boldsymbol{C}_i\boldsymbol{E}_i)^{-1}, \\ \boldsymbol{P}_i & = & \boldsymbol{I} + \boldsymbol{D}_i\boldsymbol{C}_i, \\ \boldsymbol{G}_i & = & \boldsymbol{P}_i\boldsymbol{B}_i, \\ \boldsymbol{L}_i & = & \boldsymbol{P}_i\boldsymbol{A}_i\boldsymbol{E}_i(\boldsymbol{C}_i\boldsymbol{E}_i)^{-1}, \end{array}\right\} \tag{4.72}$$

if $\boldsymbol{N}_i$ can be chosen suitably, so that:

$$\boldsymbol{L}_i\boldsymbol{C}_i - \boldsymbol{P}_i\boldsymbol{A}_i = -\boldsymbol{N}_i\boldsymbol{P}_i \tag{4.73}$$

$\boldsymbol{r}_i(t)$ will asymptotically approach zero in the absence of faults, $\boldsymbol{f}_u(t) = \boldsymbol{0}$ and $\boldsymbol{f}_y(t) = \boldsymbol{0}$.

4.8 Residual Generation via Parameter Estimation

With reference to an input-output SISO EIV model of order n in the form

$$\sum_{i=0}^{n} \alpha_i y(t-i) = \sum_{i=1}^{n} \beta_i u(t-i) \tag{4.74}$$

in which $u(t)$ represents the input, $y(t)$ the output, a Kalman filter can be used to estimate α_i and β_i model parameters.

The Kalman filter used as parameter estimator [Castaldi and Soverini, 1998] can be exploited in order to detect changes in parameters α_i and β_i due to faults which affect input and output measurements $u(t)$ and $y(t)$.

The system to design the filter is the following:

$$\left.\begin{array}{lcl} \theta(t+1) & = & \theta(t) + \omega(t) \\ y(t) & = & \theta(t)P(t) + \varepsilon(t) \end{array}\right\} \tag{4.75}$$

where the vector $\theta = [\alpha_n, \ldots, \alpha_1, \beta_n, \ldots, \beta_1]$ contains the model parameters and the measurement vector $P(t) = [y(t-n), \ldots, y(t-1), u(t-n), \ldots, u(t-1)]$. $\omega(t)$ is a white process, in order to take into account the parameter variations for non–stationary processes whilst $\varepsilon(t)$ the output error term.

Residuals can be generated, for instance, by comparing the estimate of the parameters θ given by Ordinary Least–Squares (OLS) or Recursive Least–Squares (RLS) and the one computed by the Kalman filter4.75. On the other hand, fault–free and faulty parameters $\theta(t)$ computed by Equation 4.75 in fault–free and faulty conditions can be compared.

The standard deviation of the $\varepsilon(t)$ process can be evaluated via OLS, whilst that of the $\omega(t)$ white process has to be tuned in order to obtain an accurate parameter estimate.

An application of this method is shown in Section 5.5.1.

4.9 Residual Generation via Fuzzy Models

This section exploits the approach for FDI in non–linear dynamic processes using multiple model approach. In particular, as described in Section 3.5.2, the method uses Takagi–Sugeno (TS) fuzzy models.

The non–linear dynamic process is, in fact, described as a composition of several Takagi–Sugeno models selected according to the process operating conditions.

The FDI scheme adopted to generate residuals from the measured noisy sequences $\boldsymbol{u}(t)$ and $\boldsymbol{y}(t)$ is designed by means of the non–linear TS fuzzy identified models.

In the following, it is assumed that the monitored system, depicted in Figure 4.6, can be described by a model of the type 3.97 in Section 3.5.2. $\boldsymbol{y}(t) \in \Re^m$ is the system output vector and $\boldsymbol{u}(t) \in \Re^r$ the control input vector.

In real applications the measured variables $\boldsymbol{u}^*(t)$ and $\boldsymbol{y}^*(t)$ are affected by noise. $\boldsymbol{f}_u(t)$ and $\boldsymbol{f}_y(t)$ are faults affecting system inputs and outputs.

As presented in Chapter 2, there are different approaches to generate the residuals from which it will be possible to diagnose faults associated to system inputs and outputs. In this monograph, fuzzy models are used to estimate the outputs of the system from the input-output measurements.

As depicted in Figure 4.8, residuals can be generated by the comparison of measured $\boldsymbol{y}(t)$ and estimated $\hat{\boldsymbol{y}}(t)$ outputs:

$$\boldsymbol{r}(t) = \boldsymbol{y}(t) - \hat{\boldsymbol{y}}(t). \tag{4.76}$$

The symptom evaluation is obtained by a logic device which processes the redundant signals computed by the residual generation in order to detect when a fault occurs. In such a case, faults can be detected by using a simple thresholding logic.

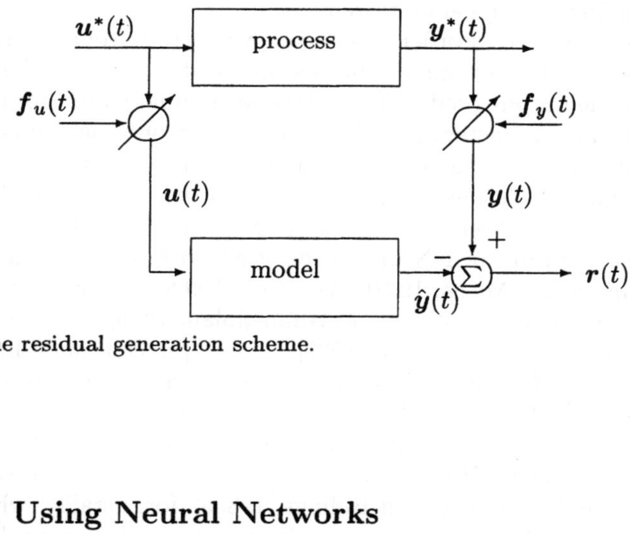

Fig. 4.8. The residual generation scheme.

4.10 FDI Using Neural Networks

Fault diagnosis and identification have been widely developed during recent years. Model–based methods, fault–tree approaches and pattern recognition techniques are among the most common methodologies used in such tasks. Neural networks have been used in fault identification problems for model approximation and pattern recognition as well. However, because of difficulties to perform Neural Network training on dynamic patterns, the second approach seems more adequate.

As described in Figure 4.9, in this section the fault diagnosis methodology consist of two stages. In the first stage, the fault has to be detected on the basis of residuals generated from a bank of output estimators, while,

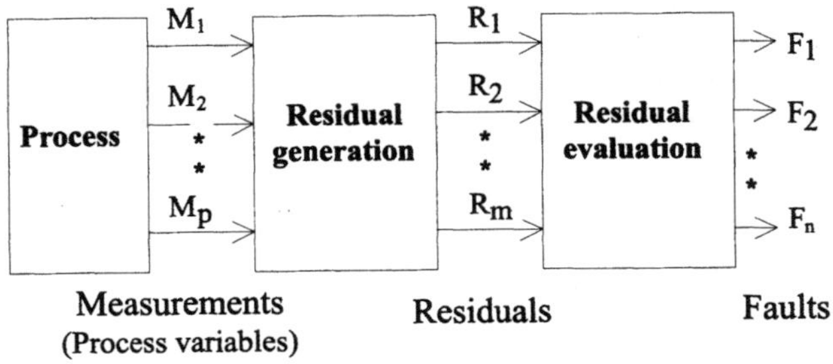

Fig. 4.9. The general structure of a diagnosis system.

in the second stage, fault identification is obtained from pattern recognition techniques implemented via Neural Networks.
Fault identification represents the problem of the estimation of the size of faults occurring in a dynamic system.

A NN is exploited in order to find the connection from a particular fault regarding system inputs and output measurements to a particular residual. In such a way the output predictor generates a residual which does not depend on the dynamic characteristics of the plant, but only on faults. Therefore, the NN classify static patterns of residuals, which are uniquely related to particular fault conditions independently from the plant dynamics.

In recent years, Neural Networks (NN) were studied when applied to fault diagnosis problem. NN have been used both as predictor of dynamic models [Marcu and Mirea, 1997] [Yu *et al.*, 1999] for fault diagnosis, and pattern classifiers [Hoskins and Himmelblau, 1988] [Dietz *et al.*, 1989] [Venkatasubramanian and Chan, 1989] [McDuff and Simpson, 1990] [Weerasinghe *et al.*, 1998] [Meneganti *et al.*, 1998] [Napolitano *et al.*, 1998] [Chowdhury and Aravena, 1998] for fault identification.

As the monitored plants have in general a dynamic behaviour, the NN should be equipped with a Non–linear Auto Regressive Moving Average structure (see [Leontaritis and Billings, 1985a] for details) when used as model estimator. However, the most frequently applied neural models are the feed-forward perceptron used in multilayer networks (see for example [Widrow and Lehr, 1990]), which have a static structure. In such a case, the introduction of explicit dynamics requires the feedback of some outputs through time delay units [Brown and Harris, 1994b].

Alternatively to static structure, NN with neurons having intrinsic dynamic properties [Werbos, 1990] can be used. However in both cases, the training procedure presents some practical difficulties as it should be performed over temporal patterns, and the neural model often presents poor approximation characteristics of the real plant. On the other hand, NN can be effectively exploited for residual signal processing, which is actually a static patter recognition problem.

On the basis of such discussion, this section addresses a methodology in which model–based approach and NN are combined to detect and identify the fault occurring in industrial processes.

Fault signals create changes in several residuals obtained by using output predictors of the process under examination. A neural network is exploited in order to find the connection from a particular fault regarding input and output measurements to a particular residual. In such a way the predictors generate residuals independent of the dynamic characteristics of the plant and dependent only on sensors faults. Therefore, the neural network evaluates static patterns of residuals, which are uniquely related to particular fault conditions independently from the plant dynamics.

The problem presented regards the detection and identification of the faults on the basis of the knowledge of the measured sequences $\boldsymbol{u}(t)$ and $\boldsymbol{y}(t)$. Moreover, it is commonly assumed that only a single fault may occur in the monitored plant.

4.10.1 Neural Network Basics

Neural networks in fault diagnosis have been usually exploited to classify measurement patterns according to the operation of the process. Unfortunately, classification of individual measurement patterns is not unique in dynamic situations.

In this section the problem is tackled by using a detection device which generates residuals independent of the dynamic characteristics of the plant and dependent on measurement faults. Static patterns can be therefore used to train the neural network.

The classification method is typically an off–line procedure where the fault mode is first defined and the data collected. In this situation, certain measurement patterns correspond to normal operation and other patterns correspond to faulty operations: the training of neural networks using this kind of information is called supervised.

A NN may be therefore classified as *supervised*, in which a teacher is used to train the network, and as *unsupervised*, in which input patterns are clustered into groups collecting similar inputs.

The MultiLayer Perceptron (MLP) and Radial Basis Function (RBF) networks are typical examples of supervised trained network architectures.

Since the diagnosis device has to represent the function mapping residuals into input sensor fault sizes, neural networks actually perform an approximation, and not properly a classification of static patterns.

The neural networks presented can be designed in the MATLAB environment by exploiting the Neural Network Toolbox [Mat, 1990].

In order to determine the network architecture which gives the best results in the noisy environment, MLP and RBF has been tested. They are both able to approximate any continuous function with an arbitrary degree of accuracy, provided with a sufficient number of neurons [Funahashi, 1989, Leshno *et al.*, 1993].

Firstly, a three–layered RBF network, namely the Generalised Regression Network, has been considered [MathWorks, 1998].

The hidden layer was composed of radial basis neurons performing a non–linear mapping of input space. Unnormalised Gaussian functions given by the equation,

$$G = e^{-\|\mathbf{x}-\mathbf{c}\|^2/\delta^2} \tag{4.77}$$

in which $\|\cdot\|$ denotes the Euclidean norm, $\mathbf{x}$ is the m–dimensional input vector, $\mathbf{c}$ the centres and δ the width of the Gaussian functions, are the most common functions in the hidden node, but several other functions have been proposed [Chen *et al.*, 1990a].

In the output layer linear neurons were used in order to perform the function approximation. The parameters of a radial basis network were obtained with the training procedure.

Centres of the Gaussian functions were the most troublesome values to tune. In particular, the network architecture was implemented by using a "non–exact" solution. An "exact" design solution requires one hidden neuron for each training pattern.

A different supervised neural network architecture was then considered, namely a so called back-propagation or multilayered feed–forward MLP network [Simani *et al.*, 1998c, Simani and Fantuzzi, 2000, Simani and Patton, 2002b, Simani and Fantuzzi, 2002].

Such a neural network consists of an input layer, one or more hidden layers and an output layer. A multilayer perceptron comprises several layers of simple computation units called *neurons*.

The mathematical description of a neuron is:

$$y_i = f_a(\mathbf{w_i}^T\mathbf{p} + \mathbf{b_i})$$

where $\mathbf{p}$ is the input pattern and $\mathbf{b}_i$, $\mathbf{w}_i$ are the parameter vectors of the neuron and y_i is neuron output. The function f_a is an activation function, generally non–linear. As the RBF, the MLP network can be designed with one hidden layer.

Moreover, since the network is used as function approximator, in the input and hidden layers sigmoidal neurons were implemented, whilst the output layer was made of a single linear neuron.

A back–propagation algorithm with adaptive learning rate was exploited to update network parameters. The aim of the algorithm is to minimise the sum of the square error (SSE) between the desired and actual network output:

$$SSE = \frac{1}{P}\sum_{p=0}^{P}\sum_{t=0}^{N}\frac{1}{2}\left(t_{t,p} - y_{t,p}\right)^2 \tag{4.78}$$

where $t_{t,p}$ and $y_{t,p}$ are the desired and actual t-th network output, respectively, regarding the p-th input pattern. N is the number of the training patterns and P, the number of the network outputs.

4.11 Fault Diagnosis of an Industrial Plant at Different Operating Points Using Neural Networks

Industrial plants often work at different operating points. However, in the literature applications of neural networks for fault detection and diagnosis usually consider only a single working condition or small changes of operating points.

A standard scheme for the design of neural networks for fault diagnosis at all operating points may be impractical due to the unavailability of suitable training data for all working conditions.

This section addresses the design of a single neural network for the diagnosis of abrupt fault (bias) in the sensors of an industrial process working at different conditions [Simani, 2000a].

Data pre-processing methods are also investigated to enhance fault classification, to reduce the complexity of the neural network and to facilitate the learning procedure [Simani, 2000a].

Results illustrating the performance of the trained neural network for sensor faults diagnosis and using simulated and real data from a single–shaft industrial gas turbine are shown in Section 5.3.6 [Simani, 2000a, Simani and Fantuzzi, 2000, Simani and Patton, 2002b, Simani and Fantuzzi, 2002].

4.11.1 Operating Point Detection and Fault Diagnosis

Industrial processes are complex and reliable diagnosis of faults can be therefore be a difficult task [Gertler, 1998]. In particular, in such a processes, sensor faults (bias) are likely to occur. Therefore, in order to prevent machine malfunctions and to determine the machine operating state, it is essential to have a sensor fault detection system as well as a fault diagnosis method.

The industrial plants here investigated and monitored are multivariable processes for which physical relationships and process coefficients are mostly unknown [Gertler, 1998]. Therefore, a FDI scheme based on a mathematical plant description [Chen and Patton, 1999] can be implemented only by identifying an accurate model for the process [Simani *et al.*, 2000a]. In such a case, a reliable FDI scheme may require an high order identified model.

Knowledge–based methods such an expert system and fuzzy logic can exploit the knowledge of an expert operator of the industrial plant [Simani, 1999a, Simani and Patton, 2002a]. The operator should formulate and design the rules for the system rule–base of a multivariable process.

The difficult task of developing a fault diagnosis technique for large-scale industrial plants can be achieved efficiently by using NNs [Simani, 2000a]. They are both pattern recognition methods as well as non–linear function approximators with arbitrary accuracy.

NNs that do not need a deep insight into the process, are robust to noise data and have the ability to generalise the relationship learnt to successfully diagnose learned faults as well as new fault conditions [Venkatasubramanian and Chan, 1989, Simani *et al.*, 1998b, Simani *et al.*, 1999d, Simani and Patton, 2002a].

In this section, the classification and approximation capabilities of a NN is to be exploited [Simani, 2000a]. The capabilities of a MLP NNs using back–propagation learning algorithm have been compared with RBF ones. The latter were found to offer better performances [Leonard and Kramer, 1991].

A basic aspect of NN design is the pre–processing of input data. Many NN applications use scaled and normalised input data before using the patterns for network training [Simani, 2000a]. Tools such as Principal Component Analysis (PCA) are also methods to exploit [Kavuri and Venkatasubramanian, 1994].

As shown in Section 5.3.6, most industrial plants operate at more than one operating point. It can be easier to obtain plant data for NN training during the main operating points and more difficult to acquire data from operating points which are not frequently used. Moreover, even if data from healthy conditions under different operating points can be found, data from fault conditions are nearly non-existent. Whilst the problem of the design of NNs for process FDI has received attention, the diagnosis of faults at different operating points has had little consideration [Simani, 2000a, Simani and Fantuzzi, 2000, Simani and Patton, 2002b, Simani and Fantuzzi, 2002].

This section describes how a NN can be successfully exploited for the diagnosis of abrupt faults on the sensors, based on a study on an industrial process [Simani, 2000a]. Fault modes can be modelled by using step functions. The process was considered to work at different operating points. Time series of real data acquired from plant sensors are available, but there was a lack of data with labelled fault classes. Due to this drawback and to the lack of a suitable and accurate mathematical model of the process, it is impossible to develop an adequate model–based fault diagnosis methodology.

Faulty sequences at different working conditions have to be generated using a simulated model of the monitored process. Then the diagnosis techniques here proposed have been applied to both the real and simulated process data [Simani, 2000a, Simani and Patton, 2002a].

The methodology involve techniques to pre–process the data by statistical scaling, algorithms to reduce the NN complexity using PCA, as well as training and testing the NN [Simani, 2000a].

The methodologies have been initially applied to design an Artificial Neural Network (ANN) to diagnose faults at the main operating point. The diagnosis of faults during start–up and and at the secondary operating point using the same NN was then investigated [Simani, 2000a].

Results are finally presented in Section 5.3.6 to illustrate the performance of the developed FDI scheme for the real plant [Simani, 2000a].

4.11.2 FDI Method Development

The method presented has been carried out in three stages. The first one consists of exploiting methods to pre–process the network input data for better classification and reduced network complexity by data scaling and PCA. The second step is the NN training and testing. Once a satisfactory network had been obtained, the third part consists of developing methods to diagnose faults at the secondary operating point using the network trained to diagnose faults at the primary operating point.

1. The magnitudes of measured process variables can span a wide range. Data conditioning is achieved by scaling the data using standard statistical normalisation methods. Time series of data are divided by the corresponding standard deviation and the mean values were subtracted. This gives all variables the same variance, brings them to comparable range. The mean and the standard deviation values used are those of the healthy condition at each operating point.
2. Since the plant is often a multivariable process, all the variables are to be used as inputs to the NN and this will result in a very complex network topology with a large number of hidden nodes. In order to reduce the input space of the NN, the well-known PCA statistical method can be used. Therefore, the number of highly correlated variables in a multivariable data set can be reduced to a smaller one of uncorrelated variables without any loss of information. Selection was carried out using methods proposed in [Jackson, 1991].
3. The data conditioned are used as inputs to the NNs. The NN training has been performed using the Neural Network Toolbox for MATLAB [Demuth and BealeDemuth, 1997]. Tests have been carried out initially on both MLP and RBF networks to compare their performances in the classification of faults.

Once the network had been trained to diagnose faults at both the primary and the secondary operating point satisfactorily, using the *simulated* process model, the next part of the study consists in developing a methodology to use this network to diagnose faults occurring under the secondary operating point of the *real* plant.

Simulated process data have to be statistically scaled, converted into principal component variables using PCA and are therefore used to train the networks.

The results of the fault diagnosis methodology presented are shown in Section 5.3.7 [Simani, 2000a].

4.12 Neuro-fuzzy in FDI

Many authors have focused on the use of neural networks in FDI applications [Marcu *et al.*, 1999, Korbicz and Obuchowitcz, 1999] for solving the specific tasks in FDI, such as fault isolation but mainly fault detection. Other authors [Koscielny *et al.*, 1999] used fuzzy logic for fault diagnosis, especially for fault isolation, but some of them even for fault detection, using for example TSK fuzzy models. In the last few years there is also an increasing number of authors [Patton *et al.*, 1999a, Calado *et al.*, 2001, Uppal and Patton, 2002, Uppal *et al.*, 2002, Palade *et al.*, 2002] who try to integrate neural networks and fuzzy logic in order to benefit of the advantages of both techniques for fault diagnosis applications.

Neural networks have been successfully applied to fault diagnosis problems due to their capabilities to cope with non–linearity, complexity, uncertainty, noisy or corrupted data. Neural networks are very good modelling tools for highly non-linear processes. Generally, it is easier to develop a non–linear neural network based model for a range of operating than to develop many linear models, each one for a particular operating point. Due to these modelling abilities, neural networks are ideal tools for generating residuals. Neural networks can also be seen as universal approximators. An usual 3 layered MLP neural network, with r inputs and m outputs, can approximate any non-linear mapping from $\Re^r$ to $\Re^m$ using an appropriate number of neurons in the hidden layer. Due to this approximation and classification ability, neural networks can also be successfully used for fault evaluation. The drawback of using neural networks for classification of faults is their lack of transparency in human understandable terms. Fuzzy techniques are more appropriate for fault isolation as it allows the integration in a natural way of human operator knowledge into the fault diagnosis process. The formulation of the decisions taken for fault isolation is done in a human understandable way such as linguistic rules.

The main drawback of neural networks is represented by their "grey box" nature, while the disadvantage of fuzzy systems is represented by the difficult and time–consuming process of knowledge acquisition. On the other hand the advantage of neural network over fuzzy systems is learning and adaptation capabilities, while the advantage of fuzzy system is the human understandable form of knowledge representation. Neural networks use an implicit way of knowledge representation while fuzzy and neuro-fuzzy systems represent knowledge in an explicit form, such as rules.

4.12.1 Methods of Neuro-fuzzy Integration

The combination of neural networks and fuzzy systems can be done in two main ways:

1. *Neural networks are the basic methodology and fuzzy logic is the second.* These hybrid systems are mainly neural networks, but the neural networks are equipped with abilities of processing fuzzy information. The systems are usually termed Fuzzy Neural Networks and they are networks where the inputs and/or the outputs and/or the weights are fuzzy sets, and they usually consist of a special type of neurons, called fuzzy neurons. Fuzzy neurons are neurons with inputs and/or outputs and weights represented by fuzzy sets, the operation performed by the neuron being a fuzzy operation.
2. *Fuzzy logic is the basic methodology and neural networks the subsequent.* These systems can be viewed as fuzzy systems augmented with neural network facilities, such as learning, adaptation, and parallelism. These systems are usually called Neuro-Fuzzy Systems. Most authors in the field of neuro–fuzzy computation understand neuro-fuzzy systems as a special way to learn fuzzy systems from data using neural network type learning algorithms. Some authors [Shann and Fu, 1995] in the field term these neuro-fuzzy systems also fuzzy neural networks, but most of them like to term them as Neuro-Fuzzy Systems. Neuro–Fuzzy Systems [Nauck and Kruse, 1998] can always be interpreted as a set of fuzzy rules and can be represented as a feed-forward network architecture.

These two previous ways of neuro–fuzzy combination can be viewed as a type of fusion systems, as it is difficult to see a clear separation between the two methodologies. One methodology is fused into the other methodology, and it is assumed that one technique is the basic technique and the other is fused into it and augments the capabilities of information processing of the first methodology. Besides these fusion forms of neuro–fuzzy systems, there is another way of hybridisation of neural networks and fuzzy systems, where each methodology maintains its own identity and the hybrid neuro–fuzzy system consists of modules structure which cooperate in solving the problem. These kind of neuro–fuzzy systems are called combination hybrid systems. The neural network based modules can work in parallel or serial configuration with fuzzy logic based modules and augments each other. In some approaches, a neural network (such as a self–organising map) can pre-process input data for a fuzzy system, performing for example data clustering or filtering noise. But, especially in FDI applications, many authors use a fuzzy system as a pre–processor for a neural network. In [Alexandru *et al.*, 2000] the residuals signals are fuzzified first and then fed into a recurrent neural network for evaluation, in order to perform fault isolation.

The most often used NF systems are fusion NF systems and the most common understanding for a Neuro–Fuzzy system is the following. A NF system is a neural network which is topologically equivalent to the structure of a fuzzy system. The network inputs/outputs and weights are real numbers, but the network nodes implement operations specific to fuzzy systems: fuzzification, fuzzy operators (conjunction, disjunction), defuzzification. In other words, a NF system can be viewed as a fuzzy system, whit its operations implemented in a parallel manner by a neural network, and that's why it is easy to establish a one–to–one correspondence between the NN and the equivalent FS. Neuro–Fuzzy systems can be used to identify fuzzy models directly from input-output relationships, but they can be also used to optimise (refine/tune) an initial fuzzy model acquired from human expert, using additional data.

4.12.2 Neuro-fuzzy Networks

In the area of neuro–fuzzy systems there are two principal types of neuro–fuzzy networks preferred by most of the authors in the field of neuro-fuzzy integration. In Sections 4.12.3 and 4.12.4, we will use kind of these structures, shortly presented bellow, for residual generation and for fault classification. The most common neuro-fuzzy network is used to develop or adjust a fuzzy model in Mamdani form given by relation 4.79, using input–output data. The network is a five layers network as shown in 4.10. A Mamdani fuzzy model consists of a set of fuzzy **IF–THEN** rules in the following form:

$$\textbf{IF}\ \ x_1\ \text{ is }\ X_{1i_1}\ \ \textit{and}\ \ x_2\ \text{is}\ X_{2i_2}\ \textit{and}\ \ \ldots\ x_n\ \text{ is }\ X_{ni_n}\ \textbf{THEN}\ \ y\ \text{ is }\ Y_j \tag{4.79}$$

where $x_1, x_2, \ldots, x_n$ are the system inputs, y is the output, X_{ki_k} with $k = 1, 2, \ldots, m$ and $i_k = 1, 2, \ldots, l_k$ are the linguistic values of the linguistic variable x_k, and Y_j, $j = 1, 2, \ldots, l_y$ are the linguistic values of the output. Every linguistic variable x_k is described by l_k linguistic values $X_{k1}, X_{k2}, \ldots, X_{kl_k}$.

Layer 1 is the input layer and each node corresponds to each input variable. Layer 2 is called *membership function layer*, the nodes from this layer mapping each input x_i to every membership function X_{ij} of the linguistic values of that input. It is possible to use, in the layer 2, a subnet of nodes to implement a desired membership function, instead of a single node. Each node in the layer 3 (called *rule layer*) performs the precondition matching – the IF part – of a fuzzy rule. The nodes from layer 4 combine the fuzzy rules with the same consequent, each node implementing a fuzzy OR operator, such as fuzzy max operator. Each node in the layer 5 corresponds to an output variable and acts as a defuzzifier. The integration and the activation functions of nodes for such a network are chosen [Shann and Fu, 1995] so that to perform specific operations in a fuzzy inference engine as described before.

Another major class of neuro–fuzzy networks are the neuro–fuzzy networks used to develop and adjust a Sugeno–type fuzzy model. The structure of such a neuro–fuzzy network is shown in Figure 4.11. The first 3 layers are the same

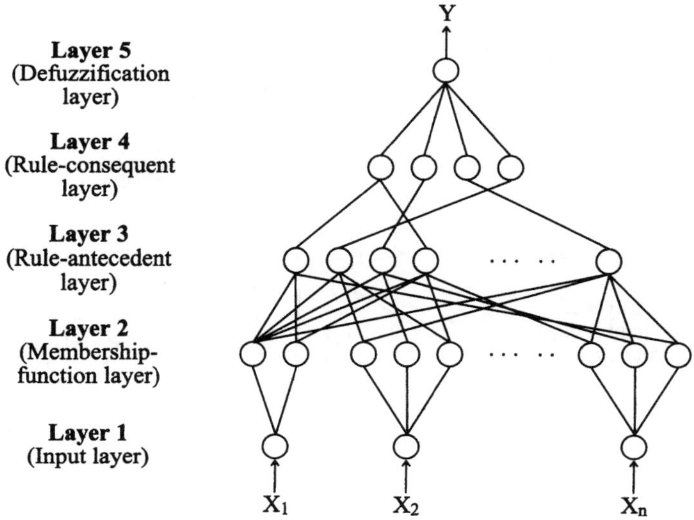

Fig. 4.10. The general structure of a Neuro–Fuzzy network for Mamdani models

with those in a neuro–fuzzy network for Mamdani models. In the rule layer, it can be used the traditional fuzzy *min* operator, but many authors prefer to use a product operator as fuzzy intersection operator. Usually, all weights of this layer are set to 1. If some prior knowledge on process functioning is available, it can be established the number of nodes in layer 3 (the number of rules or fuzzy partition regions) and the corresponding links between layer 2 and 3.

In [Zhang and Morris, 1996] the authors developed a neuro–fuzzy network for process modelling and fault diagnosis. The main shortcoming of this structure is that the user must partition the process operation into several fuzzy operating regions before training the fuzzy neural network. The partitioning is made empirically, looking to the process functioning, and it may be a very difficult task when the process has a complex nature. Different clustering techniques as well as genetic algorithms can be used to find the best fuzzy partition of the input space. Layer 4 is called the model layer, and each node implements a linear model corresponding to a rule node in the rule layer, respectively to a fuzzy operating region. The weights of a node are the parameters of the linear model and the inputs of the node are the past system inputs and outputs. Layer 5 consists of a single node, which performs the defuzzification. The most general Sugeno–type neuro–fuzzy network structure, is a network which implements a set of fuzzy rules with ARMA models of higher order in the consequence part of the rules. The rules are in the following form:

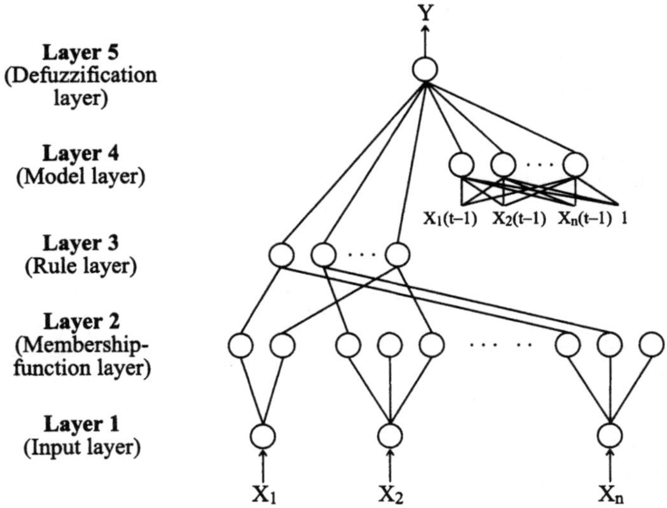

Fig. 4.11. Neuro–Fuzzy network for TSK fuzzy model implementation.

$$R_k : \textbf{IF} \ \ x_1 \ \text{is} \ X_{1i_1} \ \ \textit{and} \ \ x_2 \ \text{is} \ X_{2i_2} \ \ \textit{and} \ \ \ldots \ x_n \ \text{is} \ X_{ni_n} \ \textbf{THEN}$$
$$y_k(t) = a_{0k} + \sum_{j=1}^{n_1} a_{jk}x(t-j) + \sum_{j=1}^{n_2} b_{jk}y(t-j) \tag{4.80}$$

where $k = 1, 2, \ldots, m$, m the number of rules, and $\boldsymbol{x} = (x_1, x_2, \ldots, x_n)$ is the input vector, and $\boldsymbol{a}_{jk} = \left(a_{jk}^1, a_{jk}^2, \ldots, a_{jk}^n\right)$

When linear ARMA models of higher order are used, every node from layer 4 must be replaced by a subnet, which implements the ARMA model of the desired order. In Figure 4.12, it is shown the subnet which corresponds to node k from layer 4, when $n_1 = n_2 = 2$. The inputs of the subnet k from layer 4 are the previous inputs and outputs of the system.

4.12.3 Residual Generation Using Neuro-fuzzy Models

According to Figure 4.13, the residual signals $\boldsymbol{r}(t)$ are calculated as difference between estimated signal given by observer and the actual value of the signal. The residuals can be generated using TSK neuro–fuzzy networks. The main problem is how to find accurate neuro–fuzzy models for generating residuals, and at the same time to have as much as possible a certain degree of transparency of the models. That is why we have to find a good structure of the model and therefore a good partition of the input space, using clustering for that. There is a compromise between the interpretability and the precision of the model.

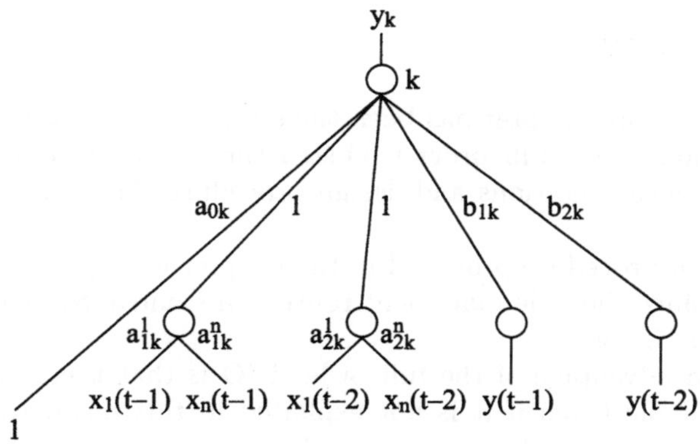

Fig. 4.12. The subnet corresponding to node k in layer 4.

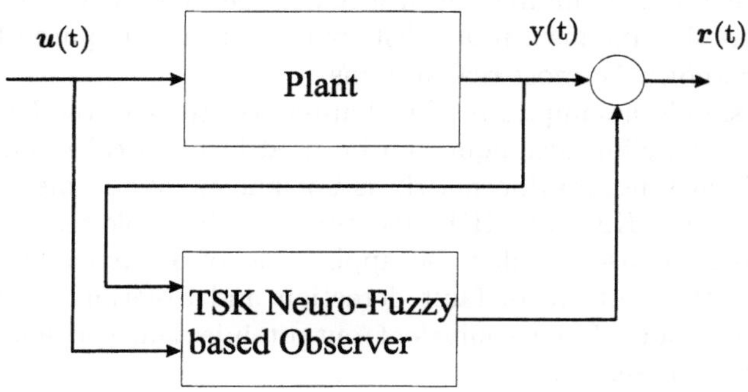

Fig. 4.13. Neuro–fuzzy based observer scheme for generating residuals.

4.12.4 Neuro-fuzzy-based Residual Evaluation

In the residual generation part of a diagnosis system the user should be more concerned on the accuracy of neuro–fuzzy models, even desirable to have interpretable models also for residual generation, such as TSK models. For the evaluation part it is more important the transparency or the interpretability of the fault classifier, in human understandable terms, such as classification rules. The main problem in neuro–fuzzy fault classification is how to obtain an interpretable fuzzy classifier, which should have few meaningful fuzzy rules with few meaningful linguistic rules for input–output variables [Babuška, 1998]. Neuro–fuzzy network for Mamdani models are appropriate tools to evaluate residuals and perform fault isolation, as the consequence of the rules contains linguistic values which are more readable than linear models in case of using TSK fuzzy models. The price paid for the interpretability of the fault classifier may be the loss of the precision of the classification task.

4.13 Summary

The purpose of this chapter has been the study of UIO–based residual generation methods and a full–order UIO structure has been recalled.

The existence conditions and design procedures for such UIO have also been presented.

The design procedure proposed in the chapter is very easy to verify and implement, since the pole placement routine in Control System Toolbox for MATLAB can be used.

The main advantage of the full–order UIO is that there is more design freedom available (even if it is not exploited in this monograph) after the unknown input de–coupling conditions have been satisfied. The remaining freedom may be used to generate directional residuals for fault isolation.

UIO–based FDI methods have been studied for many years but the number of applications is limited, even if it is increasing. The main problem is, in fact, that the unknown input distribution matrix, required for designing UIOs, is unknown for most real systems.

Under simple assumptions, the chapter has shown how UIO–based disturbance de–coupling technique can be used in practical systems, in which the disturbance distribution matrix is not known. When measurements are affected by noise, KF and UIKF structures can be exploited.

This chapter also recalled the application of a particular sliding mode observer to the problem of fault detection and isolation. The novelty lies in the application of the equivalent output injection concept to explicitly reconstruct fault signals.

A residual generation technique exploiting fuzzy models was presented while the identification of faults concerning system inputs and outputs can be performed by means of static NNs as well as Neuro Fuzzy systems.

Finally, neuro–fuzzy techniques can be applied both for residual generation and for fault classification. The combination of neural networks with fuzzy systems can produce better diagnostic results, especially when there is an interest on the transparency in human understandable terms of neuro-fuzzy models. A compromise always exists between the interpretability and the precision of the model.

CHAPTER 5
FAULT DIAGNOSIS APPLICATION STUDIES

5.1 Introduction

In this chapter several simulated and real application examples are presented in order to test the FDI techniques studied in Chapter 4 in connection with identification procedures presented in Chapter 3.

In this study complete design procedures for FDI and fault identification of actuators, components, input and output sensors of industrial processes are considered under application study according to the theory presented in Chapters 2 and 4.

The fault diagnosis can be performed using banks of dynamic observers and UIO or, when the measurement noises are not negligible, banks of Kalman Filter (KF) and of the Unknown Input Kalman Filter (UIKF) [Simani *et al.*, 2000a]. Faults affecting the actuators, components, input and output sensors are considered and simulated in the monitored systems.

As explained in Chapter 3, the FDI methods applied do not require any physical knowledge of the processes under observation since the input–output links are obtained by means of identification schemes using EE and EIV models.

In the case of noisy measurements, the identification technique (Frisch scheme) described in Chapter 3 for EIV models also gives the variances of the input–output noise signals, that are required in the design of the KF.

The identification and fault diagnosis procedure has been applied to different models of a real and simulated power plants.

In order to analyse the diagnostic effectiveness of the FDI system in the presence of abrupt changes or drifts in measurements, faults modelled by step or ramp functions have been generated.

The results obtained by this approach indicate that the minimal detectable faults on the various processes is a parameter of interest for industrial diagnostic applications.

The following processes are described.

- a Multiple-Input Multiple-Output (MIMO) SIMULINK model of a real single–shaft industrial gas turbine with variable Inlet Guided Vane (IGV) angle working in parallel with electrical mains.

- a MIMO real 120MW power plant of Pont sur Sambre. It is a double–shaft industrial gas turbine working in parallel with electrical mains.
- a MIMO SIMULINK prototype of a real single–shaft industrial gas turbine.

5.2 Physical Background and Modelling Aspects of an Industrial Gas Turbine

In this section, the dynamic non–linear model of a single–shaft industrial gas turbine has been developed as the first stage of a methodology aimed at the diagnosis of a wide range of components of an industrial process.

The model has been calibrated by means of reference steady–state condition data of a real industrial gas turbine and it has been used to simulate various machine transient.

Although the model is modular in structure and was carried out in simplified form, these features did not compromise its accuracy. Te computation time is also minimal, making the modelling methods suitable for on–line simulation.

The comparison between values of working parameters obtained by the simulation and measurements during some transients on the gas turbine in operation provided encouraging results.

The implementation of industrial gas turbine dynamic models can make it possible:

- to predict machine transient conditions due to component of different kind, volume and time constant, reducing test costs:
- to design gas turbine control systems;
- to generate time series of transient condition data.

In particular, it is possible to have a large volume of data otherwise difficulty to be available for industrial gas turbine that works mainly in steady–state conditions. In diagnostic applications, the machine dynamic model can be used to simulate operating conditions of gas turbines with faults in components, measurements and control sensors.

5.2.1 Gas Turbine Model Description

The dynamic non–linear gas turbine model has been developed by dividing the dynamic operation of the machine into elementary modules corresponding to its main components and developing a dynamic model for each of the modules.

The overall representation of a specific gas turbine is carried out by identifying the necessary modules and connecting them appropriately by means of thermodynamic and mechanic links.

The dynamic behaviour of each module is described by means of equations representing the thermodynamic transformations, the mass and momentum balance [Bettocchi *et al.*, 1996].

The mass and momentum balance equations are used in one–dimensional differential form, in the hypothesis of assimilating each block to a constant section duct [Bettocchi *et al.*, 1996].

Within these hypotheses, the the above equations take on the following forms, respectively [Shapiro, 1977]:

$$\left.\begin{array}{ll} \frac{\partial \rho}{\partial t} + \frac{\partial (\rho v)}{\partial l} = 0 & \text{mass balance} \\ \rho \frac{\partial v}{\partial t} + \rho v \frac{\partial v}{\partial l} = -\left(\frac{\partial p}{\partial l} + F + \rho g \frac{dz}{dl} \right) & \text{momentum balance} \end{array}\right\} \quad (5.1)$$

where F is the friction force, g the gravity acceleration, l the linear coordinate, p the pressure, ρ the density, v the velocity and z the altitude, with respect to time t.

Equations 5.1 are integrated considering that the change in the fluid density takes place according to an isentropic transformation.

The equations representing the thermodynamic transformation are used instead in stationary form, since the fluid thermal inertia is considered negligible in comparison to the mechanical inertia.

From a mechanical standpoint, the fluid is considered as a perfect gas and in each modules are used mean values of specific heat at constant pressure and at a constant volume depending on the temperature between modules input and output and on the fluid composition. The use of mean specific heats in dynamic simulations, where the changes in thermodynamics and performance values are analysed with reference to an initial steady-state condition, does not significantly affect the result accuracy, but considerably reduces both the model complexity and the calculation time [Bettocchi *et al.*, 1996].

The mass flow rates bled on the compressor are calculated without considering the dynamic effects on them in transient conditions, and considering that the mass flow function of air bled at the outlet of each elementary compressor module is constant in all operating conditions [Benvenuti *et al.*, 1993].

The effect on the thermodynamic cycle of turbine nozzle and blade row cooling flows, calculated starting from the mass flow rates bled on the compressor, is assessed by splitting the total cooling flow appropriately into two parts and assuming that one is mixed upstream and the other downstream from the turbine module, causing a reduction of the main flow total temperature and then, a reduction of the available enthalpy drop [Benvenuti *et al.*, 1993]

In addition to the equations describing the various modules, equations are used that represent the dynamic balance of shafts and rotating masses of the machine connected to them.

The simulation of the gas turbine working was carried out by integrating the differential equations and solving the static equations with the variable values calculated at each time instant.

Since the intention was to limit calculation and system costs in the subsequent diagnostic stage, it is necessary to be able to run the dynamic model on a PC using commercial available software. To achieve the integration of the differential equations SIMULINK® of *MATLAB* [The MathWorks Inc, 1990, The MathWorks Inc, 1991] was used, as it is a feasible (easy to use) and wide spread software.

The splitting of the gas turbine into elementary modules facilitates the modelling of other machines of any configuration with appropriate linking of various modules.

The description of how the main modules have been modelled is given below.

Duct. This term refers to the modules of the gas turbine in which no thermodynamics transformations occur, but only a mass flow. In order to carry out dynamic simulations their modelling is especially important because the "duct" modules are the site of inertial phenomena due to the mass of fluid they contain. These phenomena are described by means of mass and momentum by Equations 5.1.

In particular, for the i–section of the module, these equations take on the form [Blotenberg, 1993]:

$$\left.\begin{array}{ll} \frac{dp_i}{dt} = \frac{kRT_i}{V_i}\left(m_i - m_{(i+1)}\right) & \text{mass balance} \\ \frac{dm_i}{dt} = \frac{A_i}{L_i}\left(p_{(i-1)} - p_i\right) - \frac{\lambda_i kRT_{(i-1)} m_i^2}{A_i D_i \left(p_{(i-1)} - p_i\right)} & \text{momentum balance} \end{array}\right\} \quad (5.2)$$

where A is the area, $k = c_p/c_V$ with c_p the specific heat at constant pressure and c_V the specific heat at constant volume, D the hydraulic diameter, m the mass flow rate, p the pressure, R the gas constant, T the stagnation temperature and λ the friction coefficient.

The equations were obtained under the assumptions that the duct, whatever its geometry, may be assimilated with a constant section pipe and the change in fluid density takes place according to an isentropic transformation.

In the case of intake duct (ID), the integration of Equation 5.2 makes it possible to calculate the outlet pressure and air mass flow rate for given input conditions and duct geometry.

On the other hand, in case of the exhaust duct (ED), where the outlet pressure as well as the input conditions are known, it is sufficient to integrate the momentum balance given by Equation 5.2 alone, in order to calculate the outlet gas mass flow rate.

As an example, Figure 5.1 shows the model of the "intake duct" which is shown as "ID" in Figure 5.6, based on SIMULINK blocks.
Note that Equations 5.2, in which $i = 1$, were solved using the SIMULINK blocks and transport signal lines in place of the computer components and physical links among the various components, respectively.

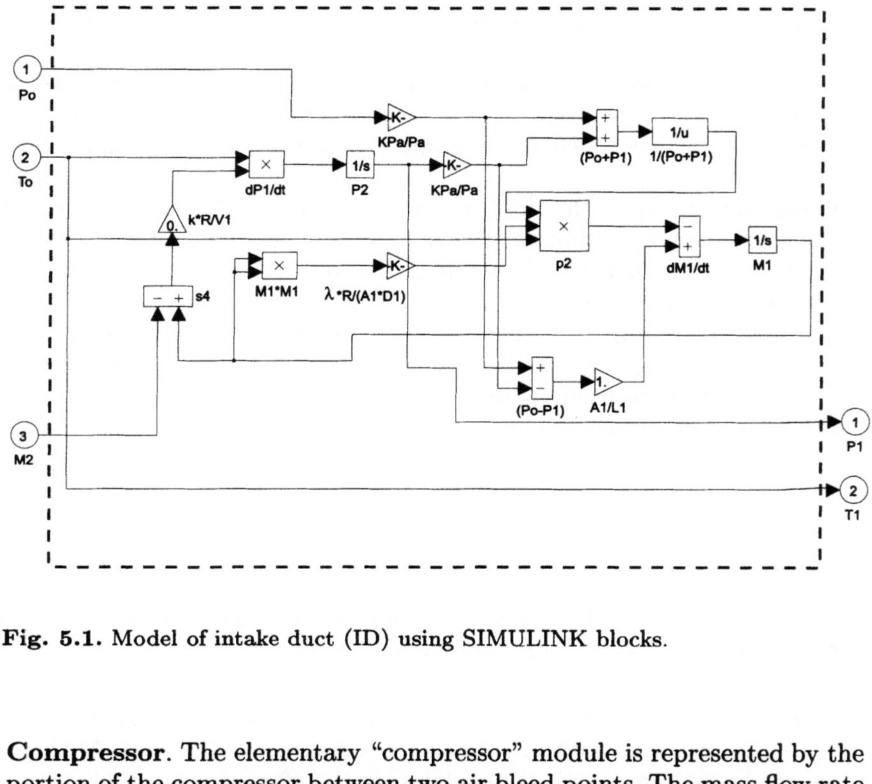

Fig. 5.1. Model of intake duct (ID) using SIMULINK blocks.

Compressor. The elementary "compressor" module is represented by the portion of the compressor between two air bleed points. The mass flow rate that passes through the module and the corresponding isentropic compression efficiency are determined using the performance maps of the particular compressor, when the pressure ratio, rotational speed function and angle of variable compressor IGV, if they exist, are known.
With reference to the i–th compressor module, Equations 5.3 makes it possible to calculate, respectively, the outlet compressor temperature T_i and the compression power, P_c, that provides shaft torque if rotational speed is known

$$\left.\begin{aligned} T_i &= T_{i-1} + T_{i-1}\left(\left(\frac{p_i}{p_{(i-1)}}\right)^{\frac{k-1}{k}} - 1\right)\frac{1}{\eta_{isc}} \\ P_c &= m_i c_p T_{i-1}\left(\left(\frac{p_i}{p_{(i-1)}}\right)^{\frac{k-1}{k}} - 1\right)\frac{1}{\eta_{isc}} \end{aligned}\right\} \tag{5.3}$$

where T is the stagnation temperature, p the pressure, $k = c_p/c_V$, η_{isc} the isentropic efficiency of compressor, m the mass flow rate, c_p the specific heat at constant pressure and c_v the specific heat at constant volume.
The outlet compressor pressure is determined by integrating the mass balance Equation 5.1 written for the compressor, where "V" represents the volume of fluid contained in the compressor and in the downstream diffuser.

Combustor. When the fluid dynamic model includes the "combustor", the mass and momentum balance Equations 5.1 are integrated to calculate the pressure and gas mass flow rate at the combustor outlet for given input conditions and geometry.
The gas temperature T_i at the combustor outlet is calculated using the following balance equation, in the hypothesis that the combustion and release of heat are instantaneous, since the thermal inertia has been neglected with respect to the mechanical inertia:

$$\begin{aligned} T_i &= \frac{m_{(i-1)} c_p T_{(i-1)} + (LHV\ \eta_{cc} + h_f) m_f}{m_i c_p} \approx \\ &\approx T_{(i-1)} + \frac{(LHV\ \eta_{cc}) m_f}{m_i c_p} \end{aligned} \tag{5.4}$$

where LHV stands for "Lower Heating Value" while η_{cc} represents the efficiency of the combustor (combustion chamber).

Turbine. In the elementary "turbine" module, the expansion is assumed to be adiabatic and with no variation in the gas mass flow rate. Mixing between the main flow and cooling flows are therefore concentrated upstream and downstream from the module.
The expansion isentropic efficiency is determined using the performance map of the particular turbine, when the expansion pressure ratio and rotational speed functions are known.
To calculate the gas mass flow rate through the turbine, it was deemed sufficiently approximate to consider the mass flow function at the turbine inlet to be constant in all operating conditions. This assumption is realistic since the transient model is used to simulate working conditions, without considering machine start-up or shut-down.
In a similar manner to the compressor, Equations 5.5 makes it possible to calculate the turbine exhaust temperature T_i and power P_t, the provided shaft torque if the rotational speed is known:

$$\left.\begin{aligned} T_i &= T_{i-1} + T_{i-1}\left(\left(\frac{p_i}{p_{(i-1)}}\right)^{\frac{k-1}{k}} - 1\right)\frac{1}{\eta_{isc}} \\ P_t &= m_i c_p T_{i-1}\left(\left(\frac{p_i}{p_{(i-1)}}\right)^{\frac{k-1}{k}} - 1\right)\frac{1}{\eta_{isc}} \end{aligned}\right\} \tag{5.5}$$

The integration of the mass balance Equation 5.2 for the turbine makes it possible to calculate the machine outlet pressure p_i.

Once the elementary module models have been set up, the overall model of the particular gas turbine was obtained by:

- appropriately linking the modules of which it is composed;
- carrying out the control logic;
- providing the values of constants that are in the various equations.

In this section, the model for simulating a single–shaft industrial gas turbine, with variable compressor IGV angle and first turbine nozzle cooled alone, working in parallel with electric mains was carried out.

Figure 5.6 shows the schematic layout and Figure 5.2 the simplified **block diagram** of the machine. These highlight boundary and control inputs and output variables, the compressor and turbine maps, direct and feedback main links among the various modules.

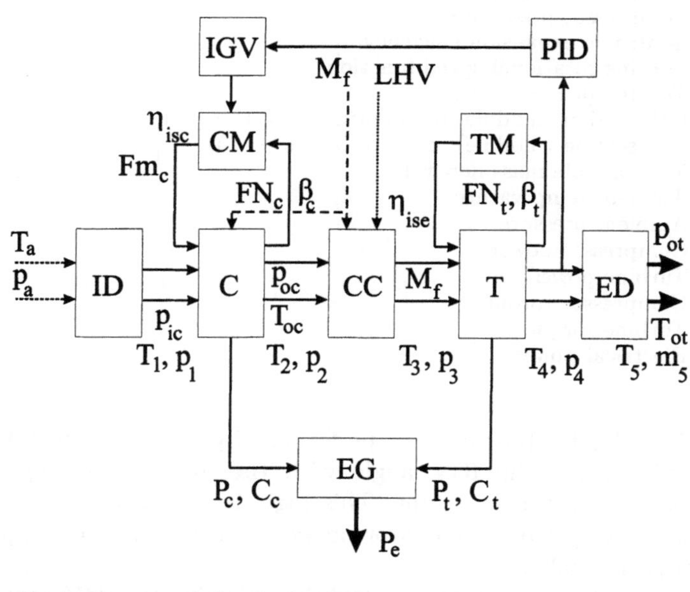

Fig. 5.2. Block diagram of the single–shaft gas turbine.

With reference to Figure 5.2, m_f is the control input (fuel mass flow rate), T_a, p_a (ambient temperature and pressure) and LHV (lower heating value) are boundary condition input, whilst P_e (electric power), T_5 and m_5 (turbine exhaust temperature and mass flow).

The nomenclature used in Figures 5.2 and 5.6 is as follow:

C	Compressor
CC	Combustor (Combustion Chamber)
CM	Compressor Map
ED	Exhaust Duct
EG	Electric Generator
ID	Intake Duct
IGV	Inlet Guide Vanes
PID	Proportional Integral Derivative Controller
T	Turbine
TM	Turbine Map

The nomenclature used in Figure 5.2 is:

M_f	Fuel mass flow rate
LHV	Lower Heating Value
η_{isc}	Isentropic compressor efficiency
Fm_c	Compressor mass flow function
FN_c	Compressor rotational speed function
β_c	Compressor pressure ratio
η_{ise}	Isentropic expansion efficiency
FN_t	Turbine rotational speed function
β_t	Turbine pressure ratio
T_i	i–th section (module) temperature ($i = 1, \cdots, 5$)
p_i	i–th section (module) pressure ($i = 1, \cdots, 5$)
m_5	5–th module mass flow rate
T_a	Ambient temperature
p_a	Ambient pressure
P_c	Compressor power
P_t	Turbine power
C_c	Compressor torque
C_t	Turbine torque
P_e	Electrical power

The machine load adjustment is performed by means of fuel flow rate control and varying the IGV angle with the logic of keeping the turbine outlet temperature constant. This logic is especially suited for optimum heat recovery steam generator operation in cogenerative applications [Bettocchi *et al.*, 1996].

To simulate this type of load control (by adjusting the IGV angle to keep the turbine outlet temperature constant) it was considered that the IGV angle at each time is obtained using a feedback PID controller applied to the turbine outlet temperature, as shown in Figure 5.2.

Since it was necessary to simulate the operation of a single–shaft gas turbine in parallel with electrical mains, it was not necessary to create a model of the rotational speed controller. In this case, the torque offered by the electric generator to the gas turbine adapts almost instantaneously to the torque delivered by the machine, thereby keeping the gas turbine rotational speed constant and equal to the synchronism speed. Therefore, the equation expressing the dynamic balance of rotating masses connected to the shaft:

$$J_g \frac{2\pi}{60} \frac{dN}{dt} = C_t - C_c - C_r \tag{5.6}$$

becomes static and makes is possible to calculate the delivered torque C_r and thus the electrical power produced. J_g represents the moment of inertia of rotating masses connected to the gas turbine shaft reduced to the shaft speed. N is the rotational speed with respect to the time t. C_t is the turbine torque, C_c the compressor torque and C_r the resisting torque.

In order to complete the overall gas turbine model, it is necessary to provide the characteristic constants of the particular machine corresponding to the appropriate equations.

The constants may be classified as

- geometric quantities, such as characteristic volumes, areas and length;
- thermodynamics and fluid-dynamics, mainly represented by the mean specific heats at constant pressure and at constant volume, and by duct friction coefficients.

Before the simulation can be run, these may be read from a startup data file and processed to calculate the constants that are in the model equations.

In addition to these constants, at the start of the simulation it is necessary to know all values in the initial steady state condition. These initial values may, for example, be calculated by a stationary program that uses the same equations of the dynamic program, and that enables the cycle to be computed in the initial state condition.

This solution basically requires the use of two programs, one static and one dynamic used one after the other. A preferred solution involved the use of a dynamic program with the initial values of a particular reference operating condition as constants. If the reference operating condition is different from the one in which the simulation must start, it is possible to go in the steady condition relative to the desired boundary conditions by means of an initial adjustment transient.

For this reason, the model depicted in Figure 5.2 may accept as inputs, in addition to the control variable represented by the fuel flow rate M_f, the variables representing the boundary conditions, such as ambient pressure and temperature (T_a, P_a) and fuel Lower Heating Value (LHV).

In order to assess the validity of the dynamic model developed, it was decided to compare results obtained from the simulation of transient conditions with measurements taken on a gas turbine working in a cogeneration plant [Bettocchi *et al.*, 1996].

Load reduction transients on a single–shaft industrial gas turbine in operation were carried out by the control system in two ways:

- reducing the fuel flow rate M_f and closing the IGV to keep the turbine outlet temperature T_ot constant;
- reducing the fuel flow rate M_f alone, after that the IGV reached the total closer position.

As an example, for the first case, the electrical power P_e, the fuel flow rate M_f and the turbine outlet temperature T_{ot} during the transient were recorded.

The measurements for the first load reduction operation are shown in Figures 5.3, 5.4 and 5.5, all values normalised with respect to the standard deviation of the corresponding signals.

In order to simulate correctly the transients caused by the different working conditions, the control system characteristics (PID constants) after the modelling stage, have to be tuned.

In the case examined, the PID control system characteristics were determined in order to reproduce, during the simulation, the electrical power P_e, the fuel flow rate M_f and the turbine outlet temperature T_{ot} experimentally recorded.

In this way, once PID constants are determined, the simulation provides the electrical power P_e, the fuel flow rate M_f and the turbine outlet temperature T_{ot} these variables are shown in Figures 5.3, 5.4 and 5.5 by using continuous lines. In the same figures, the estimated signals are then compared with the actual measurements acquired from the real process by sampling with regular time intervals (diamond symbols).
The agreement between the simulated and measured curves proves the validity of the dynamic SIMULINK model developed and therefore shows how it is possible to reproduce the real behaviour of the process under investigation by exploiting a "grey box" modelling and identification approach [Bettocchi *et al.*, 1996].

In particular, in the case of load reduction performed by the control system reducing the fuel flow rate and closing the IGV, the mean–square difference between the values obtained by the simulation and those measured experimentally are about 1.1% for the electrical power P_e, 10^{-3}% for fuel flow rate M_f and 0.4% for the turbine outlet temperature T_ot.

Similarly, in the case of load reduction performed by reducing the fuel flow rate M_f alone, the mean–square differences are about 0.8% for the electrical power P_e and turbine outlet temperature T_ot whilst 0.4% for the fuel flow rate M_f.

The percentage differences between calculated and measured transient final values, they are about 0.9% for the electrical power P_e, 0.001% for the

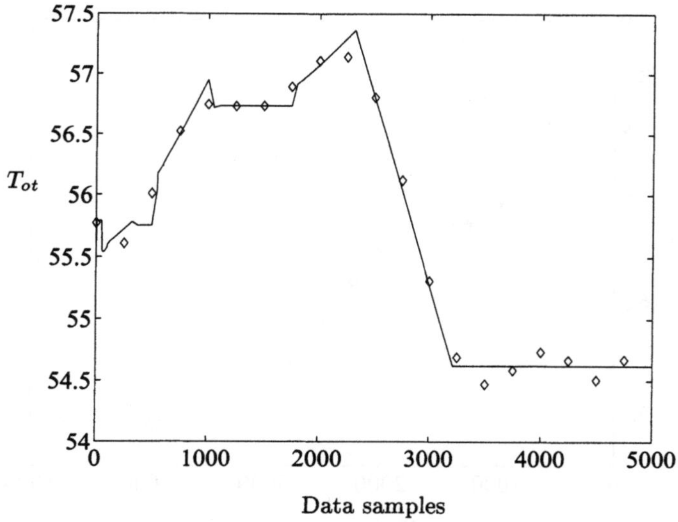

Fig. 5.3. Turbine outlet temperature T_{ot} in the case of load reduction performed reducing the fuel flow rate M_f and closing the IGV angle α.

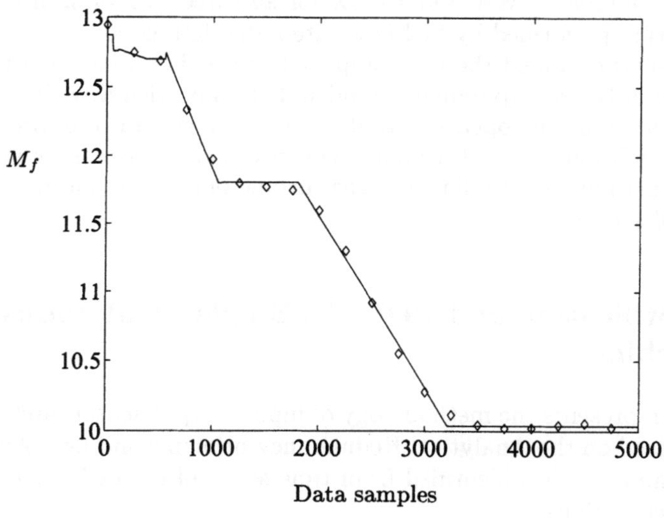

Fig. 5.4. Fuel flow rate M_f in the case of load reduction performed reducing the fuel flow rate M_f and closing the IGV angle α.

fuel flow rate M_f and 0.5% for turbine outlet temperature T_ot, in the case of load reduction performed by fuel flow rate reduction and IGV closing. The

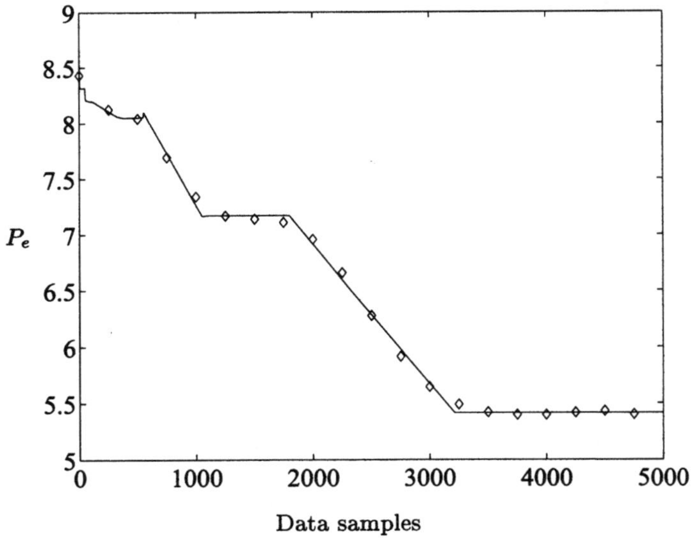

Fig. 5.5. Electrical power P_e in the case of load reduction performed reducing the fuel flow rate M_f and closing the IGV angle α.

percentage differences were about 0.6% for all three variables, in the case of load reduction performed by fuel flow rate reduction alone.

The results obtained therefore appear to provide a first confirmation of the validity of the set–up dynamic model. This is particularly the case as its simplified formulation appears suitable for use as a generator of time series of transient condition data. These data sequences can be necessary in order to develop a methodology to diagnose gas turbine operation, and measurement and control sensors.

5.3 Identification and FDI of a Single Shaft Industrial Gas Turbine

This section presents the methodology of input–output sensor fault diagnosis which is based on the Analytical Redundancy principle and uses ARX MISO linear dynamic models identified from time series of data of the gas turbine operating conditions.

Dynamic observers designed using the identified linear models allow the estimation of some measurable parameters starting from the values of other measured parameters.

The comparison between estimated and measured values of the same parameters enables a vector of residuals for the detection of a possible sensor fault, to be set up.

The application of the methodology to a single–shaft industrial gas turbine model shows the detection and isolation capabilities of faults in sensors used both in the measurements and in the machine control system feedback.

5.3.1 System Identification

The techniques of Analytical Redundancy are based on the idea that the values of all the parameters measured on the machine are functionally correlated by the same dynamic state of the machine.

In order to correlate the measured parameters among themselves, input–output linear models can be identified and therefore dynamic observers can be designed to define correlation depending on the dynamic state of the machine.

The ARX models are generated by using appropriate mathematical techniques, starting from time series of transient condition data.

The use of linear models, in particular, facilitates their set–up and implementation with low cost. The linear models, however, represent the machine only around a particular operating point, requiring a series of models to represent the overall operating field.

The technique for input-output sensor FDI presented is first applied to the model of a real single-shaft industrial gas turbine with variable IGV angle working in parallel with electrical mains in a cogeneration plant [Simani *et al.*, 1998c]. The non–linear turbine model was developed as explained in Section 5.2.

Concerning the machine layout shown in Figure 5.6, the input control sensors are used for the measurement of:

$u_1(t)$, Inlet Guide Vane (IGV) angular position (α);
$u_2(t)$, fuel mass flow rate (M_f).

The output sensors are those used for the measurement of the following variables:

$y_1(t)$, pressure at the compressor inlet (p_{ic});
$y_2(t)$, pressure at the compressor outlet (p_{oc});
$y_3(t)$, pressure at the turbine outlet (p_{ot});
$y_4(t)$, temperature at the compressor outlet (T_{oc});
$y_5(t)$, temperature at the turbine outlet (T_{ot});
$y_6(t)$, electrical power at the generator terminal (P_e).

The gas turbine main features under ISO design conditions are shown in Table 5.1.
The rotational speed sensors are not considered since the operation of the machine in parallel with electrical mains is at constant rotational speed.

The measurements of ambient temperature T_a and relative humidity were also not considered, since they are not directly used by the gas turbine control system. The ambient temperature in particular, which is an important

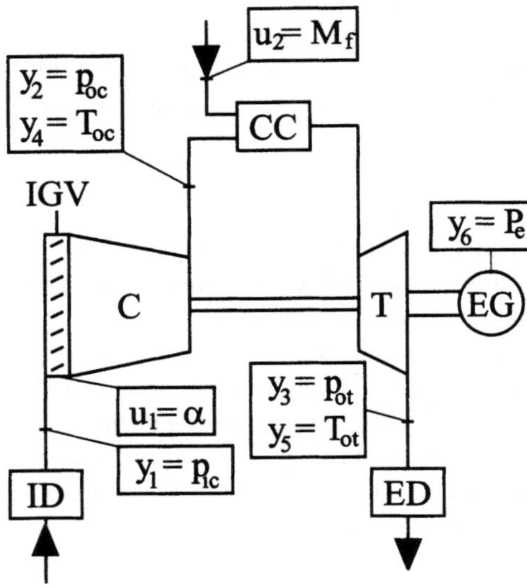

Fig. 5.6. Layout of the single–shaft industrial gas turbine with the monitored sensors highlighted.

Table 5.1. Gas turbine main cycle parameters (ISO design conditions).

Air mass flow rate [kg/s]	24.4
Cycle pressure ratio (P_{oc}/P_{ic})	9.1
Electrical power (P_e) [kW]	5220
Exhaust temperature (T_{ot})[K]	796
Fuel mass flow rate (M_f) [kg/s]	0.388
IGV angle range ($\Delta\alpha$) [deg]	17

parameter for gas turbine performance, is taken into account by the machine control system by means of the measurements of compressor outlet pressure. This pressure P_a indeed depends on the compressor mass flow rate which, in turn, depends on ambient temperature [Simani *et al.*, 1998c].

The design of the different observer configurations necessary to isolate a fault regarding one of the input-output sensors requires the knowledge of a state space model of the system under investigation.

The first step was the identification of a number of input-output models MISO equal to the number of the output variables. These models were obtained using time series of data generated with a non–linear dynamic model which simulates gas turbine operation.

The i-th model ($i = 1, \ldots, 6$) is driven by $u_1(t)$ and $u_2(t)$ and gives the prediction $\hat{y}_i(t)$ of the i-th output $y_i(t)$.

Other model input variables should be the boundary conditions (*i.e.*, ambient pressure and temperature, fuel lower heating value and composition); they were not considered as model inputs since they were assumed to be constant.

The time series data used to identify the models were generated with a non–linear dynamic model presented in Section 5.2 which simulates the gas turbine operation. The simulated process in SIMULINK® environment is shown in Figure 5.7.

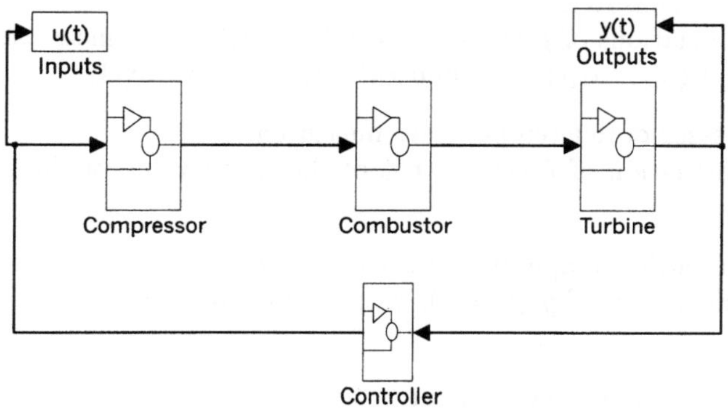

Fig. 5.7. SIMULINK block diagram of the process.

The simulator of Figure 5.7 which represents the process in Figure 5.6 provides a simulation of the power plant. As previously stated, the process consists of three major components: *the combustor*, *turbine*, and *condenser*. Furthermore, there are pumps and valves (not highlighted).

The combustor boils the water and the steam generated drives the turbine. After the turbine, the condenser cools the steam. In turn, external cooling water cools the condenser. Pumps transport the water from the condenser tank back to the combustor tank.

The user can start several simulation sequences where the measurement sensors of the power plant fail.

The non–linear model was previously developed and validated by means of measurements taken during transients on a gas turbine in operation [Simani *et al.*, 1998c] and presents an accuracy of less than 1% for all the measured variables and for a range of ambient temperature $0 \div 40^oC$ and load conditions $70 \div 100\%$.

The time series data generated with the non–linear dynamic model simulates measurements taken on the machine with a sampling rate of 0.1s. This is considered without noise due to measurement uncertainty. However, noise is usually present in the real measurement systems.

In order to simulate the measurements taken on the actual instrumentation, the following noise signals were fixed:

- the IGV angular position measurement:
  standard deviation of $\tilde{u}_1(t)$ = 1% of the mean value of the signal $u_1(t)$ (α);

- the fuel mass flow rate measurement:
  standard deviation of $\tilde{u}_2(t)$ = 2% of the mean value of the signal $u_2(t)$ (M_f);

- the pressure measurements:
  standard deviations of $\tilde{y}_1(t)$, $\tilde{y}_2(t)$, $\tilde{y}_3(t)$ = 0.4% of the mean values of the signals $y_1(t)$ (p_{ic}), $y_2(t)$ (p_{oc}) and $y_3(t)$ (p_{ot}), respectively;

- the compressor outlet temperature measurement:
  standard deviation of $\tilde{y}_4(t)$ = 0.6% of the mean value of the signal $y_4(t)$ (T_{oc});

- the turbine outlet temperature measurement:
  standard deviation of $\tilde{y}_5(t)$ = 0.7% of the mean value of the signal $y_5(t)$ (T_{ot});

- the electrical power measurement:
  standard deviation of $\tilde{y}_6(t)$ = 0.5% of the mean value of the signal $y_6(t)$ (P_e);

These noise levels are typical of the standard instrumentation of the real industrial gas turbine used to validate the non–linear dynamic model [Simani *et al.*, 2000a].

The number of samples generated by the SIMULINK model is $N = 5000$. The plots of the $r = 2$ input and $m = 6$ output measurements are shown in Figures 5.8, 5.9, 5.10 and 5.11.

The measurements depicted in Figures 5.8, 5.9 and 5.10 and 5.11 are normalised with respect to their standard deviations.
The procedure used to transform the input-output MISO model into state space representation is available in the literature [Söderström and Stoica, 1987].

Since these six state space descriptions are driven by the same two inputs, they can be easily aggregated into a single MIMO model which is the starting point for the design of the different observer configurations.

This model was tested under different operating condition and it has always provided an output reconstruction error variable in the range of $10^{-3} \div 10^{-9}$.

The parameters of each input-output model have shown remarkable robustness properties with respect to the noise variances of the corrupting data.

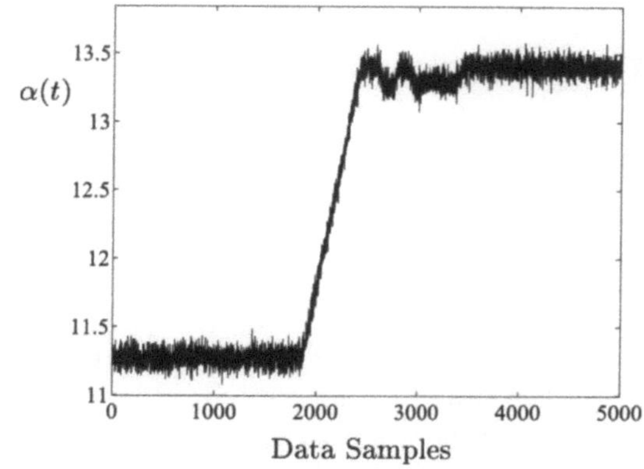

(a) First input, $\alpha(t)$

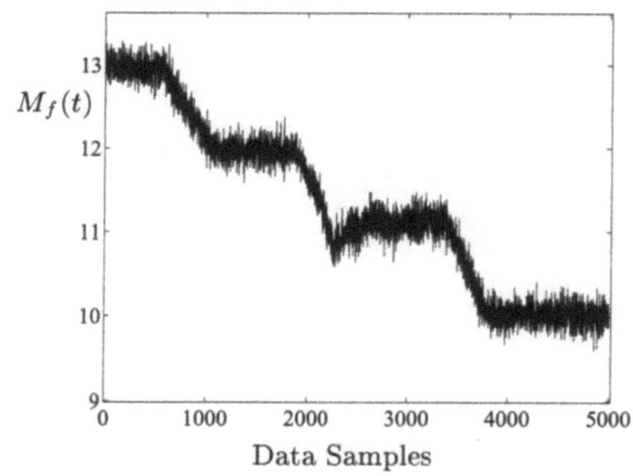

(b) Second input, $M_f(t)$

Fig. 5.8. Turbine input signals $\alpha(t)$ and $M_f(t)$.

As an example, Table 5.2 shows the parameter variations of the input–output model relative to the p_{ic} measurement versus the measurement noise. It was assumed that the measurement noise signals have identical variance and distribution.

Moreover, different time series data generated by the gas turbine non-linear model were exploited in order to identify the input-output models.

These models have always provided an output reconstruction error lower than 10^{-3}.

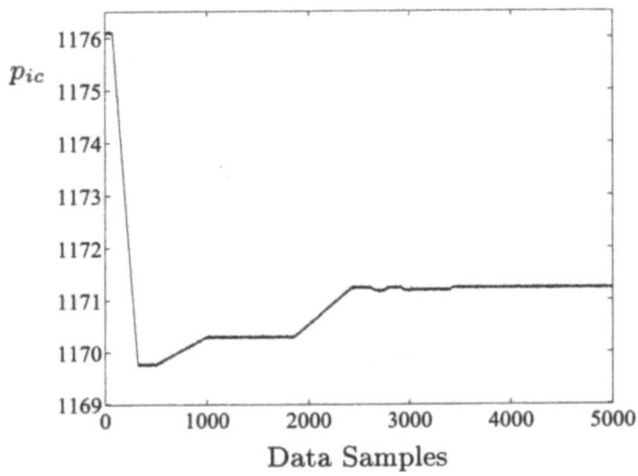

(a) First output, p_{ic}

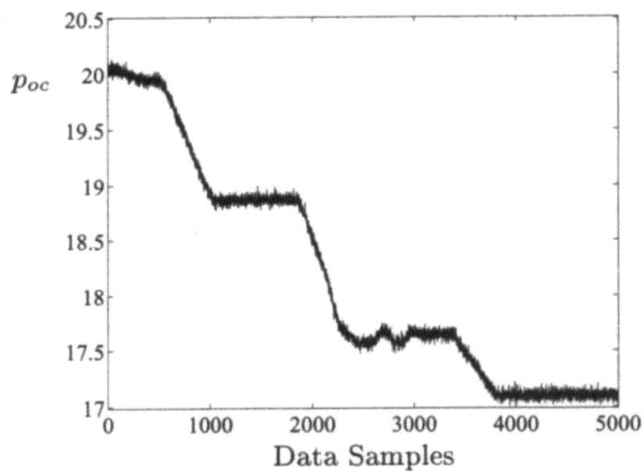

(b) Second output, p_{oc}

Fig. 5.9. Turbine first two output signals p_{ic} and p_{oc}.

Table 5.2. Parameter variation of the p_{ic} ARX model versus measurement noise.

Noise	0 %	2 %	10 %	20 %
α_2	−0.9963	−0.9941	−0.9513	−0.9325
α_1	1.9963	1.9949	1.9712	1.9486
β_{11}	0.9205	0.9368	0.9680	0.9458
β_{12}	−0.9176	−0.9455	−0.9682	−0.9864
β_{21}	0.0044	0.0178	0.0176	0.0220
β_{22}	−0.0044	−0.0092	−0.0108	−0.0197

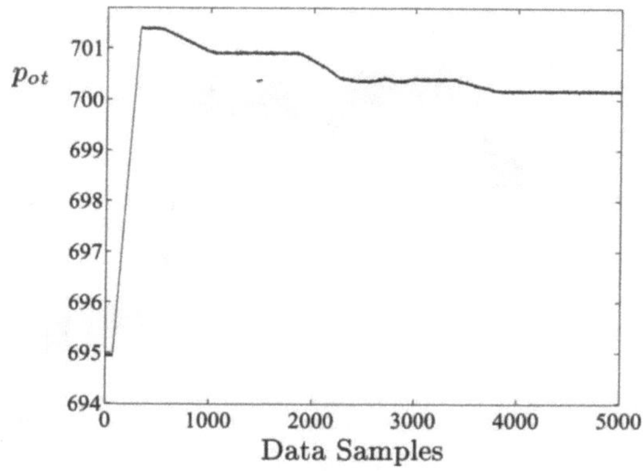

(a) Third output, p_{ot}

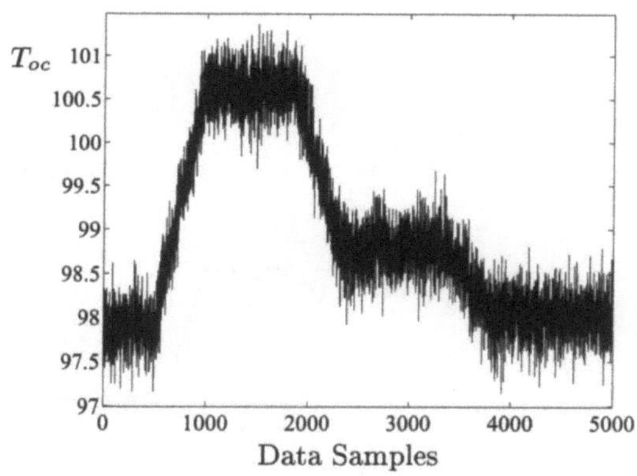

(b) Fourth output, T_{oc}

Fig. 5.10. Turbine second two output signals p_{ot} and T_{oc}.

The time series data required to generate the ARX linear models could be directly measured on the gas turbine by performing a large number of variations in operating conditions and recording data during the corresponding transients. This requires a wide campaign of experimental tests which could be compatible with the requirements of low costs typical of small and medium power size industrial gas turbines.

The time series data do not correspond to fault conditions, desirable as this may be for setting up the diagnostic algorithms.

For these reasons, a non–linear dynamic gas turbine model was used to generate the required time series data. The use of a non–linear model proves

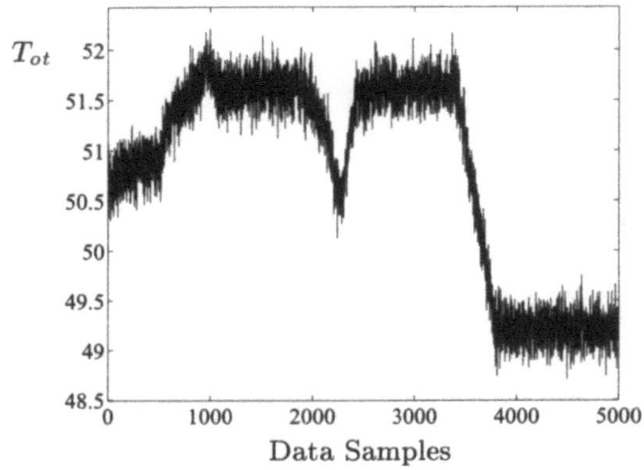

(a) Fifth output, T_{ot}

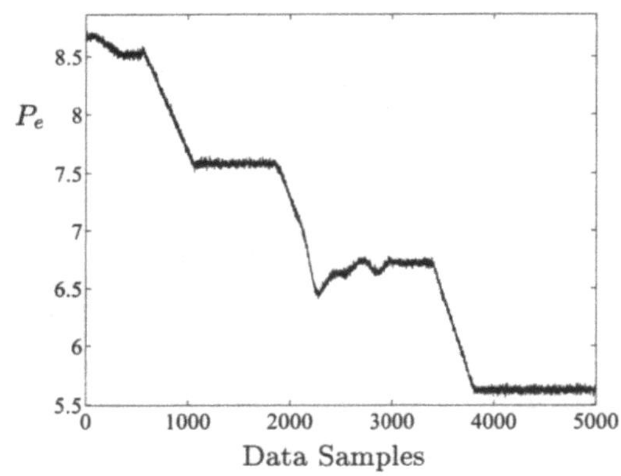

(b) Sixth output, P_e

Fig. 5.11. Turbine third two output signals T_{ot} and P_e.

to be particularly recommended in the case of simulation of gas turbine operating conditions with sensor faults in order to evaluate the effectiveness of the diagnostic tool.

5.3.2 FDI Using Dynamic Observers

To assess the technique for diagnosing sensor faults, gas turbine operating conditions with different sensor faults were simulated by using the non–linear dynamic model of the machine.

Faults in single input–output sensors were generated by producing positive and negative variations (step functions of different amplitudes) in the input-output signals. A positive and negative fault occurring at the instants of the minima and maxima values respectively of the observer residuals were chosen since these conditions represent the worst case in fault detection.

Moreover, it was decided to consider a fault during a transient since, in this case, the residual error due to model approximation is maximum (see Figures 5.12 and 5.15) and therefore it represents the most critical case.

According to the residual generation scheme developed in Section 4.4, the fault occurring on the single sensor affects the measurements of $\boldsymbol{u}(t)$ and $\boldsymbol{y}(t)$ and the observer residuals $\boldsymbol{r}(t)$. These residuals are affected (show an error) as each observer is driven by the signals $\boldsymbol{u}(t)$ and $\boldsymbol{y}(t)$. These residuals indicate fault occurrence according to whether their values are lower or higher than the thresholds fixed in fault-free conditions.

In order to determine the thresholds above which the faults are detectable, the simulation of different amplitude faults in the sensor signals was performed. Each threshold value depends on the magnitude of the residual error due to the ARX model approximation and on the real measurement noises $\tilde{\boldsymbol{u}}(t)$ and $\tilde{\boldsymbol{y}}(t)$. Table 5.3 shows the fixed values of the observer residuals.

Table 5.3. Fault detectability thresholds.

Measurement	Positive threshold	Negative threshold
T_{oc}	+0.85	-0.85
T_{ot}	+0.20	-0.22
p_{ot}	+0.022	-0.024
p_{oc}	+0.55	-0.65
p_{ic}	+0.022	-0.0225
P_e	+2.0	-2.2
M_f	+1.1	-1.1
α	+0.27	-0.41

The positive and negative thresholds correspond to fault–free residuals generated by different time series of simulated data. A margin of 10% between the positive and negative thresholds and the maximum and minimum values were respectively imposed.

Figures 5.12, 5.13 and 5.14 show an example of the residuals given by the UIO (Section 4.3) for the diagnosis of the M_f input sensor.

In particular, Figure 5.12 shows the fault-free residual generated by the input observer driven by the signal of M_f input sensor and that it is insensitive to the signal of the IGV input sensor. In this condition, it is possible to determine the thresholds above which the fault on the M_f sensor can be detected.

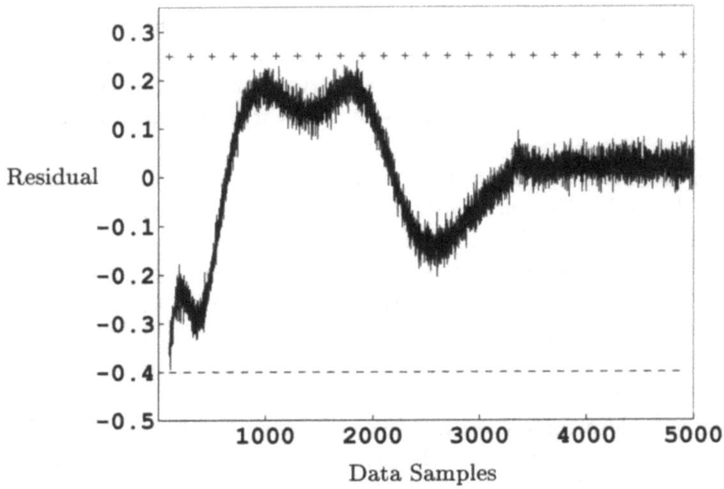

Fig. 5.12. Fault-free residual function of the UIO driven by the M_f signal with minimum positive ('+') and negative ('-') thresholds.

The eigenvalues of the state distribution matrix of the UIO are placed near to 0.2 in order to maximise the fault detection sensitivity and promptness and to minimise the occurrence of false alarms.

Figure 5.13 shows how a fault of +4% on the mean value of M_f signal at the instant of minimal residual value causes an abrupt change of the residual. In Figure 5.14 the change of the residual at the instant of its maximum is due to a fault of −4% on the mean value of the M_f signal. These fault amplitudes are those that are minimally detectable in order to identify the fault as soon as it occurs.

Figures 5.15, 5.16 and 5.17 illustrate an example of the diagnostic technique for an output sensor fault regarding the p_{ot} signal.

Figure 5.15 shows the fault-free residual obtained from the difference between the values computed by the observer (Section 4.4) of the output $y_3(t)$ (p_{ot} signal) and the one given by the sensor.

Clearly, the non–zero value of the residual is due to the identified model approximation and actual measurement noise.

The eigenvalues of the state distribution matrix of output observers are placed between 0 and 0.2 in order to maximise the fault detection sensitivity and promptness and to minimise the occurrence of false alarms.

In Figure 5.16 the abrupt change of p_{ot} residual caused by a fault of +5% on the mean value of p_{ot} signal occurring at the instant of the minimum residual value is shown.

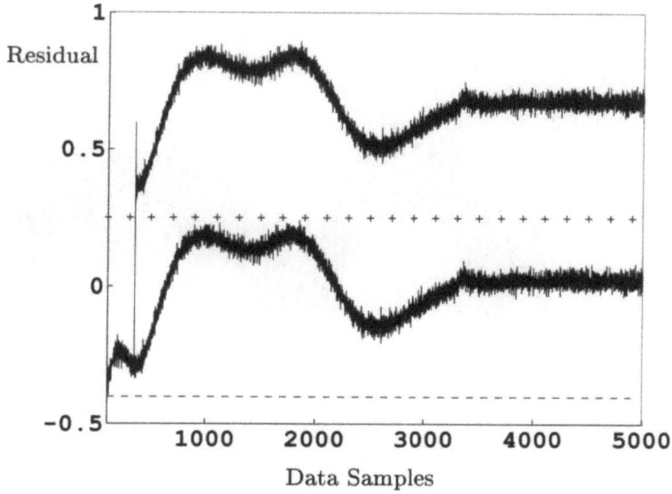

Fig. 5.13. Residual function of the UIO driven by the M_f signal in the presence of an additive positive fault signal.

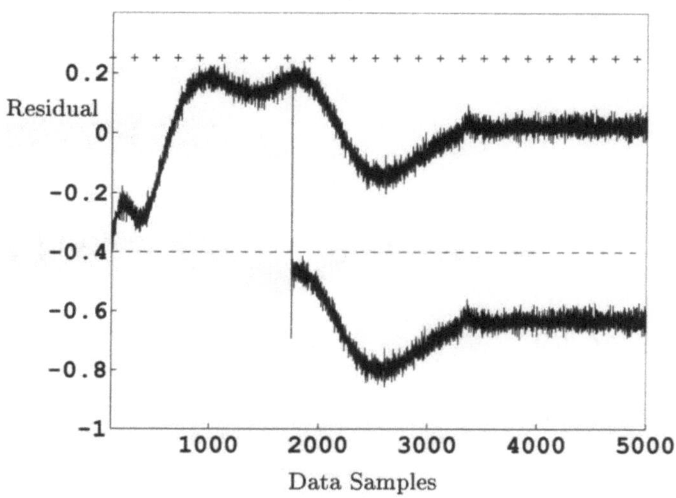

Fig. 5.14. Residual function of the UIO driven by the M_f signal in the presence of an added negative fault signal.

Figure 5.17 shows the behaviour of the residual with the same fault as the previous case (changed sign) occurring when the residual itself assumes maximal value.

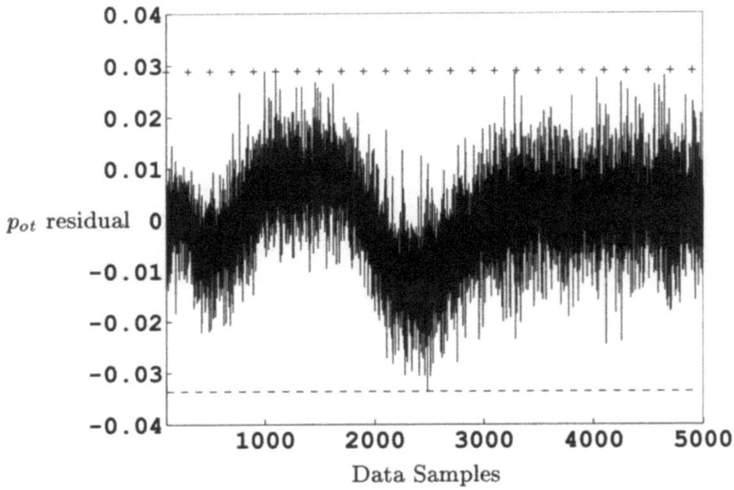

Fig. 5.15. Fault-free residual function of output observer driven by p_{ot} signal with minimum positive ('+') and negative ('-') thresholds.

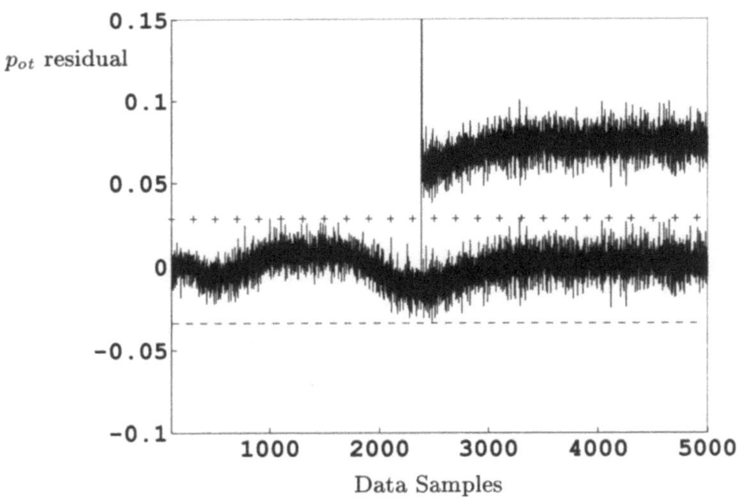

Fig. 5.16. Residual function of output observer driven by p_{ot} signal with an added positive fault signal.

The instantaneous peaks which appear in Figures 5.16 and 5.17 are generated by the abrupt change related to the fault occurrence and may be used to detect incipient anomalous sensor behaviour.

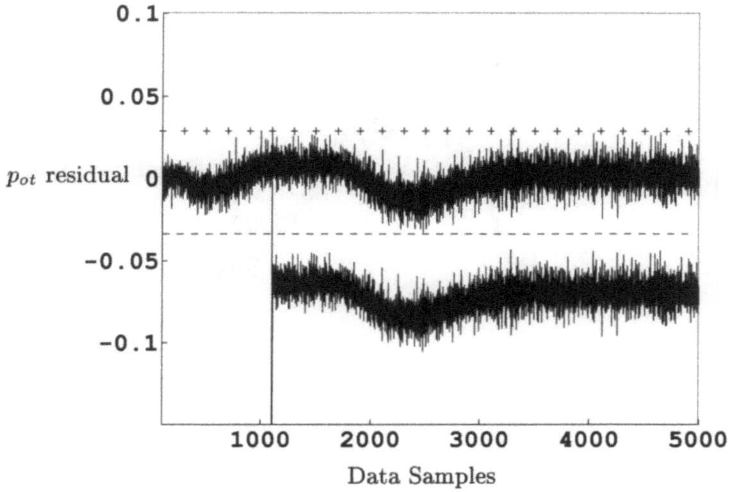

Fig. 5.17. Residual function of output observer driven by p_{ot} signal with negative failure.

In order to analyse the diagnostic effectiveness of the FDI system in the presence of measurement drifts, faults modelled by ramp functions were generated.

Figures 5.18 and 5.19 illustrate an example comprising an example based on the T_{oc} measurement signals. These also show the residual functions of the UIO observer driven by the signal α. The two ramp faults start at the sample instant 2500 and reach constant final values at the sample instant 4000. These values are equal to 4% of the mean values of α and to 5% of the mean values T_{oc}.

To summarise the performance of the FDI technique, the minimally detectable faults on the various sensors referred to the mean signal values are collected in Table 5.4, in case of step faults, and in Table 5.5, in case of ramp faults.

Table 5.4. Minimal detectable step faults.

α	M_f	p_{ic}	p_{oc}	p_{ot}	T_{oc}	T_{ot}	P_e
4%	4%	5%	7%	5%	5%	2.5%	1.7%

The minimum values shown in Table 5.4 are relative to the case in which the fault must be detected as soon as it occurs. If a detection delay is tolerable the amplitude of the minimal detectable fault is lower.

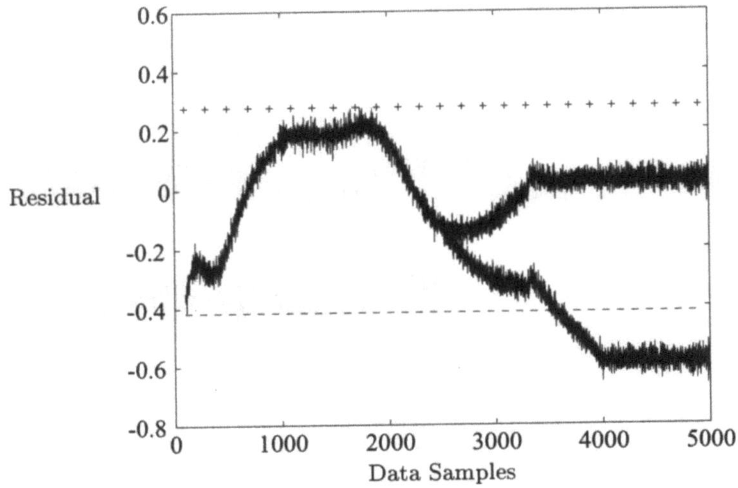

Fig. 5.18. Residual function of the UIO driven by the α signal in the presence of a drift in the α measurement.

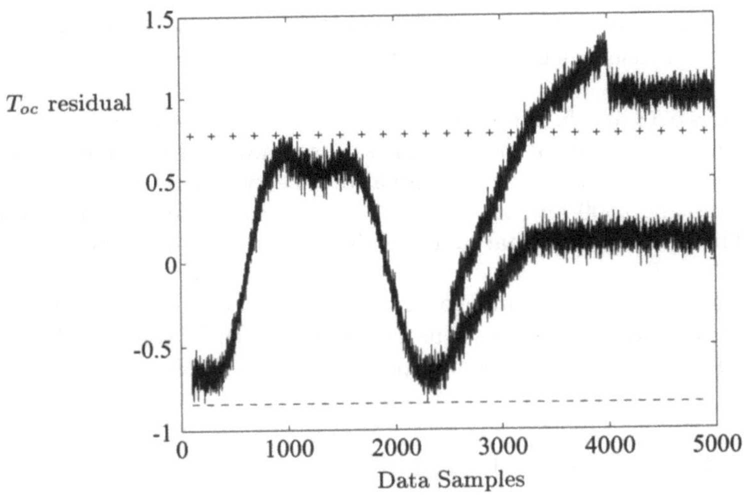

Fig. 5.19. Output observer residual signal T_{oc} corresponding to a drift in the T_{oc} measurement.

Table 5.5 shows how ramp faults can not be immediately detected, since the delay in the corresponding alarm normally depends on fault mode.

Table 5.5. Minimal detectable ramp faults.

Measurement	Fault	Detection delay [s]
T_{oc}	5 %	50
T_{ot}	3 %	100
p_{ot}	5.5 %	75
p_{oc}	7.5 %	0
p_{ic}	6 %	50
P_e	6 %	100
M_f	4 %	150
α	4 %	100

5.3.3 FDI Using Kalman Filters

According to Section 5.3.1 when signal–to–noise ratios are low, a bank of KFs can be exploited in order to diagnose malfunctions of the gas turbine sensors. This technique seems to be robust with respect to modelling uncertainty, the system parameter variations and the measurement noise, which can obscure the performance of a fault detection system by acting as a source of false alarms [Simani and Spina, 1998].

The procedure presented in this section requires the design of different KFs configurations and the basic scheme is the standard one: a set of measured variables of the system is compared with the corresponding signals estimated by filters to generate residual functions.

The diagnosis can be performed by detecting the changes of these residuals caused by a fault. The fault diagnosis of input sensors uses a number of KF equal to the number of input variables. Each filter is designed to be insensitive to a different input of the system. Output sensor faults affecting a single residual are detected by means of a classic KF, driven by a single output and all the inputs of the system.

The results and improvements obtained by using this technique are compared with the ones presented in Section 5.3.2.

Also the design of the different KF configurations necessary to isolate a fault in one of the input-output sensors requires the knowledge of a state space model of the system under investigation.

As shown in Section 5.3.1, the measurements and the noise signals $\tilde{\boldsymbol{u}}(t)$, $\tilde{\boldsymbol{y}}(t)$ with standard deviations reported in Table 5.6 were then considered as input–output time series generated by the non–linear turbine model.
As summarised in Section 4.4, the detection strategy which may be chosen in connection with KF methods for fault detection, consists of monitoring the residuals or KF innovations.

Because of the linear property of the identified model and because of the additive effect of the faults on the system, it may easily be shown that the effect of the change on the innovation is also additive.

Table 5.6. Measurement noise standard deviation.

α	M_f	p_{ic}	p_{oc}
1.08 deg	0.0076 kg/s	0.41 KPa	3.66 kPa
p_{ot}	T_{oc}	T_{ot}	P_e
0.41 kPa	3.59 K	5.59 K	23.90 kW

Any abrupt change in measurements due to a fault is reflected in a change in the mean value and in the standard deviation of innovations.

In particular, since the KF produces zero–mean and independent white residuals with the system in normal operation, a method for FDI consists of testing how much the sequence of innovations has deviated from the white noise hypothesis.

As explained in Section 2.6, the tests which are performed on the innovations $r(t)$ are the usual ones for zero–mean and variance, in the form of cumulative sum algorithms

$$\overline{r}(t) = E[r(t)] = \frac{1}{t}\sum_{j=1}^{t} r(j) \tag{5.7}$$

and

$$\sigma_r^2(t) = E[r^2(t)] = \frac{1}{t}\sum_{j=1}^{t} r^2(j) \tag{5.8}$$

and the correlation of the residuals are tested, as χ^2–type as:

$$\begin{aligned} R_r^t(\tau) &= \frac{1}{t}\sum_{j=1}^{t} r(j)r(j+\tau), \\ \zeta_r^M(t) &= \frac{t}{R_r^t(0)^2}\sum_{\tau=1}^{M} \left(R_r^t(\tau)\right)^2 \end{aligned} \tag{5.9}$$

which are computed in a growing window. The parameter $\zeta_r^M(t)$ is a chi–squared random variable with M degrees of freedom.

If a system abnormality occurs, the statistics of $r(t)$ change, so the comparison of $\overline{r}(t)$ and $\zeta_r^M(t)$ with a threshold ϵ fixed under no faults conditions, becomes the detection rule 2.17. In particular, such a threshold can be settled as in a Section 5.3.2 or, with the aid of chi–squared tables, $\epsilon = \chi_\beta^2(M)$ can be computed as a function of the false–alarms probability β and of the window size M.

As discussed in Section 5.3.2, in order to determine the thresholds above which the faults are detectable, the simulation of different fault amplitudes in the sensor signals was performed. The threshold values now depend on the residual error magnitude due to the model approximation and on the real measurement noise signals $\tilde{\boldsymbol{u}}(t)$ and $\tilde{\boldsymbol{y}}(t)$.

Figures 5.20, 5.21 and 5.22 show examples of the statistical tests 5.7, 5.8 and 5.9 respectively applied to residuals generated by the KF with unknown input. The results correspond to the fault detection of the α input sensor.

In particular, Figure 5.20 shows the mean value computed by Equation 5.7 and generated by the KF driven by the α input sensor signal. The result shows that the mean value is independent of the M_f input sensor signal. A fault of 3% on the maximum value of the α signal causes an abrupt change in the mean value of the residual computed in a growing window.

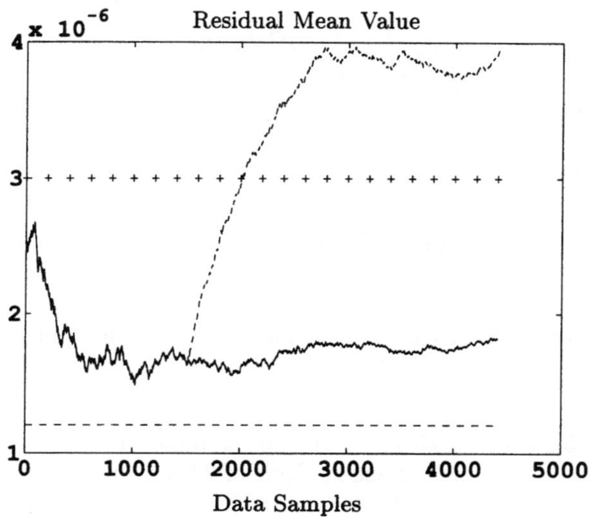

Fig. 5.20. Mean value of the residual computed by using KF with unknown input in a growing window.

This type of fault also affects the standard deviation of the same residual, as depicted in Figure 5.21. The standard deviation was computed using Equation 5.8 in a growing window. The thresholds (marked with '+' and '-') were fixed in fault-free conditions as well as by imposing an acceptable false-alarms rate.

Figure 5.22 shows how the same fault causes a change in the uncorrelation of the residual given by Eq. 5.9. The whiteness value of 20.1 was calculated by assuming that $M = 8$ and $\beta = 0.05$.

Under this condition, it is possible to determine the threshold values above which the fault on the α sensor (and also the M_f sensor) can be detected.

It is important to note that, in order to achieve the maximal input fault detection capability, the residual corresponding to the most sensitive filter to a failure on the α input was selected.

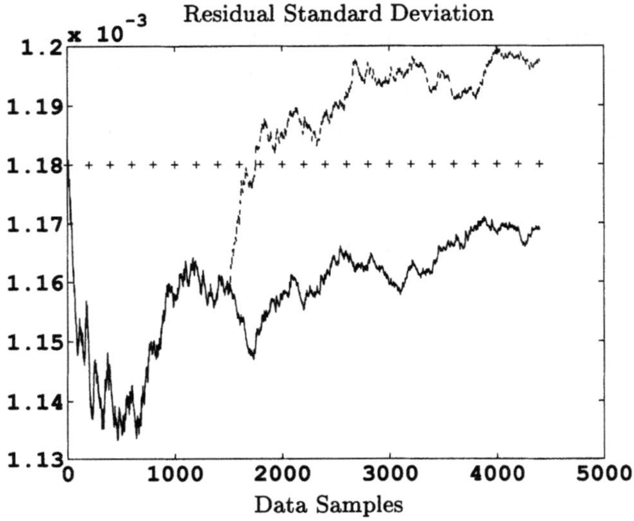

Fig. 5.21. Standard deviation of the residual computed by using a KF with unknown input in a growing window.

Figures 5.23, 5.24 and 5.25 illustrate an example of the previously shown statistical tests for the output sensor fault of 2% on the maximal value of p_{ic} signal, occurring at the sample instant 1500.

According to Eq. 5.7, Figure 5.23 shows the mean value of the residual obtained from the difference between the estimated measurements computed by the KF regarding the output $y_1(t)$ (p_{ic} signal) and the sensor measurements. Clearly, the non–zero value of the residual in the fault–free condition is due to the model approximation and to the actual measurement noise.
According to Eq. 5.8, Figure 5.24 shows the behaviour of the standard deviation of the residual with the same fault as the previous case.
Figure 5.25 shows the abrupt change in the uncorrelation of the p_{ic} residual computed by Eq. 5.9.
Tables 5.7 and 5.8 summarise the performance of the enhanced FDI technique and collect the minimal detectable fault on the various sensors, for the case the mean value and the uncorrelation of the residuals are monitored respectively.

The minimal detectable fault values in Tables 5.7 and 5.8 are expressed as percentages of the maximal signal values and are relative to the case in which the occurrence of a fault must be detected as soon as possible.

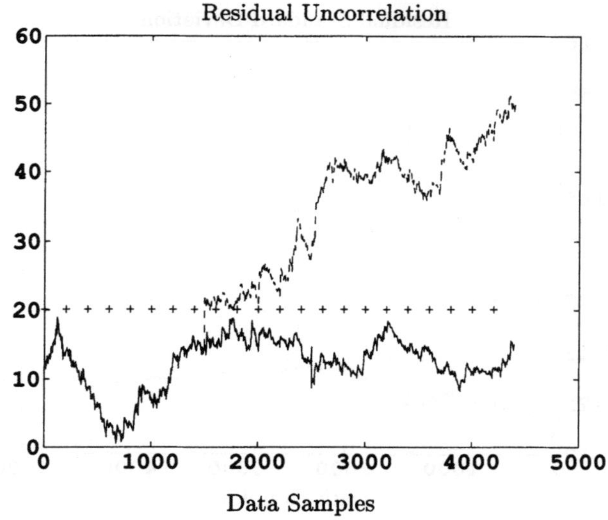

Fig. 5.22. Residual uncorrelation computed using a KF with unknown input in a growing window.

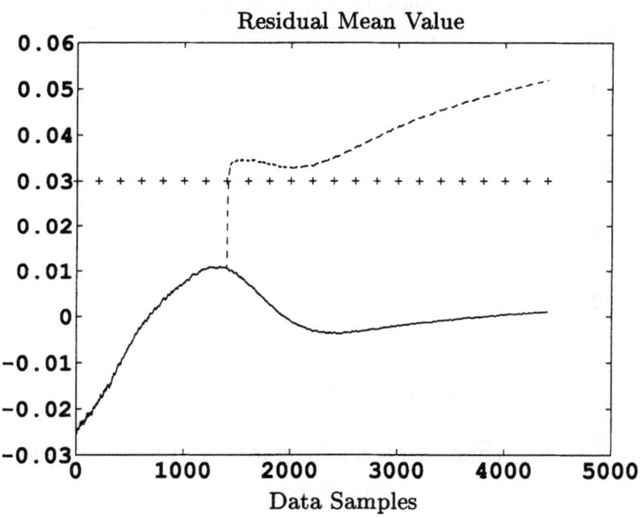

Fig. 5.23. Mean value of the p_{ic} residual computed by using a growing window.

In order to compare improvements with this FDI technique, the minimal detectable faults obtained by using observers and the geometrical analysis of residuals collected in Table 5.4 of Section 5.3.2 have to be considered.

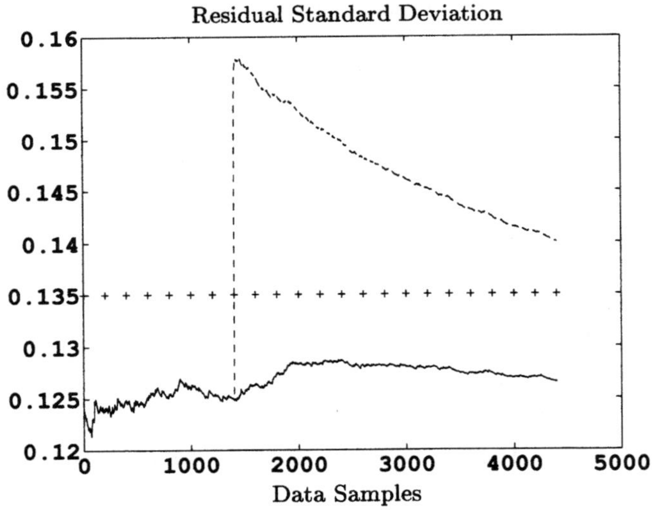

Fig. 5.24. Standard deviation of the p_{ic} residual computed by using a growing window.

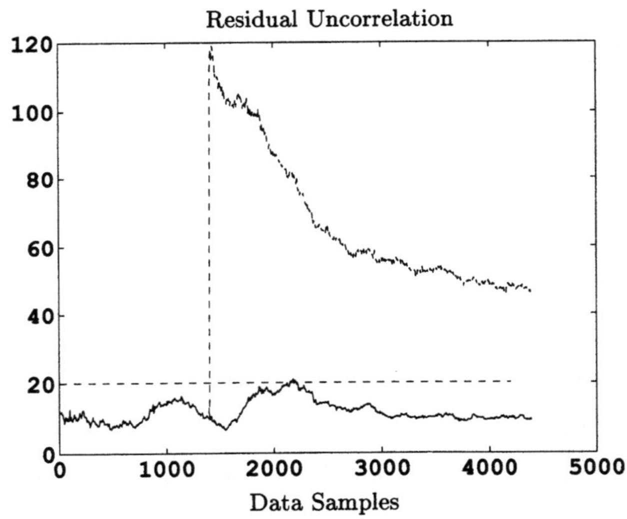

Fig. 5.25. Uncorrelation of the p_{ic} residual computed by using a growing window.

Table 5.7. Minimum detectable faults by monitoring residual mean value.

α	M_f	p_{ic}	p_{oc}	p_{ot}	T_{oc}	T_{ot}	P_e
3%	3%	2.5%	4%	1.5%	2%	2.5%	3%

Table 5.8. Minimum detectable faults by monitoring residual uncorrelation.

α	M_f	p_{ic}	p_{oc}	p_{ot}	T_{oc}	T_{ot}	P_e
2%	2.5%	0.75%	1%	0.75%	2%	0.8%	1.5%

This ensues that the fault values obtained by using statistical tests on KF innovations, collected in Tables 5.7 and 5.8, are lower than the ones reported in Table 5.4.

5.3.4 Fuzzy System Identification and FDI

This section describes some experimentations with the method for fault diagnosis of the dynamic process using the multiple–model approach. The technique presented in Section 3.5 exploits the identification of a non–linear dynamic system based on TS fuzzy models.

According to Section 3.5.2, the non–linear dynamic process can be described as a composition of several TS models selected according to process operating conditions.

In particular, the Section addresses the method for the identification and the optimal selection of the local TS models from a sequence of noisy input-output data acquired from the process.

It is assumed that the monitored system, depicted in Figure 5.6, can be described by a model of the type given by Eq. 3.97.

As presented in Section 4.9, the diagnostic scheme exploits the TS fuzzy models to generate residuals.

The problem considered here thus regards the fuzzy system identification and the sensor fault diagnosis on the basis of the knowledge of the measured noisy sequences $\boldsymbol{u}(t)$ and $\boldsymbol{y}(t)$ acquired from the input–output sensors of the industrial gas turbine (see Figure 5.6).

As stated in Section 5.2.1, the process operates mainly at steady state conditions and the 8 noisy process measurements, including temperatures, flow rates, pressures, control signals, turbine speed and torque can be acquired with a sampling rate of 0.1 s.

Because of the presence of the input and output sensors, actual measurements are affected by faults and noise.

A pressure sensor bias (abrupt fault on the p_{ot} pressure sensor signal) and an input sensor fault (abrupt faults on the $\alpha(t)$ sensor signal) were simulated to experiment with both the identification and the fault diagnosis methods.

Because of the underlying physical mechanisms and because of the modes of the control signals, the process has non–linear steady state as well as dynamic characteristics.

The GK clustering algorithm described in Section 3.5.2 was used with $M = 3$ clusters (operating conditions) and $n = 2$ the number of sample delays of the inputs and outputs. After clustering, the system parameters θ_i, with $i = 1, \cdots, M$ for each output, were estimated using the Frisch scheme. The model was then validated on a separate data set.

In fault–free conditions, Table 5.9 reports the mean–square values of the output estimation errors $r(t)$ given by classical observers using a single model (*i.e.*, with $M = 1$ and $n = 2$) for all operating conditions [Simani *et al.*, 2000a]. These values are very large and cannot be used to detect faults reliability.

A meaningful improvement has been obtained by using this identification technique where the process is described as a collection of fuzzy TS models identified using Frisch scheme method. The i-th output $y(t)$ of the plant ($i = 1, \cdots, m$ and $m = 6$) can be characterised as a TS fuzzy multiple-input single-output (MISO) model 3.73 with $r = 2$ inputs.

The mean–square errors of the output estimation errors $r(t)$, under no–fault conditions, are collected in Table 5.9.

The fuzzy multiple–model approximates the real process very accurately. The results indicate that the composite model can serve as a reliable predictor for the real process. Using this model, a model–based approach for fault diagnosis can be exploited and applied to the actual power plant.

Table 5.9. Output estimation errors with and without the multiple–model approach.

Output	p_{ic}	p_{oc}	p_{ot}	T_{oc}	T_{ot}	P_e
Classical observer	13.29	7.56	15.34	20.22	21.57	19.70
Fuzzy model	2.04	3.22	1.67	2.55	2.58	1.70

The fault occurring on the single sensor $\alpha(t)$ or $p_{ot}(t)$ causes alteration of the sensor signals $\boldsymbol{u}(t)$, $\boldsymbol{y}(t)$ and of the residuals $\boldsymbol{r}(t)$ given by the predictive model 3.73 using $\boldsymbol{u}(t)$ and $\boldsymbol{y}(t)$ as inputs. Residuals indicate the fault occurrence according to Equation 2.17 whether their values are lower or higher than the thresholds fixed under fault–free conditions.

To summarise the performance of the FDI technique, the minimal detectable faults on the various sensors, expressed as percentages of the mean values of the relative signals, are collected in Table 5.10. The minimum values shown in Table 5.10 are relative to the case in which the fault must be detected as soon as it occurs. The results were obtained by using a single model for all operating conditions.

An improvement in the FDI performance has been obtained by using the fuzzy multiple–model. Model parameters were identified under the assumptions of the Frisch scheme.

Table 5.10 summarises the performance of the enhanced FDI technique and shows the minimal detectable fault size for the various sensors. The fault sizes are expressed as percentages of the signal mean values.

Table 5.10. Minimal detectable step faults with and without the multiple–model approach.

Sensor	α	M_f	p_{ic}	p_{oc}
Classical observer	4%	4%	5%	7%
Fuzzy model	1.8%	2.3%	0.60%	0.8%
Sensor	p_{ot}	T_{oc}	T_{ot}	P_e
Classical observer	5%	5%	2.5%	1.7%
Fuzzy model	0.65%	1.7%	0.65%	1.2%

The residuals obtained by using the multiple–model approach are more sensitive to a fault occurring on the sensors, since the corresponding output estimation errors are smaller. Noise rejection is, in fact, achieved by means of the dynamic Frisch identification scheme. Moreover, smaller thresholds can be placed on the residual signals to declare the occurrence of faults.

As an example, fault–free and faulty residuals regarding the $\alpha(t)$ sensor signal are reported in Figures 5.26(a) and 5.26(b). These were generated by using a classical observer designed and the identified fuzzy system, respectively. Fault-free thresholds were marked by using "−" and "+".
The consequence is that the values of the faults, reported in Table 5.10, obtained by using the fuzzy multiple–model approach are lower than the ones corresponding to classical observers. Moreover, the minimal detectable faults on the various sensors seem to be adequate for the industrial diagnostic applications.

However, these improvements are not free of charge: they have been obtained with a procedure of greater complexity and, consequently, with a growing computational cost.

5.3.5 Sensor Fault Identification Using Neural Networks

In this section, the problem of the *identification* of faults regarding control sensors of the single shaft industrial gas turbine is studied [Simani *et al.*, 1998b, Simani *et al.*, 1999d, Simani and Fantuzzi, 2000].

Faults modelled by step functions create changes in several residuals obtained by using dynamic observers of the process under examination.

A Neural Network (NN) is exploited in order to find the connection from a particular fault regarding input and output sensors to a particular residual. In such a way the observers generate residuals that do not depend on the dynamic characteristics of the plant, but only on sensors faults. Therefore, the NN classifies static patterns of residuals, which are uniquely related to particular fault conditions independently from the plant dynamics.

A number of residuals equal to the number of the outputs of the process is obtained by the difference between the estimated measurements computed by observers and the real measurements.

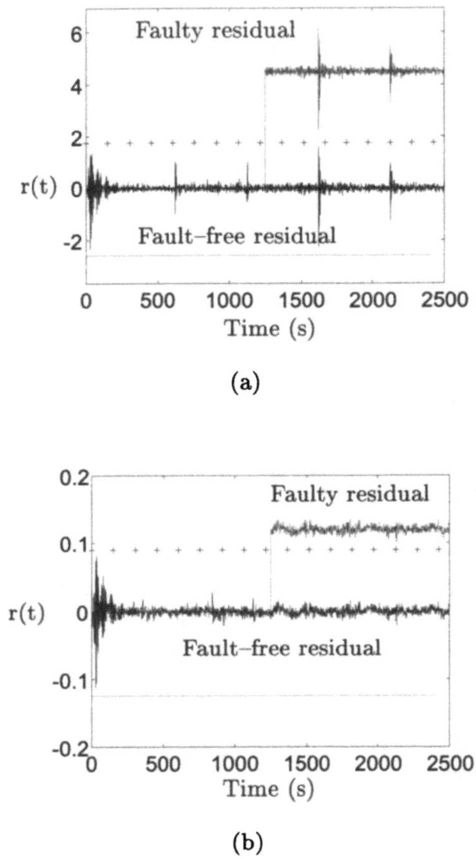

Fig. 5.26. (a) single model and (b) fuzzy model residuals $r(t)$ for the signal $\alpha(t)$.

The identification of output sensor faults is indeed very easy, since each output measurement is directly connected to a single residual generator. This situation does not hold for the inputs, and the relation between input faults and residuals should be determined.

The solution to this problem was obtained either by monitoring changes in residuals by means of a geometrical analysis or using special testing methods, *e.g.* a whiteness and a chi–squared test of the residual of the KF. An alternative solution is presented exploiting the learning capabilities of a NN.

In order to find the relationships that exist between input sensor faults and residuals, the NN is to classify the residual computed by observers according to the operation of the process. In this latter approach the process dynamics are not required.

The classification method is typically an off–line procedure in which the fault mode is first defined and the data (residuals) are then collected.

The classification of process residuals can be carried out in accordance with the information about different faults. Then, it is known that certain residual patterns correspond to the normal operation and other patterns correspond to the faulty operation. With this kind of data the training of the NN is performed.

The NN implemented by the Neural Network Toolbox for MATLAB are *Multilayer Perceptron* and *Radial Basis Function* NN described in Section 4.10. They are both able to approximate any continuous function with an arbitrary degree of accuracy, provided with a sufficient number of neurons.

The technique for the input–output sensor fault identification presented here was applied to the gas turbine simulated model of Figure 5.6 introduced in Section 5.3.2.

The first type of NN considered is the Radial Basis Function (RBF) network.

The simulations basically concern two aspects, namely the generation of pattern for the NN training and the fault diagnosis validation. The first step regarded the generation of pattern of residuals and fault signals.

The training set includes simulated faults on the sensors of variables M_f and IGV. An RBF network with a number of inputs equal to the number of output residuals and a number of outputs equal to the number of fault functions has been considered. Therefore, a six inputs–one output RBF network has been trained by using steady-state residual sequences comprising 1100 samples as shown in Figures 5.27 and 5.28.
Figure 5.27 shows the six steady–state residuals used as inputs for the training of the network whilst Figure 5.28 corresponds to the output target.
The sequences considered comprise 11 fault conditions, namely no fault and faults varying from 5%, 10% to 90% of the maximum value of input measurements. Each fault condition is composed of 100 samples.

The network training is performed with a trial and error procedure to arrange the number of hidden neurons in respect to the network output error. Even if an output error goal (SSE) of less than 0.1 was reached (sometimes with more than 100 hidden neurons), generalisation properties were unsatisfactory.

A different supervised NN architecture was then considered, namely a feed–forward MLP network [Simani *et al.*, 1998b, Simani and Fantuzzi, 2000].

Such a NN consists of an input layer, one or more hidden layers and an output layer. A six inputs–one output MLP network was designed with one hidden layer. Since the network is used as a function approximator, sigmoidal neurons were implemented in the input and hidden layer, whilst the output

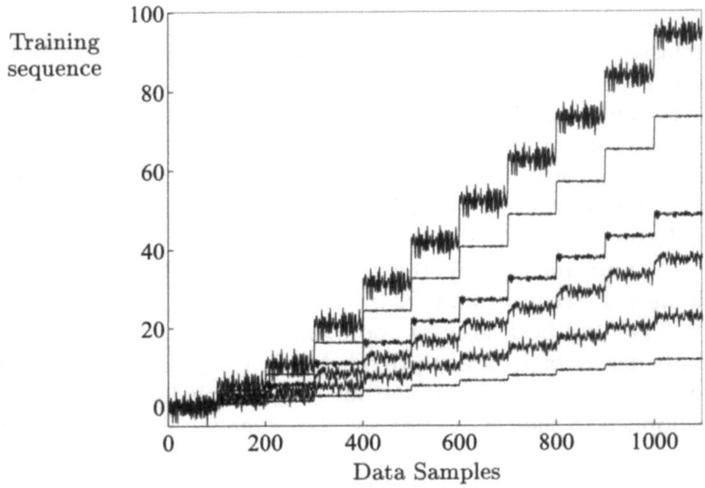

Fig. 5.27. NN input pattern.

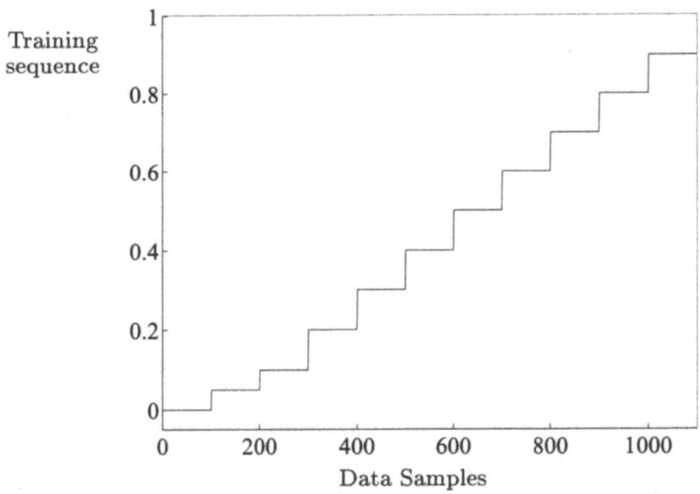

Fig. 5.28. Output pattern of the NN.

layer was made of a single linear neuron. A back–propagation algorithm with adaptive learning rate was exploited to update network parameters.

The training patterns were the ones used for the RBF network. The selection of training parameters in the back–propagation algorithm as well as the tuning of the number of hidden neurons of the network were difficult. In particular the convergence of the network depends on the number of neurons in the hidden layer. The momentum term is varied between 0.7 and 0.9.

In the Tables 5.11 and 5.12, the results of training sessions regarding the inputs M_f and IGV are shown, respectively, for different values of neurons and training epochs.

Table 5.11. Training results concerning the M_f sensor.

Input layer	Hidden layer	SSE after 70000 epochs
15	15	0.27
15	20	0.264
20	50	**0.127**

Table 5.12. Training results concerning the IGV sensor.

Input layer	Hidden layer	SSE after 70000 epochs
15	15	0.17
15	20	0.24
20	30	**0.108**

Even if the SSE value is usually fixed in a range between 0.01 and 0.001, due to the noisy environment, the network architectures providing the lowest SSE were chosen. These values allow estimating the input sensor fault amplitude with an accuracy of at least 1%.

NN minimal fault values concerning both input sensors are shown in Table 5.13. These minimum detectable faults can be compared with the ones obtained by using statistical tests on KF innovations as well as geometrical analysis of residuals generated by means of output dynamic observers.

Table 5.13. Minimal detectable step faults.

Method	M_f	IGV
(NN)	3%	2.5%

The fault sizes are expressed as percentages of the mean signal values.

One should note how the values of the faults obtained by using statistical tests on KF innovations are lower than the ones obtained with geometrical analysis of dynamic observer residuals and they appear comparable to the ones estimated by the NN. However, the minimal detectable faults on the various input sensors seem to be adequate for industrial diagnostic applications. The improvements achieved have been obtained with a procedure of greater complexity and consequently, with a growing computational cost.

5.3.6 Multiple Working Conditions FDI Using Neural Networks

The process under investigation is the single–shaft industrial gas turbine presented in Section 5.3 [Simani *et al.*, 1998c].

As stated in Section 5.2.1, the monitored process operates mainly at steady state and 8 noisy process measurements, including temperatures, flow rates, pressures, control signals, turbine speed and torque can be acquired.

In this application study, data for two abrupt faults and the healthy conditions were extracted from measurements and were used to obtain the results. Although an additional two faults were present in the available data, they were not included here [Simani *et al.*, 1998c, Simani and Spina, 1998].

- *Fault 1.* **Pressure sensor bias**: abrupt faults on the p_{ot} pressure sensor signal.
- *Fault 2.* **Actuator failure**: abrupt faults on the $\alpha(t)$ actuator signal.

Several sets of process data from the gas turbine were available for investigation. The data sets have an average length of 5000 samples acquired every 0.1s. for the 8 variables. These include some data sets that were not suitable for the investigation, due to the turbine start–up and shut–down during that period (*i.e.* because of transient conditions).

Data acquired at two working conditions were available, both for analysis and for the development of the NN fault diagnosis scheme.

There was considerably more data from the primary operating point (shaft speed $2 \times 10^4 \frac{rad}{s}$) than data from the secondary condition (shaft speed $1 \times 10^4 \frac{rad}{s}$). The data available from the secondary operating point consisted mainly of healthy operating conditions, with little fault data.

For the development of the method it was necessary to obtain enough labelled fault data during the different working conditions. This was achieved using a non–linear simulation of the gas turbine system in SIMULINK® environment. The model used for this purpose is described in Section 5.2. The simulations were performed under different (but realistic) operating conditions.

5.3.7 FDI Method Development

As stated in Section 4.11.2, the method presented was carried out in three stages. The first consisted of exploiting methods to pre–process the network input data. The second step was the NN training and testing, whilst the third part consisted of developing methods to diagnose faults at the secondary operating point using the network trained to diagnose faults at the primary operating point.

1. The magnitudes of measured process variables can span a wide range. Data conditioning was achieved by scaling the data using standard statistical normalisation methods. Data time series were divided by the corresponding standard deviation and the mean values were subtracted. This

gives all variables the same variance and brings them to comparable range. The mean and the standard deviation values used are those of the healthy condition at each operating point.
2. As the plant is a multivariable process, all the variables are to be used as inputs to the NN and this will result in a very complex network topology with a large number of hidden nodes. In order to reduce the input space of the NN, the well-known PCA statistical method can be used. Therefore, the number of highly correlated variables in a multivariable data set can be reduced to a smaller one of uncorrelated variables without any loss of information.
3. The conditioned data were used as inputs to the NNs. The NN training was performed using the Neural Network Toolbox for MATLAB [Demuth and BealeDemuth, 1997]. Tests were initially carried out on both MLP and RBF networks to compare their performances in the classification of faults. RBF NN, giving the best results, were used for further development in the FDI technique.

Once the network had been trained to recognise faults at both the primary and the secondary operating point satisfactorily, using the *simulated* turbine model 5.2, the next part of the work consisted of developing a methodology to use this network to diagnose faults occurring under the secondary operating point of the *real* plant (see Section 5.2).

Simulated turbine data were scaled statistically, converted into principal component variables using PCA and used to train the networks.

5.3.8 Multiple Operating Point Simulation Results

The simulated process was run at the primary and secondary points, and steady state data were collected from 8 variables, for the healthy condition and two faults. These data were used to develop the FDI techniques mentioned in Section 5.3.7, involving data scaling, input reduction and NN training.

In order to reduce the dimensionality of the data set, it was decided to use the first 4 principal components that accounted for a variance of 95% of the data set. This resulted in a reduction of dimensionality, from 8 process variables to 4 principal component variables.

RBF networks were trained with the principal component converted data as inputs, and the final network was selected for the simulated process with 4 inputs, one for each principal component, 8 centres and 3 outputs, one for a healthy condition and one for each fault. The root mean–square (RMS) value for the network output error on the data set was 0.001.

Figures 5.29 and 5.30 show the faulty residuals compared with fault–free ones.

In this case, the residual was defined as the difference between the measured output and its estimate, given by the NN.

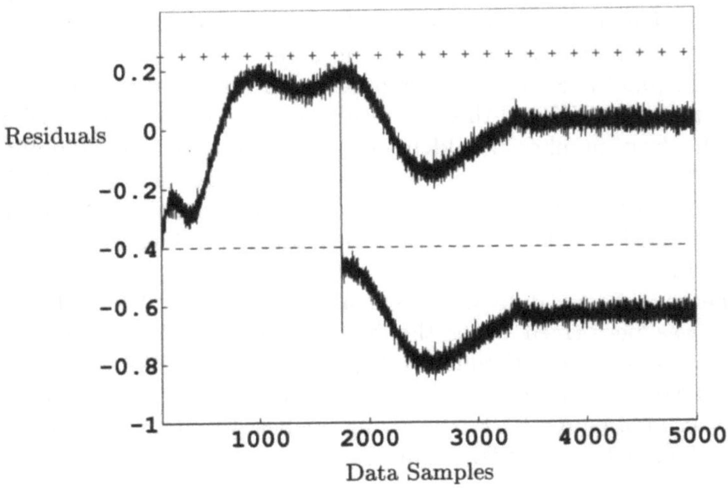

Fig. 5.29. $\alpha(t)$ fault signal.

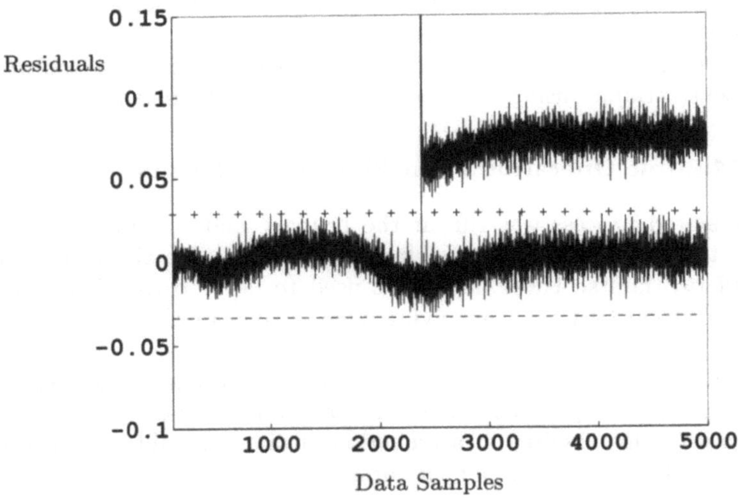

Fig. 5.30. $p_{ot}(t)$ fault signal.

A successful classification from simulated data was obtained and no information was lost reducing the input dimensions using PCA.

The trained network was then applied to fault classification of the real plant sensor and actuator. The RMS error of the network output applied to real data was 0.06. The output nodes correctly classified faults occurring on the sensor and actuator of the real plant at both the primary and the secondary working points.

The classification results demonstrate that for the secondary (and primary) operating points for the real process, these two faults can be detected and isolated successfully using the same NN trained to diagnose faults at the primary and secondary operating point of the corresponding simulated model.

5.4 Identification and FDI of Double Shaft Industrial Gas Turbine

The technique for robust input–output sensor FDI introduced in Section 4.7 was applied to real data from the 120MW power plant of Pont–sur–Sambre [Guidorzi, 1996, Simani, 1999b].

It consists of a double–shaft industrial gas turbine working in parallel with the electrical mains.

5.4.1 Process Description

The block-diagram of the plant is shown in Figure 5.31 where the numbers refer to:

1. super heater (radiation);
2. super heater (convection);
3. super heater;
4. reheater;
5. dampers;
6. condenser;
7. drum;
8. water pump;
9. burner.

The available data from the control inputs $u_i(t)$ ($i = 1, \cdots, r$, with $r = 5$) were $N = 2200$ samples from normal operating records of:

$u_1(t)$	C_b	gas flow
$u_2(t)$	O_s	turbine valves opening
$u_3(t)$	Q_d	super heater spray flow
$u_4(t)$	R_y	gas dampers
$u_4(t)$	Q_a	air flow

The data from the output sensors $y_i(t)$ ($i = 1, \cdots, m$, $m = 3$) were the corresponding values of:

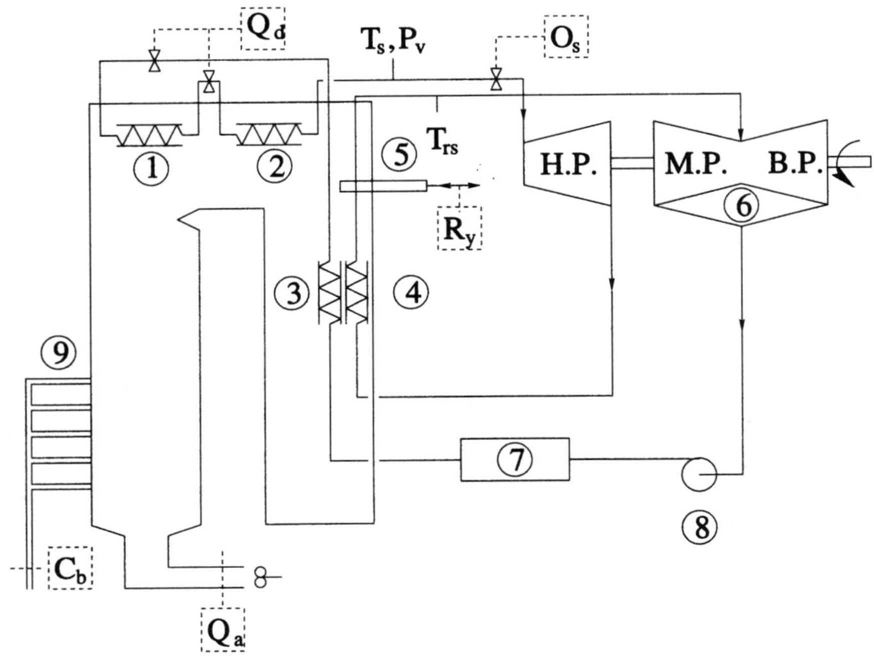

Fig. 5.31. The structure of the power plant.

$y_1(t)$	P_v	steam pressure
$y_2(t)$	T_s	main steam temperature
$y_i(t)$	T_{rs}	reheat steam temperature

The sampling time was 10 seconds and as this is small compared with the time constants of the plant, it has been increased to about 60 seconds.

The number of samples has thus been reduced to $N = 367$. Their plots are shown in Figures 5.32, 5.33, 5.34 and 5.35.

The process depicted in Figure 5.31 provides an example of application to a real power plant. This industrial process consists mainly of three major components: the *reactor*, *turbine*, and *condenser*. Furthermore, there are several pumps, valves (not highlighted) and one turbine. The boiler boils the water and the steam generated drives the turbine. After the turbine, the condenser cools the steam. In turn, external cooling water cools the condenser. The cooling pumps transport the water from the condenser tank back to the boiler tank.

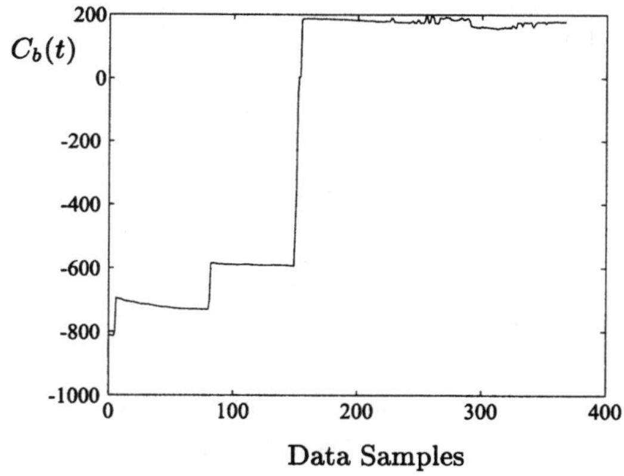

(a) First input, C_b

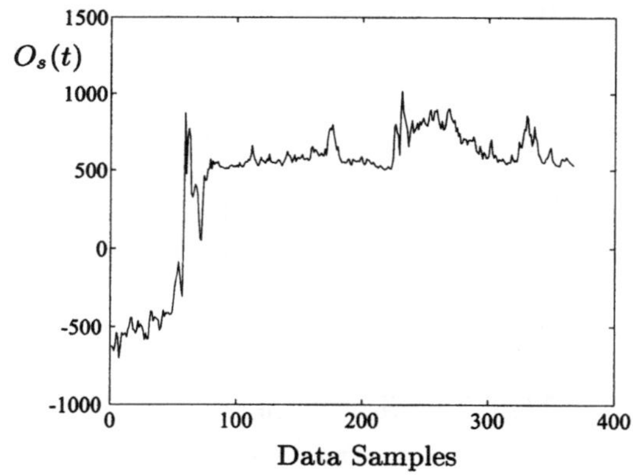

(b) Second input, O_s

Fig. 5.32. First two inputs of the power plant.

5.4.2 System Identification

The computational procedure which has been performed on the data is the identification of the triple (A_i,B_i,C_i) and disturbance distribution matrix E_i (see Equation 4.70) from the equation error model ($i = 1, \ldots, m$) corresponding to the MISO subsystem described by Eq. 4.70 that links each output with the five ($r = 5$) inputs (see Chapter 3).

Moreover, the triple (A,B,C) from the EIV model and the estimation of the input–output noise variances were obtained. The matrices A, B and C

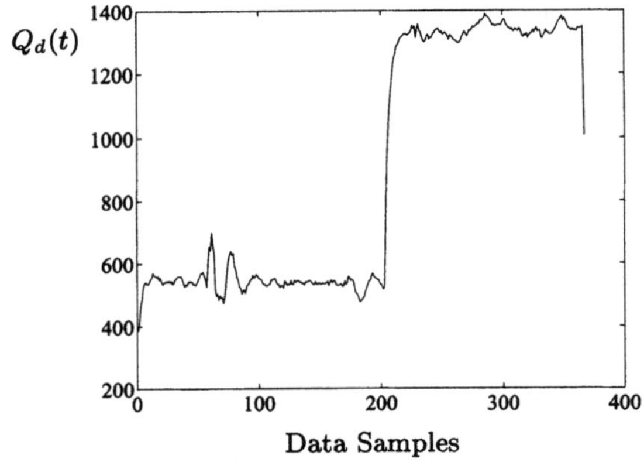

(a) Third input, Q_d

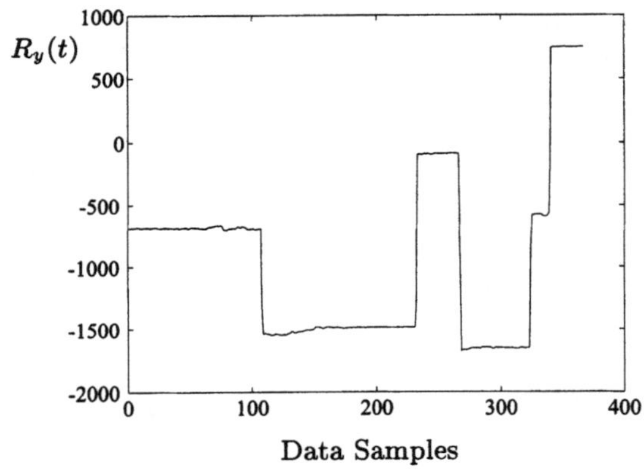

(b) Fourth input, R_y

Fig. 5.33. Second two inputs of the power plant.

were obtained by grouping the A_i, B_i and C_i $(i = 1, \ldots, m)$ corresponding to the MISO subsystem which links each output with the five $(r = 5)$ inputs. Three subsystems $(m = 3)$ with order two have thus been considered.

The design of the UIO described by Eq. 4.71 requires the knowledge of a minimal form model (A, B, C) for the system under investigation.

The determination of the order of every subsystem has been performed by considering the Final Prediction Error (FPE), Akaike's Information Criterion (AIC) and Minimum Description Length (MDL) identification criteria [Söderström and Stoica, 1987].

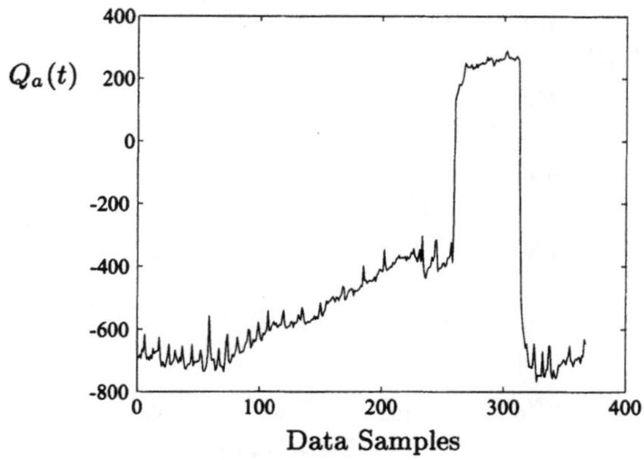

(a) Fifth input, Q_a

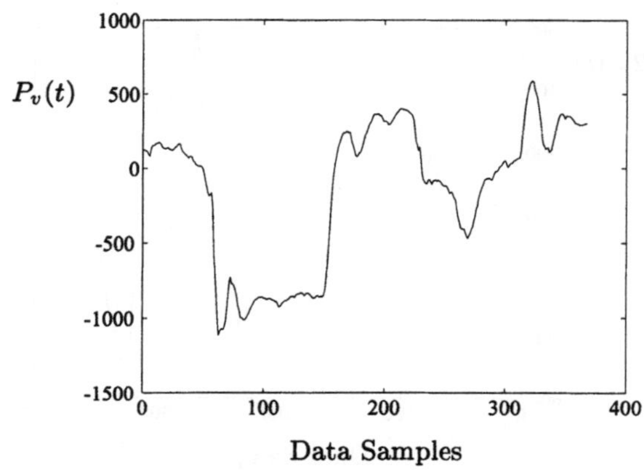

(b) First output, P_v

Fig. 5.34. Last input and first output of the power plant.

5.4.3 FDI Using Unknown Input Observers

Faults in a single output sensor were generated by producing positive and negative variations (step and ramp functions of different amplitudes) in the output signals.

A positive and negative fault occurring respectively at the instant of the minimum and maximum values of the observer were chosen since these conditions represent the worst case in fault detection.

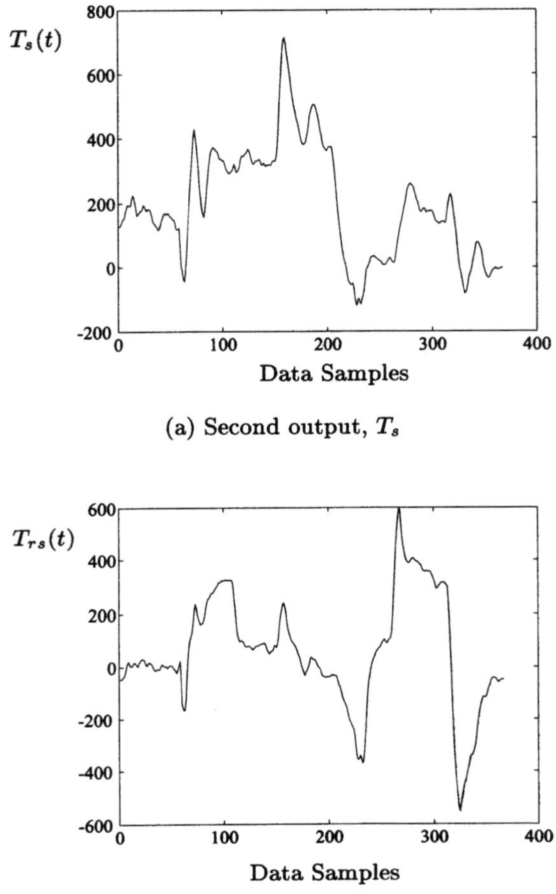

(a) Second output, T_s

(b) Third output, T_{rs}

Fig. 5.35. Last two outputs of the power plant.

Moreover, it was decided to consider a fault during a transient since, in this case, the residual error due to model approximation is maximum and therefore it represents the most critical case.

The fault occurring on the single sensor causes alteration of the sensor signal and of the residuals given by observers and filters using this signal as input. These residuals indicate that faults have occurred according to whether their values are lower or higher than the thresholds that have been fixed under fault–free conditions.

In order to determine the thresholds above which the faults are detectable, the simulation of different amplitude faults in the sensor signals was performed. The threshold value depends on the residual error amount due to the model approximation. These thresholds were settled on the basis of fault-free residuals. A margin of 10% between the thresholds and the residual values was imposed.

In Figures 5.36 and 5.37 an example of the residuals given by UIO 4.71 for the diagnosis of O_s input sensor is shown.

In particular, Figure 5.36 shows the fault-free residual generated by the input observer driven by the signal of O_s input sensor $u_2(t)$ and insensitive to the signal of C_b input sensor $u_1(t)$. In this condition, it is possible to determine the thresholds above which the fault on the O_s sensor can be detected.

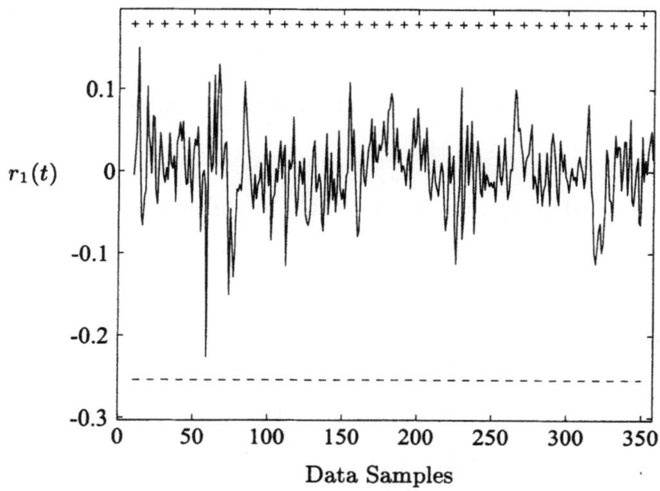

Fig. 5.36. The fault–free residual function $r_1(t)$ of the UIO driven by the O_s signal with minimum positive ('+') and negative ('-') thresholds.

The eigenvalues of the UIO state distribution matrix (Equations 4.26 with $i = 1$) of the input observer are placed near to 0.2 in order to maximise the fault detection sensibility and promptness and to minimise the occurrence of false alarms.

Figure 5.37 shows how a fault of 25% on the mean value of O_s signal at the sample $T = 150$ causes an abrupt change of the residual.
Figures 5.38 and 5.39 illustrate an example of the diagnostic technique for output sensor fault regarding the T_{rs} signal.

Figure 5.38 shows the fault-free residual (Equation 4.18) obtained from the difference between the values computed by the observer of the output $y_3(t)$ (T_{rs} signal) and the one given by the sensor $y_3(t)$. Clearly, the non–zero

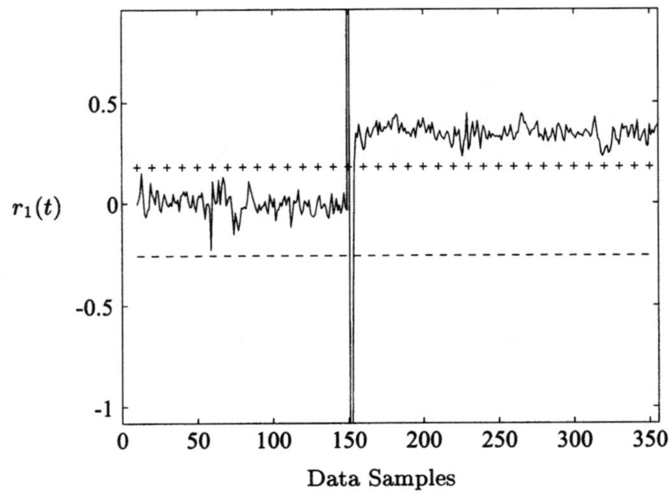

Fig. 5.37. Residual function $r_1(t)$ of the UIO driven by the O_s signal in the presence of a fault.

value of the residual is due to the ARX model approximation and actual measurement noise.

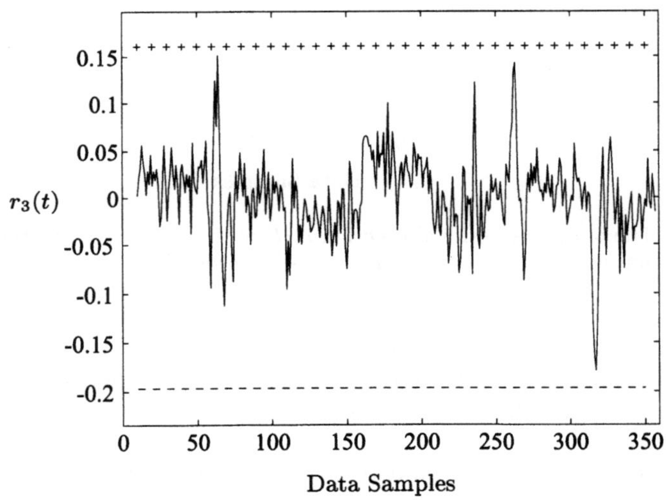

Fig. 5.38. The fault–free residual function $r_3(t)$ of output observer driven by T_{rs} signal with minimum positive ('+') and negative ('-') thresholds.

The eigenvalues of the state distribution matrix (matrix $(A^i - K^iC^i)$ in Equation 4.18 with $i = 3$) of the output state observer are placed between 0 and 0.2 in order to maximise the fault detection sensitivity and promptness and to minimise the occurrence of false alarms.

In Figure 5.39 the abrupt change of the T_{rs} residual caused by a fault of 10% on the mean value of T_{rs} signal occurring at the instant of $T = 150$ is shown.

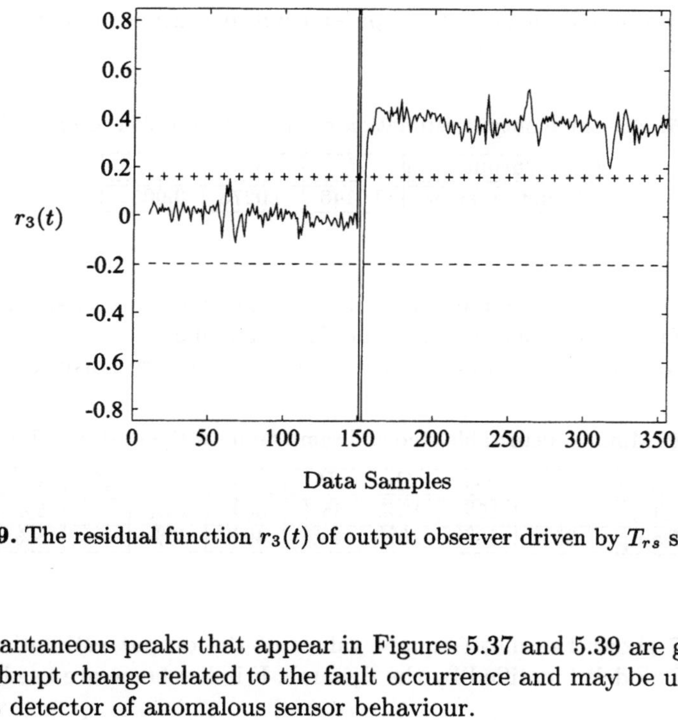

Fig. 5.39. The residual function $r_3(t)$ of output observer driven by T_{rs} signal with a fault.

The instantaneous peaks that appear in Figures 5.37 and 5.39 are generated by the abrupt change related to the fault occurrence and may be used as an incipient detector of anomalous sensor behaviour.

To summarise the performance of the FDI technique using classical observers and UIO, the minimal detectable faults on the various sensors referred to the mean signal values are collected in Table 5.14, in case of step and ramp faults.

Table 5.14. Minimal detectable step and ramp faults with classical observers and UIO.

Sensor	C_b	O_s	Q_d	R_y	Q_a	P_v	T_s	T_{rs}
Step	30%	25%	20%	40%	45%	15%	5%	10%
Ramp	40%	30%	35%	55%	50%	40%	20%	30%

Finally, Table 5.15 shows the mean–square values of the output estimation errors corresponding to the state space systems obtained by the equation errors models in deterministic case.

5.4.4 FDI Using Kalman Filters

An improvement on the performance of the FDI device was obtained by using both the classical KF and the UIKF.

The noise signals affecting the input–output measurements were identified using the Frisch scheme method.

Table 5.15. The three output estimation errors with equation error models.

Output	P_v	T_s	T_{rs}
Equation error	0.0146	0.0273	0.0051

Also in this case, the comparison of the residuals with the thresholds (fixed under no fault conditions) remains the detection rule.

Table 5.16 shows the minimal detectable faults in the noisy case.

Table 5.16. Minimal detectable step and ramp faults with classical KF and UIKF.

Sensor	C_b	O_s	Q_d	R_y	Q_a	P_v	T_s	T_{rs}
Step	25%	15%	12%	35%	35%	10%	3%	5%
Ramp	35%	20%	20%	45%	40%	30%	5%	8%

Table 5.17 shows the mean–square values of the output estimation errors when EIV models identified by the dynamic Frisch scheme are used.

Table 5.17. The three output estimation errors with EIV models.

Output	P_v	T_s	T_{rs}
EIV	0.0026	0.0018	0.0012

When comparing the deterministic estimation errors with those of the EIV models, the latter are smaller in magnitude because the noise rejection is achieved using the dynamic Frisch scheme. One must recall that this scheme includes a mechanism for estimating the noise variances. Consequently, the residuals generated via the KF are more sensitive to a fault occurring on the sensors. Moreover, smaller thresholds can be placed on the residual signals to declare the occurrence of faults.

5.4.5 Disturbance Decoupled Observers for Sensor FDI

Under the hypothesis that the system under investigation can be described as an equation error model, this section presents the method of obtaining the disturbance distribution matrix from the fault–free system data, by taking into account the equation error term.

The UIO performing the disturbance decoupling can be designed from the equation error model [Simani *et al.*, 1999a].

The identification scheme exploited to extract the disturbance distribution matrix from input-output data was illustrated in Section 4.7.

In the previous section the characteristics of the industrial process, such as the 120MW power plant of Pont sur Sambre, used to illustrate the method proposed in this work, were shown.

The results obtained by using UIO which perform the diagnosis of faults regarding output sensors are shown below. These results can be compared with the ones obtained without disturbance decoupling recalled in Section 5.4.2.

Table 5.18 reports the mean–square values of the output estimation errors given by the FDI observers without disturbance decoupling. These values are very large and they cannot be used to detect faults reliability.

Slightly better results than the previous ones have been obtained by using a technique presented in [Simani and Spina, 1998] where the process was described as an errors–in–variables model and the Frisch scheme dynamic system identification was performed (Section 3.3.2).

The KFs were exploited to generate residuals in connection with step and ramp faults.

Table 5.18. The three output estimation errors without disturbance decoupling.

P_v	T_s	T_{rs}
581.25	51.46	55.88

The mean–square errors of the output estimation errors obtained by using the KF are collected in Table 5.19.

Table 5.19. The three output estimation errors with KF.

Output	P_v	T_s	T_{rs}
KF	181.92	28.42	33.69

A meaningful improvement on the performance of the FDI device was obtained by using the UIO exploiting the disturbance decoupling technique presented in Section 4.7.

Table 5.20 shows the minimal detectable faults concerning system outputs in case of disturbance decoupling.

Table 5.20. Minimal detectable step and ramp faults with UIO.

Sensor	P_v	T_s	T_{rs}
Step	5%	1%	1.7%
Ramp	20%	4.5%	4.7%

Table 5.21 shows the mean–square values of the output estimation errors when UIO is used.

Compared with the ones concerning classical observers, the residuals are very small because disturbance decoupling is achieved, and consequently, their increase can be significantly detected when a fault occurs on the sensors. Moreover, smaller thresholds can be placed on the residual signals to declare the occurrence of faults.

This demonstrates the improved efficiency of the FDI technique when decoupling of disturbances is performed.

Table 5.21. The three output estimation errors with disturbance decoupling.

Output	P_v	T_s	T_{rs}
UIO	20.45	12.24	15.55

5.4.6 Fuzzy Models for Fault Diagnosis.

This section proposes an approach for FDI in the power plant of Pont sur Sambre using the multiple–model approach presented in Section 3.5.2.

This technique concerns the identification and design of a fuzzy system based on Takagi–Sugeno fuzzy models.

The non–linear dynamic process is described as a composition of several TS models selected according to the process operating conditions.

The FDI scheme adopted to generate residuals exploits the non–linear TS fuzzy model [Simani, 1999a].

With reference to the fuzzy identification method presented in Section 3.5.2 and implemented using the Fuzzy Modelling and Identification Toolbox for MATLAB [Babuška, 1998] the GK clustering algorithm was used with $M = 4$ clusters for each output (operating conditions) and $n = 3$ the number of shifts of inputs and outputs.

After clustering, the system parameters θ_i, with $i = 1, \cdots, M$ for each output, were estimated using the dynamic Frisch scheme identification method. The model was then validated on a separate data set.

Table 5.18 shows the mean–square values of the fault–free output estimation errors $\boldsymbol{r}(t)$ given by classical observers, using a single model for all operating conditions. These values are very large and consequently cannot be used to detect faults reliability.

A meaningful improvement has been obtained by using the identification technique presented in Section 3.5.2 where the process is described as a collection of fuzzy TS models identified using Frisch scheme method.

The i-th output $y_i(t)$ of the plant ($i = 1, \cdots, m$ and $m = 3$) can be characterised as a TS fuzzy multiple-input single-output (MISO) model 3.73 with $r = 5$ inputs.

The mean–square errors of the fault–free output estimation errors $\boldsymbol{r}(t)$ are collected in Table 5.22.

Table 5.22. The three output estimation errors with fuzzy multiple–model.

Output	P_v	T_s	T_{rs}
Multiple–model approach	10.46	8.90	6.91

The corresponding results are shown in Figures 5.4.6, 5.4.6 and 5.4.6.

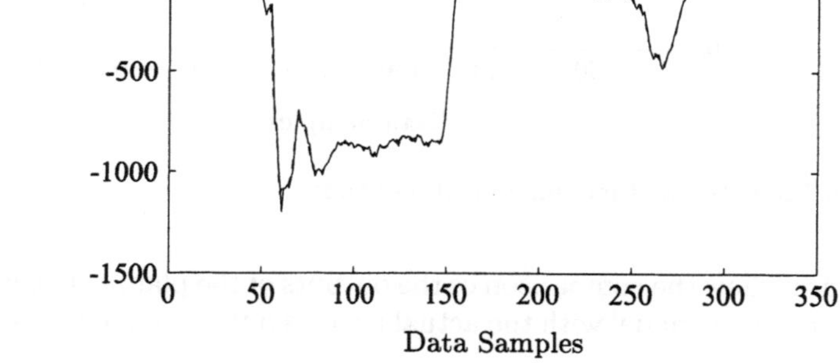

Fig. 5.40. Predicted and measured $P_v(t)$ output.

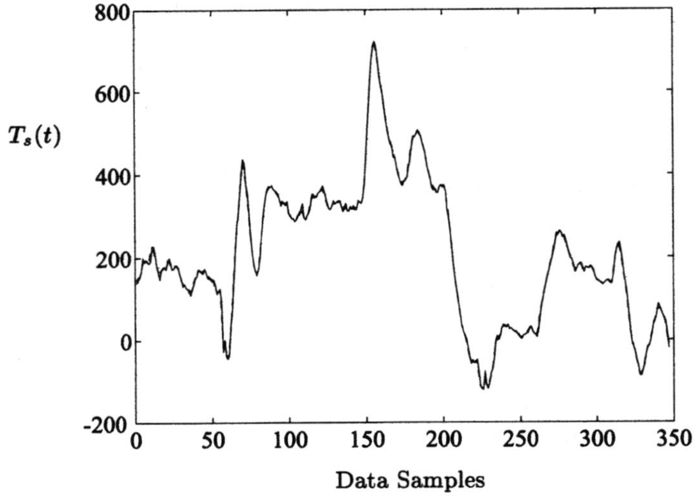

Fig. 5.41. Predicted and measured $T_s(t)$ output.

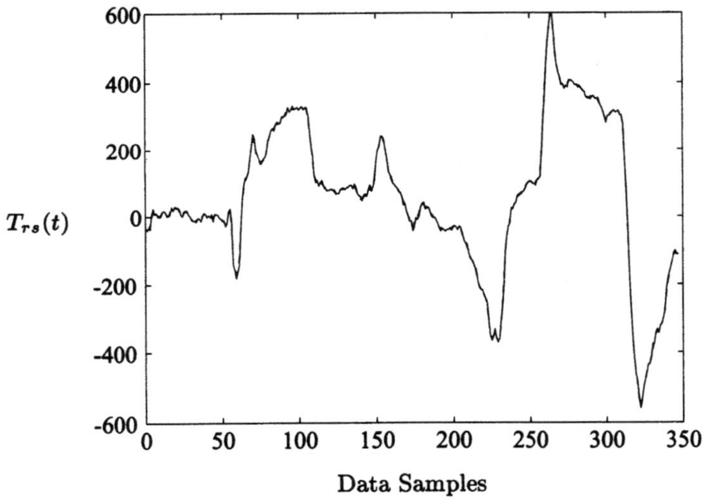

Fig. 5.42. Predicted and measured $T_{rs}(t)$ output.

These figures show the comparison of the outputs of the plant calculated using the fuzzy multiple–model with the actual process outputs on a validation data set.

Therefore, as depicted in Figure 5.43, residuals can be generated by the comparison between the measured and the estimated outputs.

$$\boldsymbol{r}(t) = \hat{\boldsymbol{y}}(t) - \boldsymbol{y}(t). \tag{5.10}$$

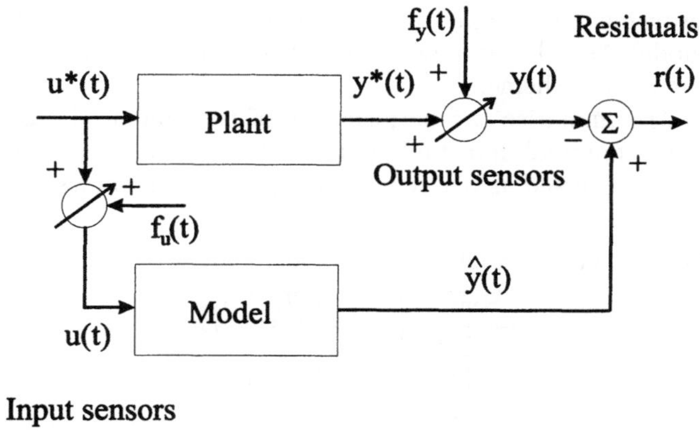

Fig. 5.43. The residual generation scheme.

The dashed line corresponds to the i-th predicted output ($i = 1, \cdots, 3$), $\hat{y}_i(t)$, and the solid line to the measured output, $y_i(t)$. The fuzzy multiple–model approximates the real process very accurately.

The results indicate that the composite model can serve as a reliable predictor for the real process. Using this model, a model–based approach for fault diagnosis can be exploited and applied to the actual power plant.

Single faults were generated by adding step and ramp signals in the input and output measurements. It was decided to consider fault occurrences during a transient since, in this case, the residual error due to model approximation is maximum and therefore it represents the most critical case in failure detection.

The fault occurring on the system output causes alteration of the signal $\boldsymbol{y}(t)$ and of the residuals $\boldsymbol{r}(t)$ given by the predictive model 3.73 using $\boldsymbol{u}(t)$ as input. Residuals indicate fault occurrence according to 2.17 whether their values are lower or higher than the thresholds fixed in fault-free conditions.

To summarise the performance of the FDI technique, the minimal detectable faults on the various outputs, expressed as percentages of the mean values of the relative signals, are collected in Table 5.16, in case of step and ramp faults.

The minimum values shown in Table 5.16 are relative to the case in which the fault must be detected as soon as it occurs.

The results were obtained by using a single model for all operating conditions. If a detection delay is tolerable the amplitude of the minimal detectable fault is lower.

One should note how faults modelled by ramp functions may not be immediately detected, since the delay in the corresponding alarm normally depends on the fault mode.

An improvement in FDI performance has been obtained by using the fuzzy multiple–model. Model parameters were identified under the assumptions of the dynamic Frisch scheme.

Table 5.23 summarise the performance of the enhanced FDI technique and collect the minimal detectable fault on the various output signals. The fault sizes are expressed as per cent of the signal mean values.

Table 5.23. Minimal detectable step and ramp faults with multiple–model.

Sensor	P_v	T_s	T_{rs}
Step	3%	1%	2%
Ramp	10%	8%	6%

The values shown in Table 5.23 are relative to the case in which the occurrence of a fault must be detected as soon as possible.

The residuals obtained by using multiple–model approach are more sensitive to a fault occurring on the system outputs, since the corresponding output estimation errors are smaller. Noise rejection is, in fact, achieved by means of the dynamic Frisch Scheme identification method. Moreover, smaller thresholds can be placed on the residual signals to declare the occurrence of faults.

The result is that the values of the faults obtained by using fuzzy multiple–model approach, collected in Table 5.23, are lower than the ones reported in Table 5.16.

Moreover, the minimal detectable faults on the various sensors seem to be adequate for the industrial diagnostic applications, by also considering that the minimal detectable faults can be reduced if a delay in detection promptness is tolerable. However, these improvements are not free of charge: they have been obtained with a procedure of greater complexity and, consequently, with a growing computational cost.

5.5 Modelling and FDI of a Turbine Prototype

This section shows a complete design procedure of a model–based fault diagnosis system, starting from system identification, both in the deterministic and stochastic environment, to residual generation, fault detection and isolation [Simani and Patton, 1999, Simani *et al.*, 2000c, Simani *et al.*, 2000b].

The procedure is applied to a model of a real industrial plant (a single shaft gas turbine) [Simani and Patton, 1999, Simani *et al.*, 2000c, Simani *et al.*, 2000b].

Linear state space models have been obtained for principal working points of the plant since state space descriptions provide general and mathematically rigorous tools for system modelling and residual generation that may be used successfully in fault detection. Residuals should then be processed to detect an actual fault condition, rejecting any false alarms caused by noise or spurious signals.

In particular, this work addresses output estimation approach for fault diagnosis [Simani *et al.*, 2000a] of actuators, components and input-output sensors, mainly in conjunction with residual processing schemes which include a simple threshold detection [Chen and Patton, 1999] as well as residual statistical analysis.

One of the main aspects of the proposed methodology should be stressed. Linear prototypes for the design of linear output estimators [Simani *et al.*, 1999a, Simani, 1999b, Simani *et al.*, 2000a] have been developed instead of complicated non–linear models obtained by modelling techniques in connection with non–linear observers. In fact, even if the number of studies addressing non–linear fault diagnosis theory steadily increases over the years, in some cases, the linear approach is still advantageous in terms of solution complexity and performance. Moreover linear system methods are still very valid since the feature of the system supervision is to monitor the operation and performance of the system with respect to an expected point of operation. It must be realised that, of course, a change in point of operation can be indicative of a fault in the process.

5.5.1 System Modelling and Identification

The identification procedure presented in Chapter 3 has been applied to a model of a single–shaft industrial gas turbine prototype developed in MATLAB–SIMULINK environment [Simani *et al.*, 2000b].

It is a strongly non–linear model since it is mainly based on non–linear functions and look–up tables that model the thermodynamic relations among the variables involved.

Figure 5.44 shows the block schematic diagram of the gas turbine including its inputs and outputs.
Air flows (ambient air and pressure, p_a and t_a) via an inlet duct to the compressor ("compressor" block), high pressure air from the compressor is heated in combustion chambers ("combustor" block) and expands through a single stage compressor turbine ("turbine" block). A butterfly valve (valve angle, a_v) provides a means of controlling the speed of the turbine (first control input, $u_1(t)$). Cooling air is bled from the compressor outlet to cool the turbine stator and rotor.

A non–linear regulator ("controller" block) regulates the combustor fuel flow (f_f) to maintain the compressor speed (N_t) at a set–point value. Under steady state conditions, the power generated by the turbine is balanced by that absorbed by the compressor and losses since there is no power turbine present in the model.

The process inputs $u_i(t)$ are the ambient air temperature t_a and pressure p_a, fuel flow f_f ($u_2(t)$) and the butterfly valve opening angle ($a_v = u_1(t)$).

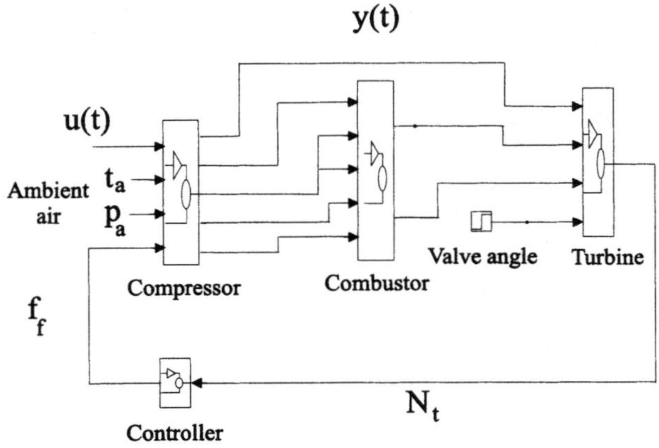

Fig. 5.44. The monitored system.

In particular, the input signals $a_v(t)$ and $f_f(t)$ are shown in Figures 5.45(a) and 5.45(b).
The process outputs $y_i(t)$ consist of all the 28 measurements that can be acquired from each block of the simulated system, *e.g.* mass flow (m_j), temperature (t_k), pressure (p_h), torque (q_l) and speed (N_t) signals.

The SIMULINK prototype, depicted in Figure 5.44, can be described by the closed–loop scheme in Figure 5.46, in which the faults $\boldsymbol{f}_u$, $\boldsymbol{f}_s$, $\boldsymbol{f}_c$ and $\boldsymbol{f}_y$ are likely to occur in the real plant.
They represent actuator, system, controller component and output sensor faults, respectively. In particular, they are modelled as ramp functions [Simani *et al.*, 2000b].

The time series of data $(u(t), y(t))$ used to identify the models were generated with a non–linear dynamic model in SIMULINK© environment and they simulate measurements taken on the actual machine with a sampling rate of 0.08 s.

The non–linear SIMULINK© model of the gas turbine was validated in steady state conditions against engine measurements when they were available, and against the prediction of a more rigorous steady state gas turbine model when measurements were not available. The accuracy of variables from identified linear model was found to be within 5% of the reference (real measurement and reference model) values. For the majority of variables the accuracy was within 1%.

Table 5.24 shows the input measurement accuracy, when orders and output reconstruction errors of each ARX model are shown in Table 5.25.

The i–th model (with $i = 1, \cdots, m$ and $m = 28$) is driven by $\boldsymbol{u} = [a_v(t),\ f_f(t)]$ and gives the prediction of the i–th output $y_i(t)$.

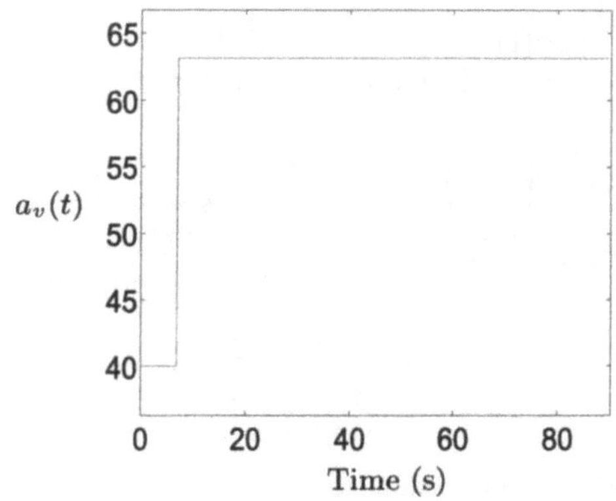

(a) $u_1(t) = a_v(t)$

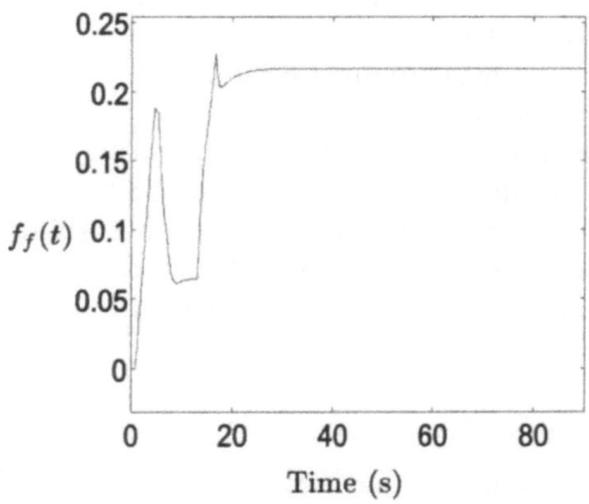

(b) $u_2(t) = f_f(t)$

Fig. 5.45. Gas turbine input signals: (a) valve angle and (b) fuel flow.

Table 5.24. Dynamic model identification: turbine inputs.

Variable	Name	Accuracy
t_a	amb. air temp.	$\pm 0.4^o C$
p_a	amb. air press.	$\pm 1\%$
f_f	fuel flow	$\pm 5\%$
a_v	valve angle	$\pm 2\%$

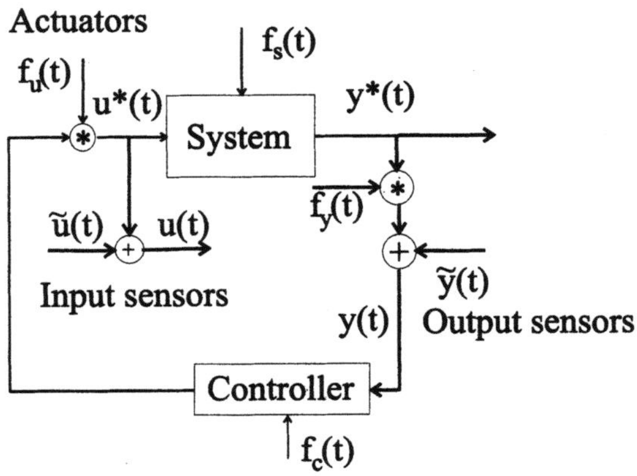

Fig. 5.46. Turbine closed–loop scheme.

In the model of the monitored system shown in Figure 5.44 the ambient pressure and temperature (p_a and t_a) are not considered as inputs as they are considered constant at all times.

Table 5.25 also shows measurement accuracy of the output variables $y_i(t)$, with $i = 1, \cdots, m$ and $m = 28$.

Each model was tested under different operating conditions and it has always provided an output reconstruction error SSE lower than 0.5%. Moreover, two time series of data generated by the gas turbine non–linear model were exploited in order to validate the ARX models (see Table 5.26 in the following). These models have always provided in full simulation an output reconstruction error SSE lower than 1%.

Table 5.25. Turbine output signals and MISO ARX model characteristics.

Variable label	Variable name	Model order	SSE	Accuracy
m_j	mass flow	2	$< 10^{-3}$	$\pm 5\%$
p_h	pressure	2	$< 10^{-4}$	$\pm 1\%$
q_l	torque	2	$< 10^{-4}$	$\pm 5\%$
t_k	temperature	2	$< 10^{-4}$	$\pm 1.5^o C$
w_t	speed	2	$< 10^{-5}$	$\pm 1\%$

A very effective way of evaluating the adequacy and flexibility of identified models consists in their use for performing complete simulations (*i.e.*, using only the initial samples of the observed outputs) and in comparing the obtained predictions with observed output samples.

This procedure, which can be applied when a single set of data is available, gives the best results when applied to sequences different from those used to identify the model. The mean–square prediction error SSE between the observed outputs and the ones obtained by simulation can be used to compare models with different orders.

The reconstruction errors of each ARX model are summarised in Table 5.26. The SSE prediction errors are also reported with respect to three different sequences of data. In Table 5.26, the first SSE column refers to the model prediction errors (see Equation 3.17), whilst the second and the third ones correspond to the SSE values for two validation sequences.

Table 5.26. Dynamic ARX model validation.

Variable	Model order	SSE identif.	SSE 1st valid.	SSE 2nd valid.
m_j	2	$< 10^{-3}$	$< 10^{-3}$	< 0.01
p_h	2	$< 10^{-4}$	$< 10^{-3}$	< 0.01
q_l	2	$< 10^{-4}$	$< 10^{-3}$	< 0.1
t_k	2	$< 10^{-4}$	$< 10^{-3}$	< 0.1
w_t	2	$< 10^{-5}$	$< 10^{-5}$	< 0.1

Regarding the identification procedure for noisy data introduced in Chapter 3, the Frisch scheme can be applied to perform the dynamic system identification of the plant. Such a scheme facilitates the determination of a linear discrete–time dynamic model that generates the noisy sequences as well as the variances of the noises $\tilde{\boldsymbol{u}}(t)$ and $\tilde{\boldsymbol{y}}(t)$ corrupting the data.

In the ideal Frisch scheme these signals are assumed to be white noise, mutually uncorrelated and uncorrelated with every component of real measurements $\boldsymbol{u}^*(t)$ and $\boldsymbol{y}^*(t)$.

The Table 5.27 summarises the reconstruction errors concerning the MISO models in the form of Equation 3.23 with two inputs ($\alpha(t)$ and $IGV(t)$) and each monitored output variable, as output.

It is worthwhile observing that only four output measurements (p_{h*}, m_{j*}, q_{l*}, t_{k*}) were considered in Table 5.27, corresponding to the residual signals that will be used in the fault detection and isolation procedures treated in the following sections [Simani *et al.*, 2002].

Table 5.27. Frisch scheme model reconstruction errors.

Variable	Name	Model order	$J(\theta)$	Accuracy
p_{h*}	Pressure	2	0.0054	$\pm$ 1%
m_{j*}	Mass flow	2	0.0049	$\pm$ 5%
q_{l*}	Torque	2	0.0042	$\pm$ 5%
t_{k*}	Temperature	2	0.0031	1.5°

Table 5.28 collects parameters of second order models ($n = 2$) as well as the input and output noise signals.

Table 5.28. Frisch 2-nd order model parameters and noise variances.

Variable	Model parameters θ	
p_{h*}	$[-0.0295,\ 1.0054,\ 0.1369,\ -0.1328,\ 0.0402,\ -0.0232]$	
m_{j*}	$[0.6655,\ 0.2885,\ -0.0579,\ 0.0651,\ 0.2408,\ -0.2065]$	
q_{l*}	$[-0.9920,\ 1.9904,\ -0.0179,\ 0.0181,\ 0.0111,\ -0.0100]$	
t_{k*}	$[-1.1760,\ 2.1882,\ 0.0283,\ -0.0311,\ -0.3202,\ 0.3133]$	
Variable	Input noises $\tilde{\sigma}_u$	Output noise $\tilde{\sigma}_y$
p_{h*}	[0.0004, 0.0023]	0.0026
m_{j*}	[0.0004, 0.0023]	0.0026
q_{l*}	[0.0004, 0.0023]	0.0015
t_{k*}	[0.0004, 0.0023]	0.0024

On the basis of the data collected in Table 5.28, four Kalman filters with two inputs ($r = 2$) and one output ($m = 1$) can be designed for residual generation in the noisy case. The residual generation problem will also be considered in Sections 5.6 and 5.6.5 [Simani *et al.*, 2002].

The residual generator is implemented by means of dynamic observers or KFs, in order to produce a set of signals from which it will be possible to isolate faults associated to actuators, components and sensors.

5.6 Turbine FDI Using Output Observers

Model–based FDI methodology has been applied to detect faults in a single–shaft industrial gas turbine prototype.

Test and measurements are simulated using the plant model developed in MATLAB–SIMULINK environment. Details on system and linear modelling procedure were described in the previous sections [Simani *et al.*, 2002].

In particular, four fault cases have been considered, namely:

1. Compressor contamination (system fault), $f_s(t)$;
2. Thermocouple sensor fault (output sensor fault), $f_y(t)$;
3. Turbine damage (system fault), $f_s(t)$;
4. Controller actuator fault (actuator fault), $f_c(t)$.

Note that in real industrial applications it is commonplace for each of the above faults to develop slowly over a period of months. For the purpose of this simulation, in order to avoid excessively long duration simulations, the fault development rate will be increased so that significant effects are present after one hour. However, this is still considerably longer than the duration of the gas turbine dynamics which occur over periods of seconds, a factor which must be taken account of in any FDI algorithm design.

In the presence of a fault condition, the challenge for the designer of the FDI algorithm may be summarised as follows:

1. Detect that a fault condition exists with minimum delay from the initial occurrence of the fault.
2. Identify the nature, magnitude and location of the fault, again with minimum delay from the initial occurrence of the fault.

Note that it is desirable to avoid introducing perturbation signals onto the model variables. In the first instance an FDI design should be based upon data which is available from the normal day to day operation of the plant, for example during transient and over prolonged periods of steady state operations.

The rate of development and magnitude of faults have been set to nominal values in this case study. It will be of interest to know how small the fault parameters can be made whilst still maintaining good FDI performance.

Moreover, it is assumed that only a single fault may occur in the actuators, components or output sensors of the plant.

5.6.1 Case 1: Compressor Failure (*Component Fault*)

Fault "case 1" represents fouling of the surfaces of the compressor blades, this reduces air flow, changes the blade aerodynamics and consequently changes the surface roughness. The failure is modelled as a gradual decrease in mass flow rate for a given pressure ratio.

The maximum decrease in mass flow rate is set nominally at 5% while the fault development rate is set to (5% decrease of normal flow rate)/hour.

In order to design the system component FDI scheme ($f_u(t) = 0$, $f_y(t) = 0$, $f_c(t) = 0$ and $f_s(t) \neq 0$) the subsystem depicted in Figure 5.47 was considered.

The inputs for the system are $\boldsymbol{u}(t)$ while $\boldsymbol{y}(t)$ are the outputs which could be affected by the fault $f_s(t)$.

The detection of a compressor fault was performed by using the classical output observer configuration exploited for the FDI of output sensor faults, as depicted in Figure 5.48.

The inputs $a_v(t)$, $f_f(t)$ and the output $y_i(t)$ feed the observer to estimate the signal $y_i(t)$ itself, $\hat{y}_i(t)$, and to generate the residual $r(t)$. In fact, $y_i(t)$ represents the output measurement which is the most sensitive signal to a fault affecting the compressor $f_s(t)$. Under this assumption, $y_i(t)$ consists of a torque measurement q_{l*} directly acquired from the compressor.

The observer is obtained from a second order ($n = 2$) ARX MISO ($r = 2$, $m = 1$) model, that was identified with an output reconstruction error $J(\theta) = 6.03 \times 10^{-5}$ [Simani *et al.*, 2002].

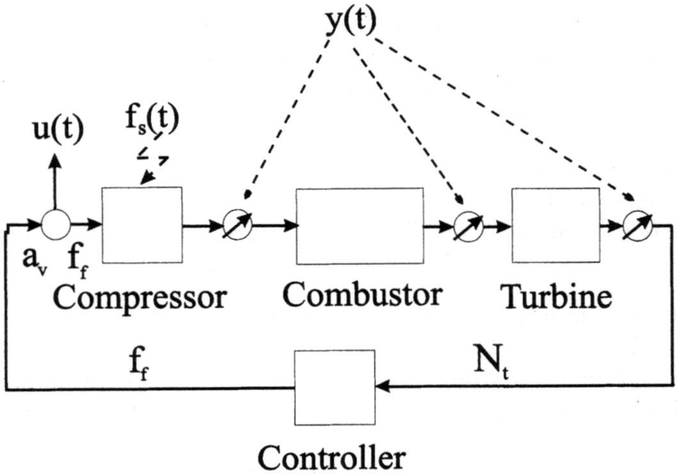

Fig. 5.47. The monitored subsystem.

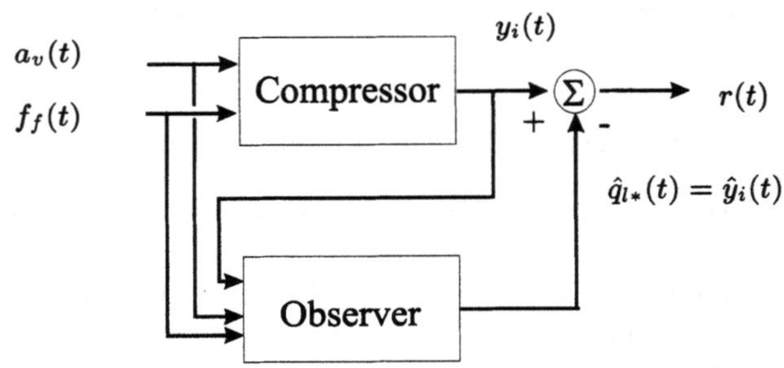

Fig. 5.48. Scheme of the $y_i(t) = q_{l*}(t)$ residual generator.

The parameters of such a model, driven by $a_v(t)$ and $f_f(t)$ signals, are represented by the vector $\theta = [-0.9246, 1.9238, -0.0009, 0.0010, -0.0353, 0.0359]$.

The diagnosis of the $q_{l*}(t)$ torque signal (linked to the faulty compressor component $f_s(t)$) requires the knowledge of the triple $(\boldsymbol{A}_i, \boldsymbol{B}_i, \boldsymbol{C}_i)$ and the identification of an ARX model with two inputs which gives the prediction of the output $y_i(t) = q_{l*}(t)$.

The poles $\boldsymbol{p}$ of the output observer for the signal $q_{l*}(t)$ were chosen near 0.5 according to the minimisation of the function $V(\boldsymbol{p})$, shown in Figure 5.49 and presented in Section 4.2.
The output signal $y_i(t) = q_{l*}(t)$ is depicted in Figure 5.50(a), whilst Figure 5.50(b) shows the ramp fault $f_s(t)$.

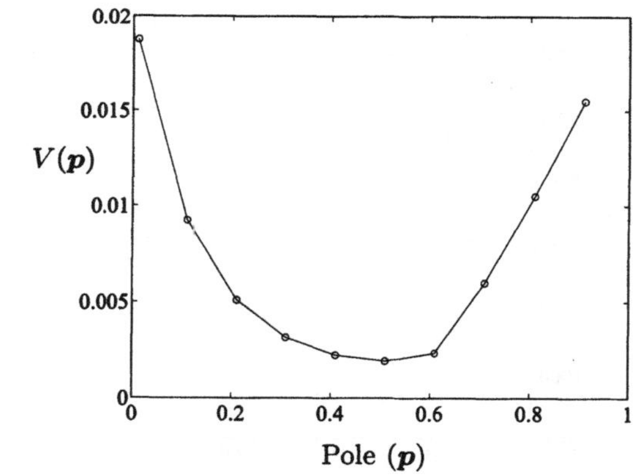

Fig. 5.49. Pole assignment cost function $V(\boldsymbol{p})$.

It is worth noting how the shape of transient of the measured variable $q_{l*}(t)$ between 0 to 20s. is determined by the input variation and is not related to the incipient compressor fault.
On the other hand, Figure 5.51(a) shows the estimate of the fault $f_s(t)$ obtained by computing the difference between the fault–free (solid line) and the faulty residual (dotted line), depicted in Figure 5.51(b).

5.6.2 Case 2: Fault Diagnosis of the Output Sensor

The "case 2" fault represents the malfunctioning of a thermocouple in the turbine gas path which leads to a slowly increasing or decreasing reading over time.

There is no limit placed on the error magnitude while the fault development rate is set to (5% error in measuring actual temperature)/hour.

As in the previous case, in order to diagnose a single $f_y(t)$ fault on the i–th output sensor ($\boldsymbol{f}_u(t) = \mathbf{0}$, $\boldsymbol{f}_s(t) = \mathbf{0}$, $\boldsymbol{f}_c(t) = \mathbf{0}$) when the measurement noise signals are negligible ($\tilde{\boldsymbol{u}}(t) \cong \mathbf{0}$ and, $\tilde{\boldsymbol{y}}(t) \cong \mathbf{0}$) the model of the i–th output observer ($i = 1, \cdots, m$) has been used [Simani *et al.*, 2002].

The construction of the observer for the diagnosis of the output sensor fault (thermocouple fault) affecting the measurement of the temperature t_{k*} requires the knowledge of the triple ($\boldsymbol{A}_i$,$\boldsymbol{B}_i$,$\boldsymbol{C}_i$) and therefore the identification of an ARX model with two inputs which gives the prediction of the turbine output t_{k*}.

A second order ARX MISO model ($r = 2$ and $m = 1$), driven by $a_v(t)$ and $f_f(t)$ input signals, was identified. Such a model gives an output reconstruction error equal to 1.13×10^{-5}. The parameters of the ARX model are described by the vector $\theta = [-0.0244, 1.0295, -0.0020, -0.0014, -0.3180, 0.3140]$.

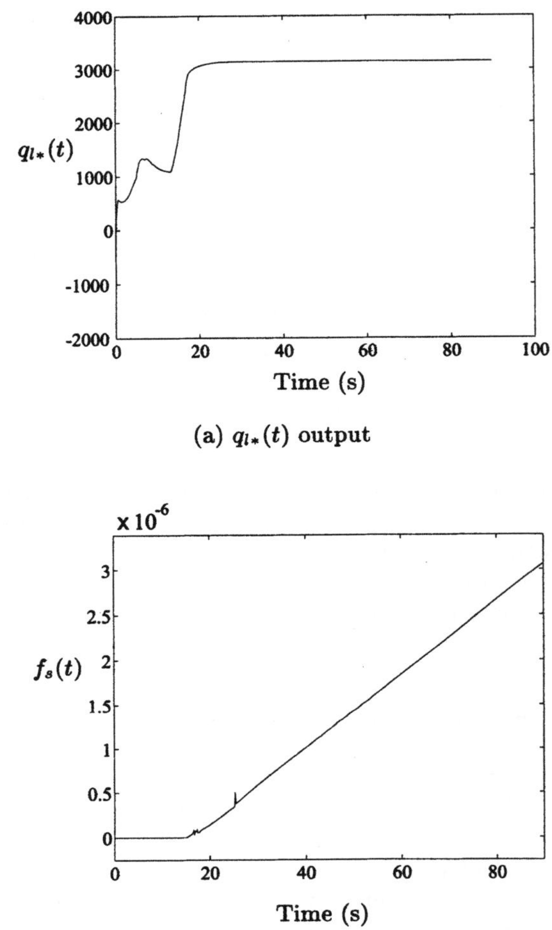

(a) $q_{l*}(t)$ output

(b) The simulated fault $f_s(t)$

Fig. 5.50. The monitored signal versus the the component fault mode.

The poles of the output observer, whose scheme is depicted in Figure 5.52, were chosen near 0.3 in order to minimise the function $V(\boldsymbol{p})$.
As shown in Figure 5.53, an incipient fault (drift) was generated in the output sensor of the SIMULINK© model by adding a ramp function with a slope of $0.008\frac{^oC}{s}$ to the $y_i(t) = t_{k*}$ output signal.

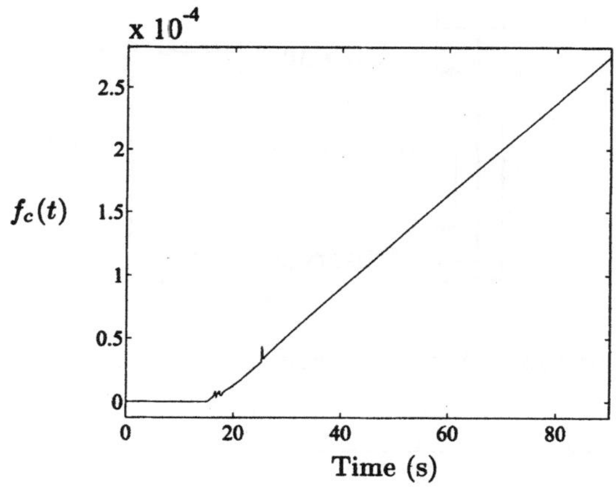

(a) $f_c(t)$ fault estimate

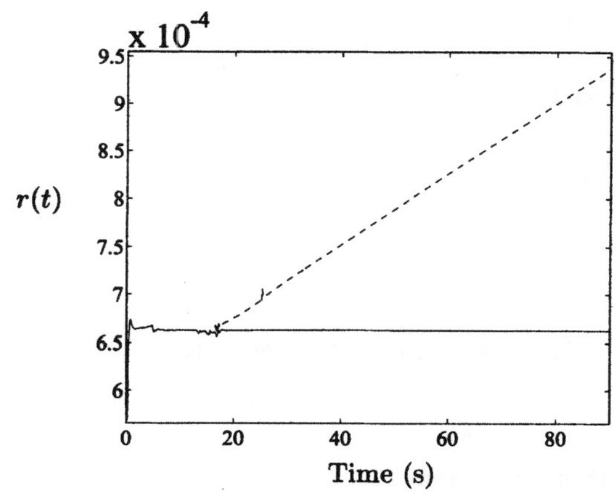

(b) Fault-free (solid line) and faulty (dashed line) residual $r(t)$

Fig. 5.51. Results from the residual generation.

Moreover, it was decided to consider a fault during a transient since, in this case, the residual error due to ARX model approximation is maximum and therefore it represents the most critical case.
The fault occurring on the single sensor causes alteration of the sensor signal $y_i(t) = t_{k*}$ and of the residuals given by the output observer using this signal as input. These residuals indicate a fault occurrence when their values are lower or higher than the thresholds fixed in fault-free conditions.

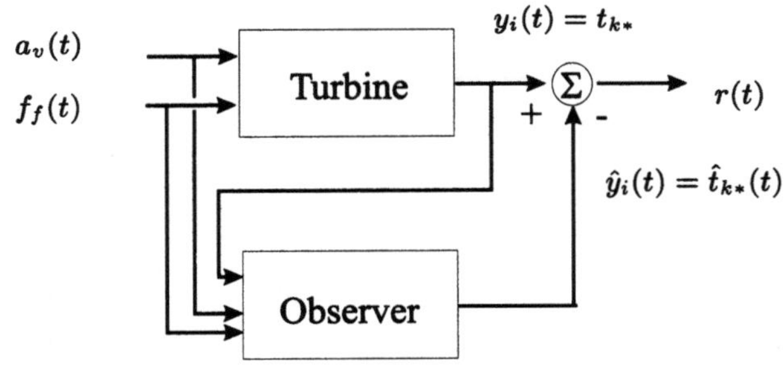

Fig. 5.52. Output sensor observer scheme.

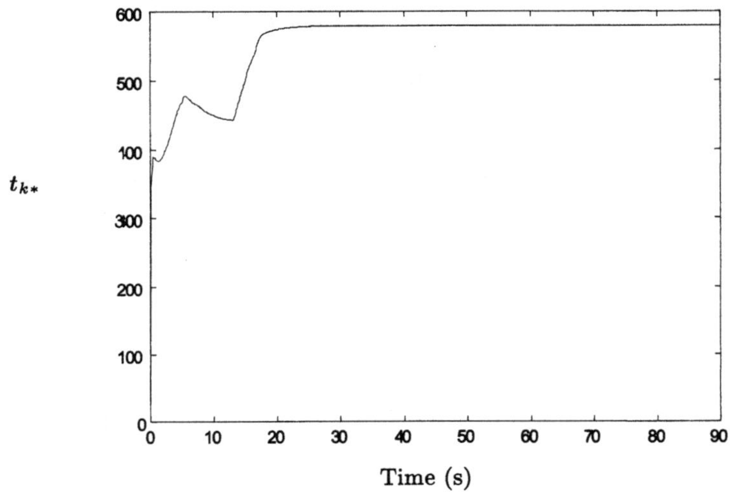

Fig. 5.53. t_{k*} output measurement.

Figure 5.54(a) shows the fault-free $y_i^*(t) - \hat{y}_i(t)$ (continuous line) and faulty $y_i(t) - \hat{y}_i(t)$ (dotted line) residual obtained from the difference between the values computed by the observer related to the output $y_i(t) = t_{k*}$ and the ones given by the sensor. Obviously, the non–zero value of the residual is due to the ARX model approximation. The drift (ramp fault) in Figure 5.54(b) starts at the instant $t = 15$s.

Since the observer gives the estimate $\hat{y}_i(t)$ of $y_i(t)$ at the instant t by using measurements available from the instant $t = 0$ to $t = n - 1$, a fault occurring at the instant t affects only $y_i(t)$. This change produces the instantaneous peak which appears in Figure 5.54(b).

In this case, the peaks are not due to instantaneous changes in the input signals, *e.g.* fuel flow $f_f(t)$ or valve position $a_v(t)$. Thus, they may be used as incipient detector of anomalous behaviour of the output sensors.

Figure 5.54(b) shows the behaviour of the residual with the same fault as the previous case occurring at the instant $t = 35$s in different operating conditions of the plant.

The fault–free residual, $y_i^*(t) - \hat{y}_i(t)$, is depicted by the continuous line, whilst the residual corresponding to the fault, $y_i(t) - \hat{y}_i(t)$, is shown with the dotted line.

The peak that appears in the Figure 5.54(b) is generated by the change related to the fault occurrence at the same instant.

Figure 5.55(a) depicts the dynamics of the drift $f_y(t)$ affecting the t_{k*} output sensor, whilst Figure 5.55(b) shows the fault estimate obtained from the difference between the fault-free and the faulty residual.

The peak that appears in the Figure 5.55(b) is generated by the instantaneous difference between measured $y_i(t)$ and estimated output $\hat{y}_i(t)$ at the instant t related to the fault occurrence.

It is worth noting how, because of the links between fault and symptom signals, the failure estimates may have different scales from the real ones. The estimates of the faults can, in fact, only capture the shape (ramp nature) of the fault and not the precise magnitude.

5.6.3 Case 3: Turbine Damage (*Turbine Component Fault*)

Fault "case 3" represents the fault $f_s(t)$ of the turbine. This results in a reduction in turbine efficiency.

The fault $f_s(t)$ is modelled as a gradual reduction in turbine efficiency over time. The maximum decrease in turbine efficiency is set nominally at 5% while the fault development rate is set to (5% reduction of normal efficiency)/hour.

An output observer fed by the inputs $\alpha(t)$, $M_f(t)$ and by the output measurement $p_{h*}(t)$ of the pressure of the turbine (see Figure 5.56) has been designed in order to detect such kind of failure [Simani *et al.*, 2002]. Noise–free conditions ($\tilde{\boldsymbol{u}}(t) = \mathbf{0}$, $\tilde{\boldsymbol{y}}(t) = \mathbf{0}$) have been assumed.

The corresponding MISO ARX model, having parameters $\theta = [0.4234, -1.7905, 2.3658, 0.0002, 0.0008, 0.0933, -0.2035, 0.1113]$, gives a mean–square reconstruction error equal to 1.8013×10^{-6}. Observer eigenvalues were chosen near 0.3 to minimise the cost function $V(\boldsymbol{p})$.

The component fault dynamics $f_s(t)$ and its estimate $\hat{f}_s(t)$ obtained by the output observer are shown in Figures 5.57(a) and 5.57(b), respectively.

The scheme used to generate the redundant residual regarding the $p_{h*}(t)$ output signal is depicted in Figure 5.58(a). The fault–free and the faulty residual are also shown in the Figure 5.58(b).

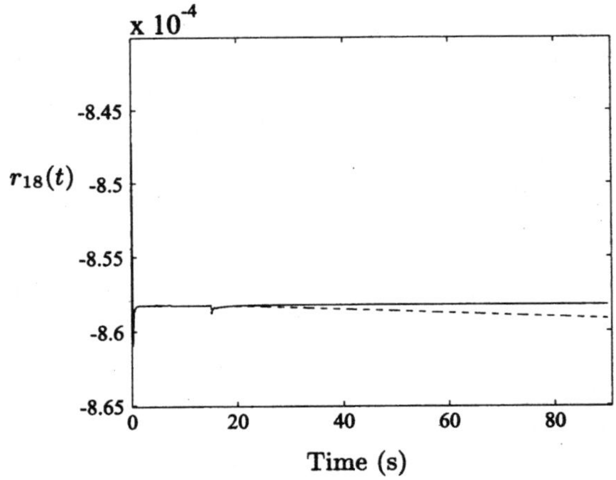

(a)

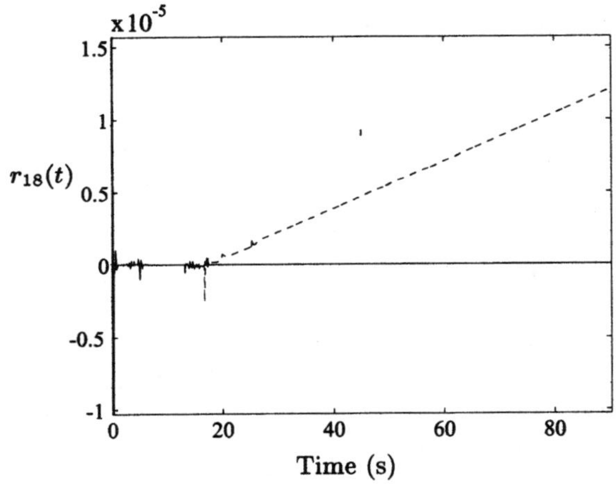

(b)

Fig. 5.54. Residual function in different operating points.

5.6.4 Case 4: Actuator Fault (*Controller Malfunctioning*)

As depicted in Figure 5.47 and in the related work [Simani *et al.*, 2002], fault "case 4", $f_c(t)$ affects the actuator of the turbine controller .

Under the assumption that there are no actuator dynamics in the current turbine model, the fault $f_c(t)$ of the actuator causes a slower response to

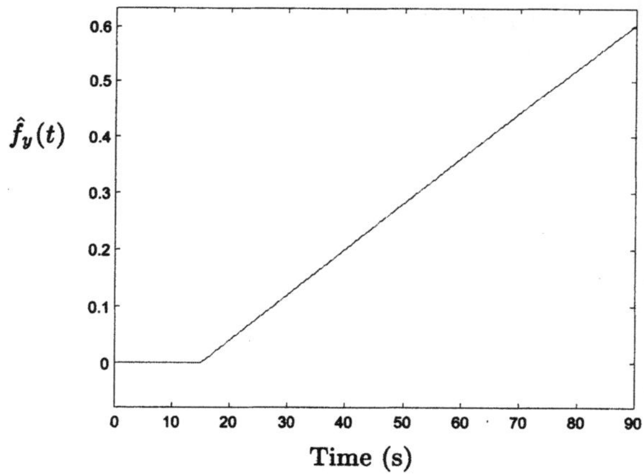

(a) Measured fault function

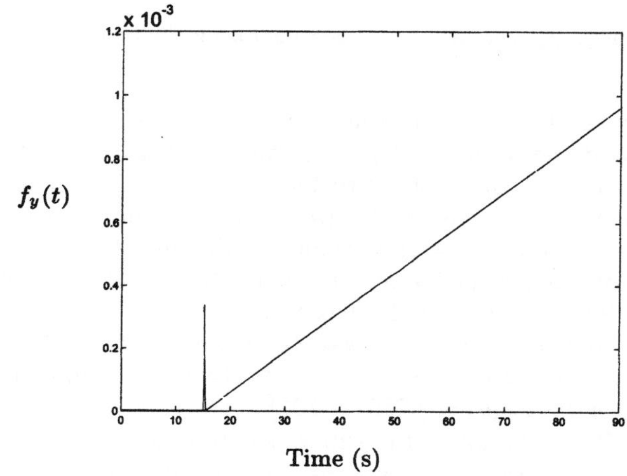

(b) Estimated fault signal

Fig. 5.55. Real and estimated fault function.

demanded flow rates. Its effect is modelled as a simple first order lag on the resulting fuel flow. The actuator response time constant increases linearly with the time in order to represent a progressive damage to the actuator.

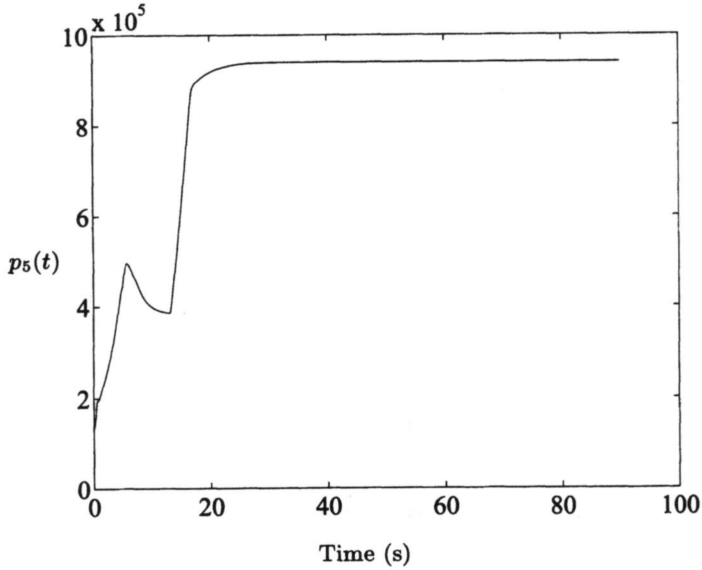

Fig. 5.56. The $p_h(t)$ measured pressure signal from the turbine.

In particular, the inputs of the turbine, the fuel flow, $M_f(t)$, the valve angle, $\alpha(t)$ and the outputs $\boldsymbol{y}(t)$ were considered. In particular, the speed demand, N_t, and the speed of the turbine, ω_t, were taken in account.

For each output, a third order ($n = 3$) ARX models with two inputs and one output ($m = 1$, $r = 2$) were identified. The ARX model parameters are collected in the parameter vector $\boldsymbol{\theta} = [0.2018, -1.3242, 2.1207, -0.0069, 0.0632, -0.0560, 0.0187, -0.0464, 0.0286]$.

A single fault $f_c(t)$ was simulated by means of the SIMULINK model, and $m_{j*}(t)$ was determined as the most sensitive output to a fault regarding the actuator, with a $J(\theta) = 4.7857 \times 10^{-5}$.

Figure 5.59(a) illustrates the subsystem considered in this case. Figure 5.59(b) shows the observer scheme used to generate the residual signal used to detect the fault $f_c(t)$.
The observer eigenvalues $\boldsymbol{\lambda}$ were chosen close to 0.4 to minimise the cost function $V(\boldsymbol{p})$.

Figure 5.60 shows the function $V(\boldsymbol{p})$ is depicted, with $p_1 = \ldots p_{n-1} = 0.4$ and $\boldsymbol{p} \equiv p_n$, since $V(\boldsymbol{p})$ represents a four–dimensional function ($n = 3$).
Figure 5.61(a) illustrates the dynamic behaviour of the $m_{j*}(t)$ signal, a measurement of turbine mass flow, while the effects of the fault on the symptom signal $r(t)$ are shown in Figure 5.61(b).

Because of the closed–loop configuration of the subsystem considered in Figure 5.59(a), the fault shape cannot be described by using a ramp function.

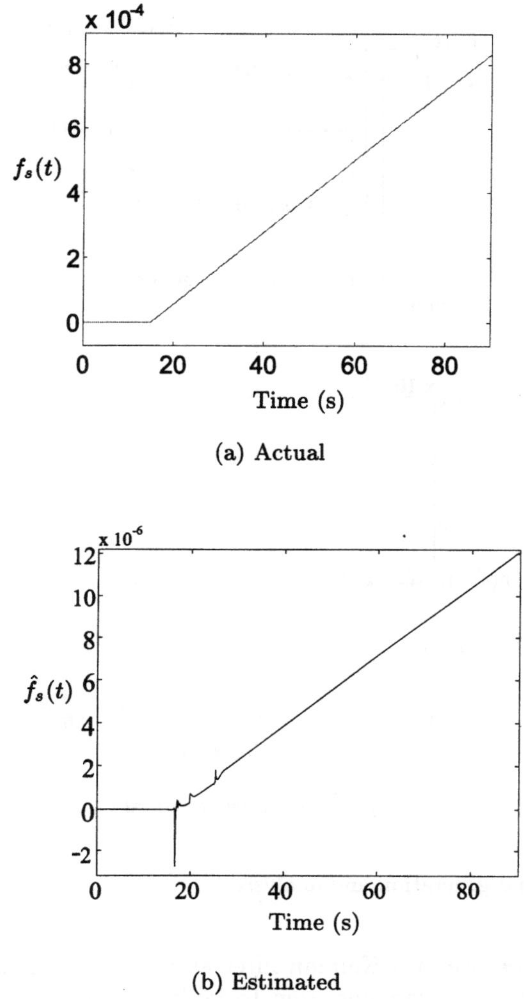

Fig. 5.57. "Case 3" seal fault $f_s(t)$ dynamics.

Figure 5.62 shows the fault-free (see Figure 5.62(a)) and faulty (see Figure 5.62(b)) residual $r(t)$ obtained from the difference between the values computed by the observer related to the output $m_{j*}(t)$ and the ones given by the sensor. These residuals indicate a fault occurrence when their values are lower or higher than the thresholds fixed in fault-free conditions.

In order to improve the fault detection capabilities of the proposed method regarding the "case 4", the technique presented in Section 4.8 is exploited.

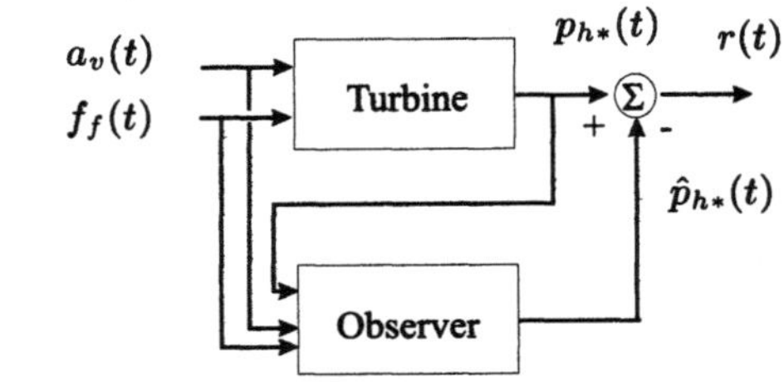

(a) $p_{h*}(t)$ residual generation scheme

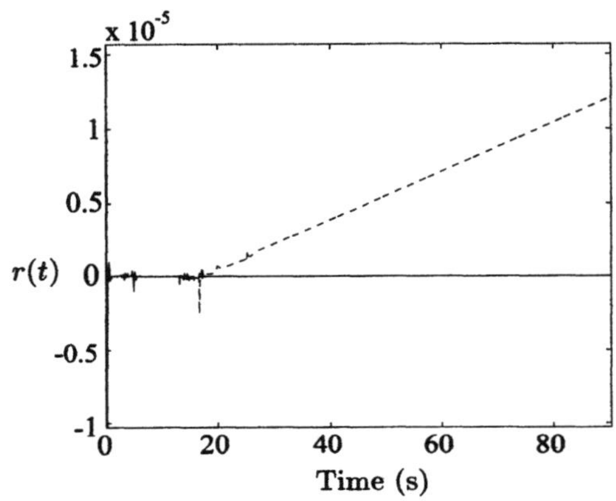

(b) $p_{h*}(t)$ observer residual

Fig. 5.58. Residual generation and analysis.

It concerns the use of a Kalman filter as parameter estimator, in order to detect changes in parameters due to faults affecting input and output measurements.

Figure 5.63(a) depicts the recursive estimation of one entry of the parameter of the MISO ARX model for the $m_{j*}(t)$ output $y_i(t)$ given by the Kalman filter (solid line) and the estimate computed by the OLS (Ordinary Least–Squares) method (dotted line) [Ljung, 1999].

Note how the real process with $\alpha(t)$ and $M_f(t)$ as inputs and $m_{j*}(t)$ as output $y_i(t)$ is non–stationary and the estimates are different.
Figure 5.63(b) shows the change of the most sensitive parameter $\theta_i(t)$ of $\theta(t)$ due a fault, by using the Kalman filter for a third order ARX model ($n = 3$), with a covariance matrix for $\varepsilon(t)$ and $\omega(t)$ processes estimated from the OLS.

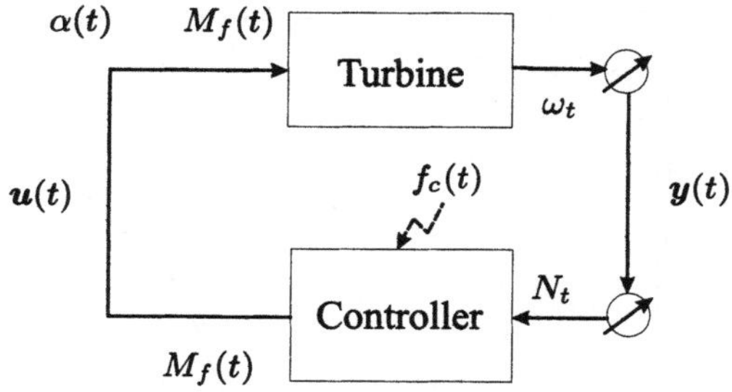

(a) The diagnosis subsystem

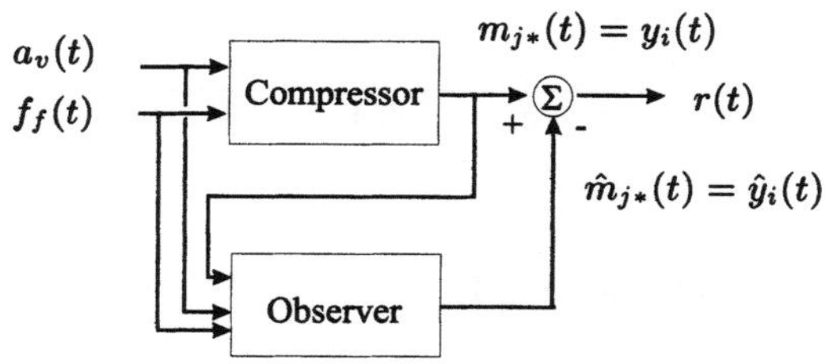

(b) The observer scheme

Fig. 5.59. Schemes for the fuel actuator (controller) fault $f_c(t)$.

The $f_c(t)$ actuator fault occurs as a ramp function when $t \geq 15$s, and it is injected into the feedback controller system. The fault effects on output measurement are different from a ramp mode.

In particular, the "case 4" fault mode is depicted in Figure 5.61(b) and the non–linear effect on $\theta_i(t)$ of the $f_c(t)$ signal is very similar to a step change, as shown in Figure 5.63(b).

5.6.5 FDI in Noisy Environment Using Kalman Filters

Under the assumption of noisy measurements $\boldsymbol{u}(t)$ and $\boldsymbol{y}(t)$, Figures 5.64, 5.65, 5.66 and 5.67 show results from the application of model–based FDI techniques exploiting Kalman filter for residual generation [Simani *et al.*, 2000a].

In particular, Figure 5.64(a) shows the value of the fault $f_s(t)$ affecting the $r(t)$ residual concerning the torque measurement $q_{l*}(t)$ ("case 1"), whilst Figure 5.64(b) depicts fault-free and faulty residuals generated by the Kalman filter having two inputs ($\alpha(t)$, $M_f(t)$) and one output $y_i(t) = q_{l*}(t)$.

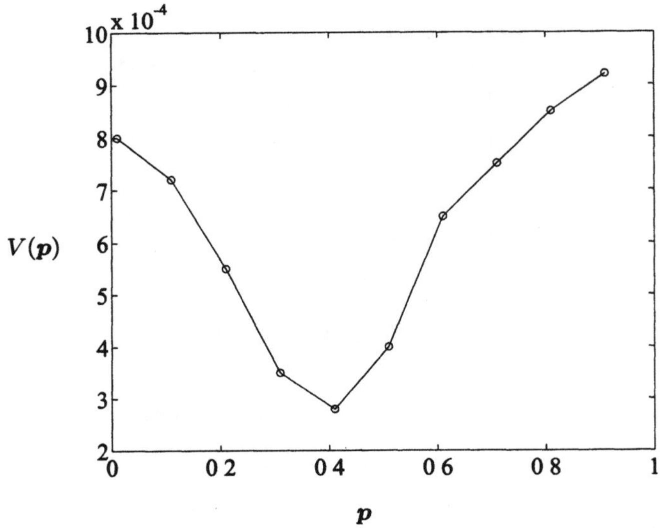

Fig. 5.60. The cost function $V(\mathbf{p})$.

It is important to note that, in order to achieve the maximal fault detection capability, the residual corresponding to the most sensitive filter to a failure on the $q_{l*}(t) = y_i(t)$ measurement was selected.

Figure 5.65(a) shows the simulated fault $f_y(t)$ affecting the output sensor $y_i(t)$ for the measurement of the turbine temperature t_{k*} concerning "case 2".

In Figure 5.65(b) fault–free and faulty residuals regarding the $t_{k*} = y_i(t)$ signal are shown. The residuals are obtained from the difference between the values $y^i(t|t)$ computed by the Kalman filter 4.31 and the ones measured by the temperature sensor t_{k*} .

It is worth noting that the non–zero value of the residual in fault–free conditions is due to the ARX model approximation and to the actual measurement noise.
Figure 5.66 shows simulated fault 5.66(a) and residuals 5.66(b) corresponding to component fault ("case 3").

According to results from the identification steps exposed in previous sections, the residual is computed monitoring the $y_i(t)$ pressure signal $p_{h*}(t)$. Finally, Figure 5.67 shows the actuator fault $f_c(t)$ 5.67(b) and the residuals 5.67(b) concerning $y_i(t) = m_{j*}(t)$ measurement due to a ramped incipient actuator fault ("case 4").
Because of the nature of the incipient ramp fault $f_c(t)$ affecting the regulator in the feedback control loop, the output measurements affected by the fault itself are different from ramp signals, as depicted in Figure 5.67(a).

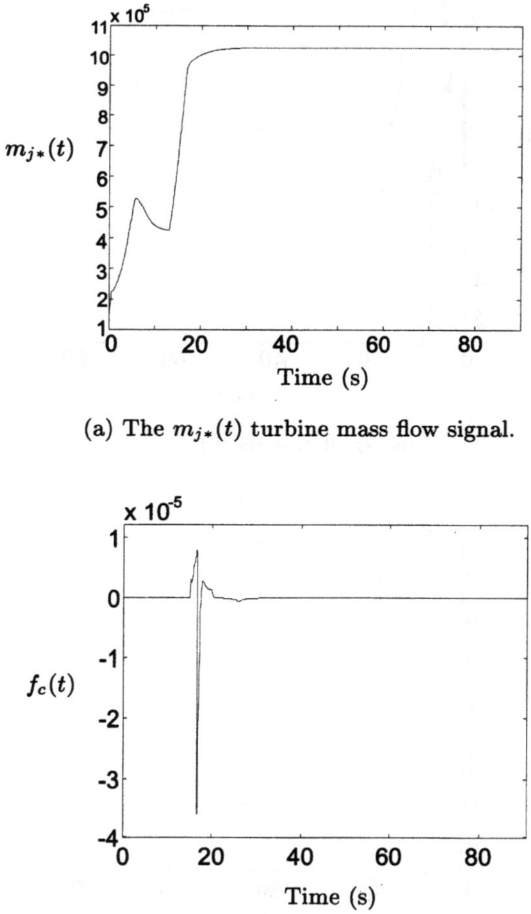

(a) The $m_{j*}(t)$ turbine mass flow signal.

(b) The fault $f_c(t)$ concerning $m_{j*}(t)$

Fig. 5.61. Diagnosis of the $m_{j*}(t)$ mass flow signal.

5.6.6 Fault Isolation

In order to summarise the FDI capabilities of the presented schemes, Table 5.29 shows the "fault signatures" in case of a single fault in each actuator, component and sensor.

Table 5.29 was obtained by performing residual sensitivity analysis, *i.e.* by selecting the most sensitive residuals to the faults.

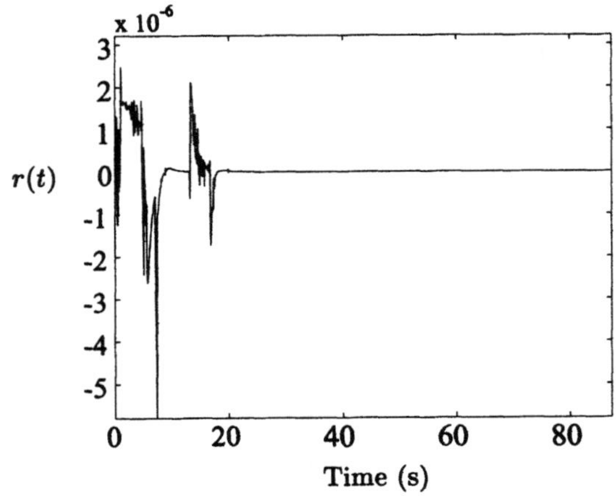

(a) Fault-free residual

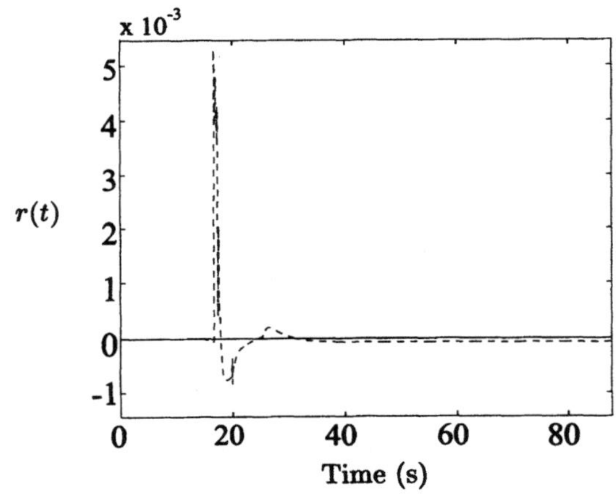

(b) Residual in the presence of $f_c(t)$ fault

Fig. 5.62. The fault–free and faulty residual signals.

The residuals that are affected by faults are denoted by a '1' in the corresponding table entry, while an entry '0' means that the fault does not affect the correspondent residual.

Under these conditions, the entries '1's in Table 5.29 represent distinguishable residuals: it means that their magnitude is greater than a fixed threshold. On the other hand, a '0' entry means that the residual is lower than the fixed threshold.

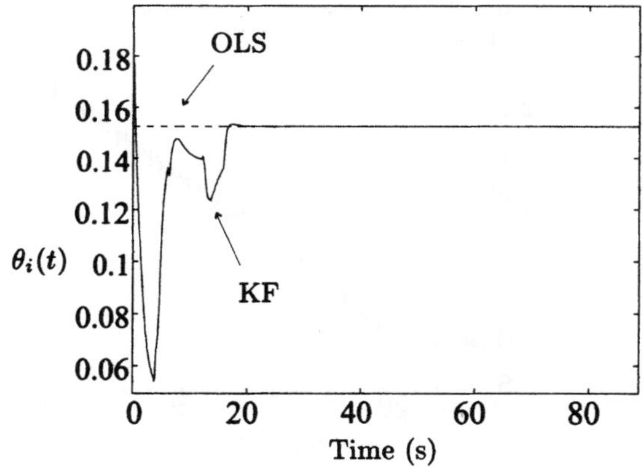

(a) $\theta_i(t)$ parameter variation due to a fault

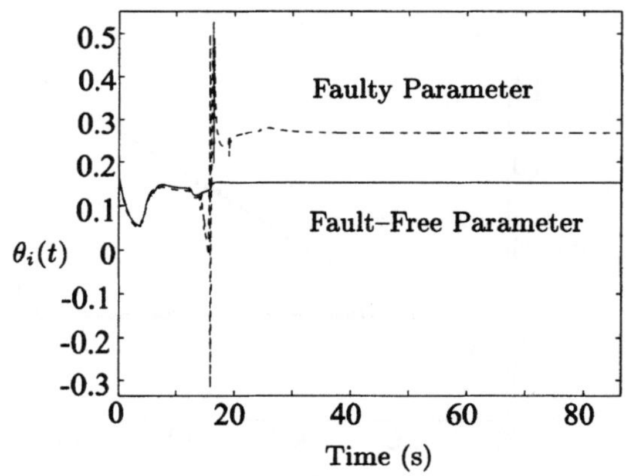

(b) Fault–free and faulty parameter

Fig. 5.63. Kalman filter parameter variations due to the $f_c(t)$ fault.

Table 5.29. Fault signature.

Fault/$r(t)$	q_{l*}	t_{k*}	p_{h*}	m_{j*}
Case 1	1	0	0	0
Case 2	0	1	0	0
Case 3	0	0	1	0
Case 4	0	0	0	1

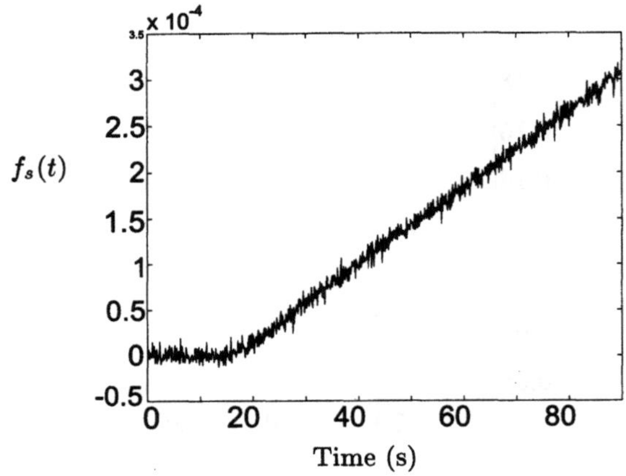

(a) System fault $f_s(t)$

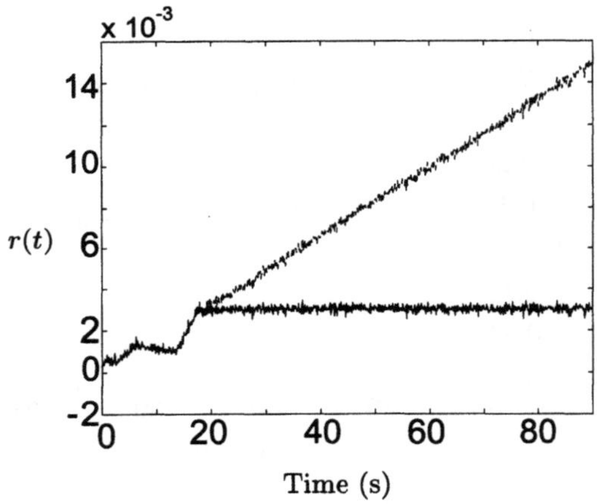

(b) Kalman filter residuals in fault–free (black line) and faulty (gray line) cases.

Fig. 5.64. Fault and residual signal for component fault ("case 1"). The fault is simulated by a ramp signal commencing at $t = 15$s.

Note how faults occurring at the same time in actuator, components and sensor can be isolated since each fault affects only the residual function of the observer driven by the same output.

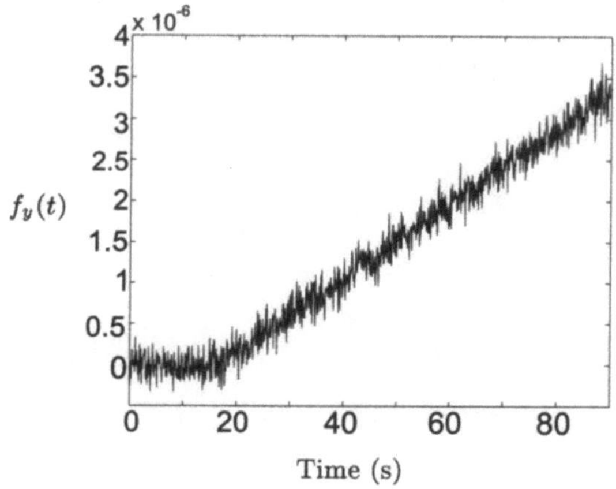

(a) Output sensor fault $f_y(t)$

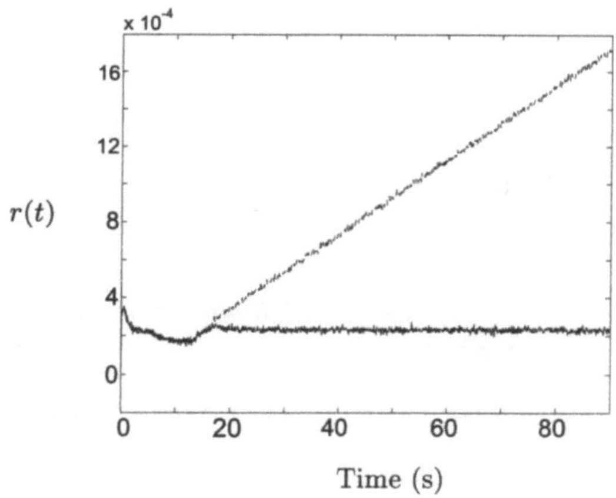

(b) Kalman filter residuals in fault–free (black line) and faulty (gray line) cases.

Fig. 5.65. Fault and residual signal for output sensor fault ("case 2"). The fault is simulated by a ramp signal commencing at $t = 15$ s.

5.6.7 Minimal Detectable Faults

Table 5.30 summarises the performance of the FDI technique both in noise–free and noisy environments.

The minimal detectable fault values are expressed as percentage of the signal values and are relative to the case in which the occurrence of a fault must be detected as soon as possible.

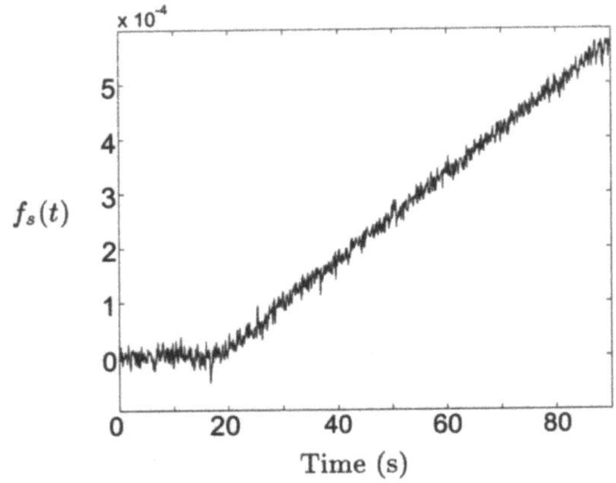

(a) System fault $f_s(t)$

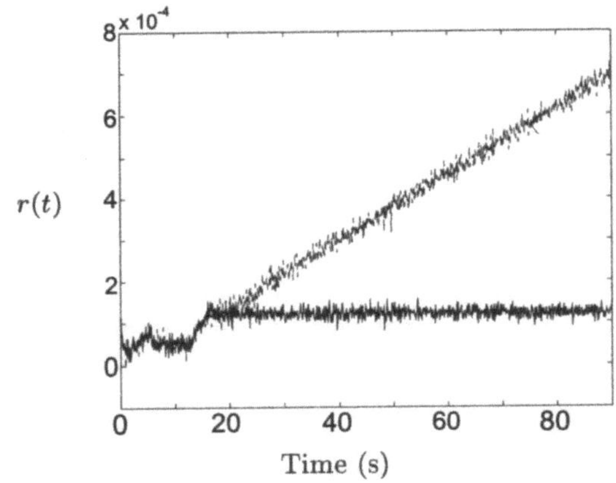

(b) Kalman filter residuals in fault–free (black line) and faulty (gray line) cases.

Fig. 5.66. Fault and residual signal for component fault ("case 3"). The fault is simulated by a ramp signal commencing at $t = 15$ sec.

The values of the faults obtained by using geometrical analysis on Kalman filter residuals are different from the ones computed in the deterministic environment exploiting classical observers. It is worth noting how faults modelled by ramp functions may not be immediately detected, since the delay in the corresponding alarm normally depends on the fault mode.

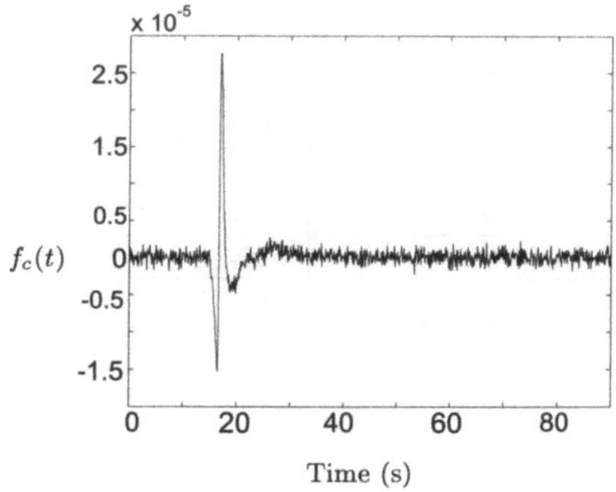

(a) Actuator fault $f_c(t)$

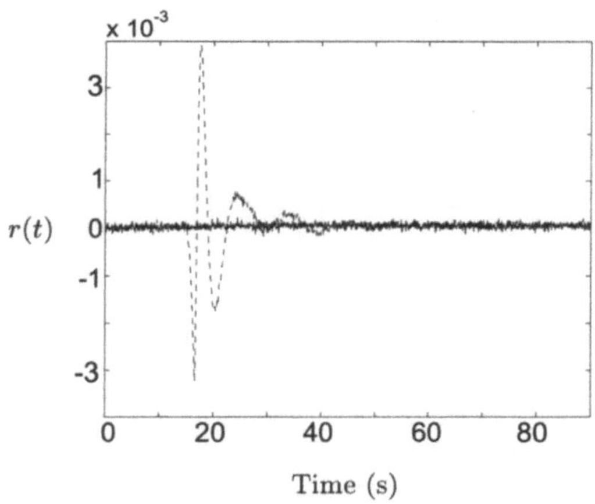

(b) Kalman filter residual

Fig. 5.67. Fault and residual signals for actuator fault ("case 4"). The $f_c(t)$ fault commences at the instant $t = 15$ sec.

The minimal detectable fault can be found by fixing a detection delay, defined in Figure 5.68. If a detection delay is tolerable, the amplitude of the minimal detectable fault is lower.

The minimal detectable faults on the various sensors seem to be adequate for the industrial diagnostic applications, by considering also that the minimal detectable faults can be reduced if a delay in detection promptness is tolerable.

Table 5.30. Minimum detectable faults by monitoring residual and innovation values.

Fault Case	Monitored signal	Noise free	Noisy case	Detection delay
Case 1 (Compressor fault)	$q_{l*}(t)$	0.5%	1%	30s
Case 2 (Thermocouple sensor fault)	$t_{k*}(t)$	10%	12%	30s
Case 3 (Turbine failure)	$p_{h*}(t)$	5%	7%	60s
Case 4 (Actuator fault)	$m_{j*}(t)$	1%	3%	10s

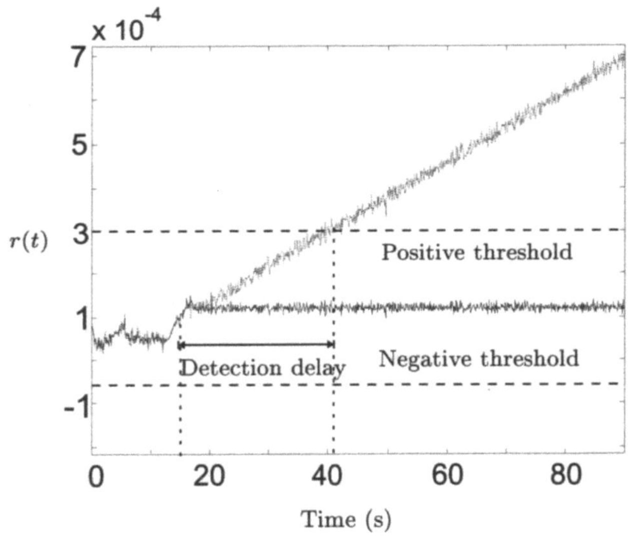

Fig. 5.68. Detection delay definition.

5.7 FDI with Eigenstructure Assignment

This section presents some results concerning robust fault diagnosis of the dynamic processes using a parametric identification technique.

As stated in Section 4.7, the first step of the considered approach estimates an equation error model by means of the input–output data acquired from the monitored system. In particular, the equation error term of the model takes into account disturbances (non measurable inputs), non–linear and time–variant terms, measurement errors, etc.

Section 4.7 proved that the method requires a state space realization of the input–output equation error model which allows leads to define an equivalent disturbance distribution matrix related to the error term. Therefore, the eigenstructure assignment results for robust fault diagnosis can be successfully applied.

The procedure proposed in Section 4.7 has been tested by means of the industrial gas turbine process simulator. In such a manner, sensor, component and actuator faults can be simulated on an single shaft gas turbine. Results from this simulator were also reported.

5.7.1 Robust Fault Diagnosis of the Industrial Process

The proposed robust diagnosis procedure has been applied for the detection and the isolation of faults regarding actuators, components and sensors of the single–shaft industrial gas turbine system.

The process and the fault simulator was developed in SIMULINK© environment [Simani *et al.*, 2000b].

The process has strong non–linear behaviour since it is mainly based on non–linear functions and look-up tables which simulate thermodynamic relations among the variables involved.

The fault diagnosis problem has been approached by using both a bank of classical observers, Kalman filters (Sections 4.4 and 4.6) and the residual generation scheme with eigenstructure assignment, presented in Section 4.7.

In both cases, the design of the residual generators requires the identification of a number of equation error MISO ($r = 2$ and $m = 3$) models (see Equation 4.70) equal to the number m of the output variables $y_i(t)$.

The i–th model ($i = 1, \ldots, m$, $m = 3$) is driven by $u_1(t)$ and $u_2(t)$ and gives the prediction of the i–th output $y_i(t)$.

Each model was tested in different operating conditions and it has always provided an output reconstruction error lower than 0.1%. Moreover, as previously stated, the design of the residual generator (Section 4.4) requires the knowledge of a state space representation (A, B, C, D, E) (Section 3.2).

It should be noted that each residual signal corresponds to a system output in the traditional Luenberger scheme (three residual signals in our example), but in the second scheme the number of residual depends on structure of Q matrix (Section 4.7).

In order to show the effectiveness of the proposed robust fault detection procedure, the comparison between the two residual generation schemes has been performed.

Residuals computed in fault–free conditions by means of the traditional observer scheme (as presented in [Simani *et al.*, 2000b]) and by using the eigenstructure assignment method are shown in Figures 5.69, 5.70, 5.71 and 5.72, 5.73, respectively.
From the comparison among Figures 5.69, 5.70, 5.71 and 5.72, 5.73 it is worth noting how in the second simulation case (eigenstructure assignment method), the residual magnitudes are lower than the ones of the residuals generated using classical dynamic observers.

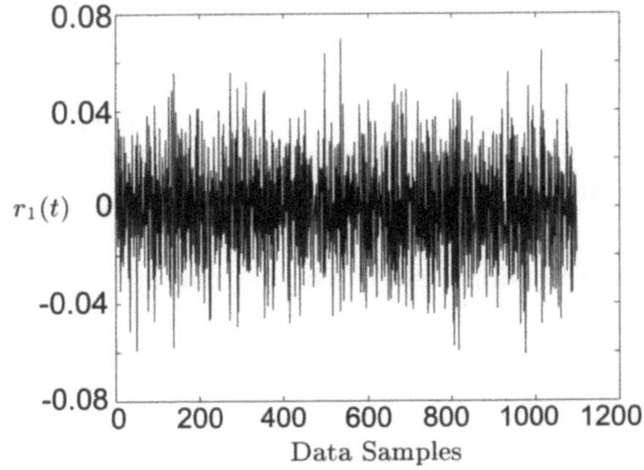

Fig. 5.69. The residual $r_1(t)$ for $y_1(t)$ generated by using Luenberger dynamic observer in fault–free conditions.

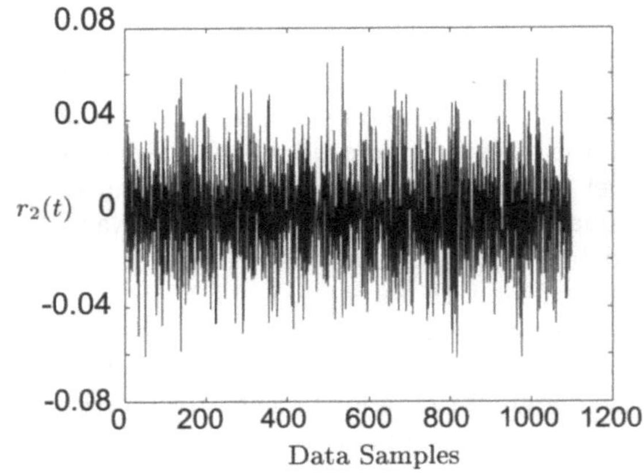

Fig. 5.70. The residual $r_2(t)$ for $y_2(t)$ generated by using Luenberger dynamic observer in fault–free conditions.

This confirms the effectiveness of the proposed eigenstructure assignment procedure for residual disturbance de–coupling. The residuals are therefore robust with respect to system uncertainties. The next step consists of showing how the developed diagnosis scheme allows to increase residual sensitivity to faults in the presence of disturbances.

Three fault cases can be simulated using the SIMULINK simulator of Figure 5.44, [Simani *et al.*, 2000b]. According to Section 5.5.1, these fault situations are

– Compressor contamination (component fault);
– Thermocouple sensor failure (sensor fault);

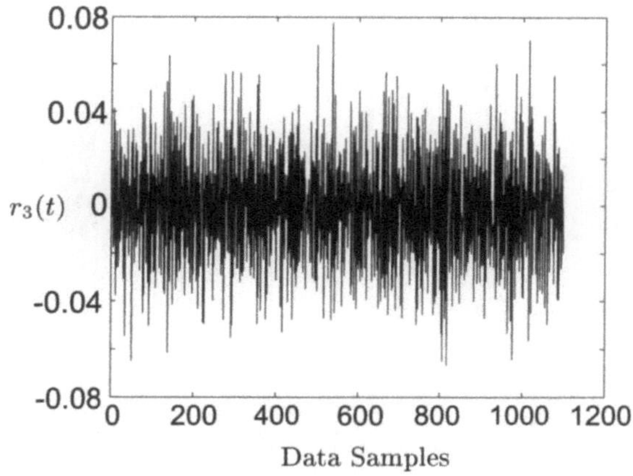

Fig. 5.71. The residual $r_3(t)$ for $y_3(t)$ generated by using Luenberger dynamic observer in fault–free conditions.

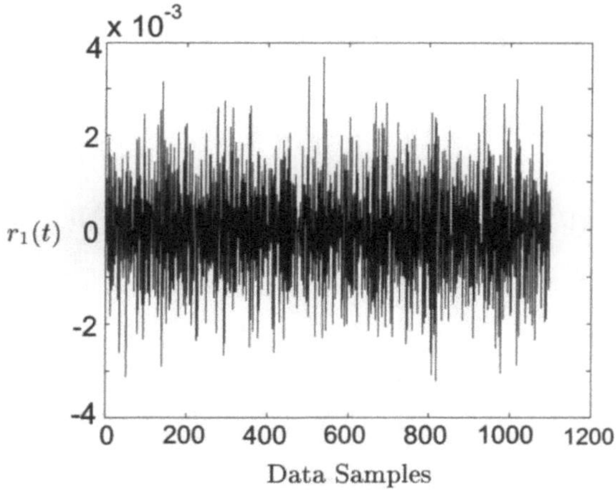

Fig. 5.72. The residual $r_1(t)$ computed by the observer with eigenstructure assignment procedure in the fault–free case.

– Actuator failure (actuator fault).

As an example, the fault on the compressor (component fault) has been simulated. The corresponding simulation results are shown by Figures 5.74, 5.75 and 5.76.
Because of the fault size and the model uncertainties, all residuals seem not to be sensitive to this fault. Their behaviours are similar to the fault–free case and the fault detection is not possible.

On the other hand, Figures 5.77 and 5.78 show residuals computed by the observer with the eigenstructure assignment approach. In this case, the

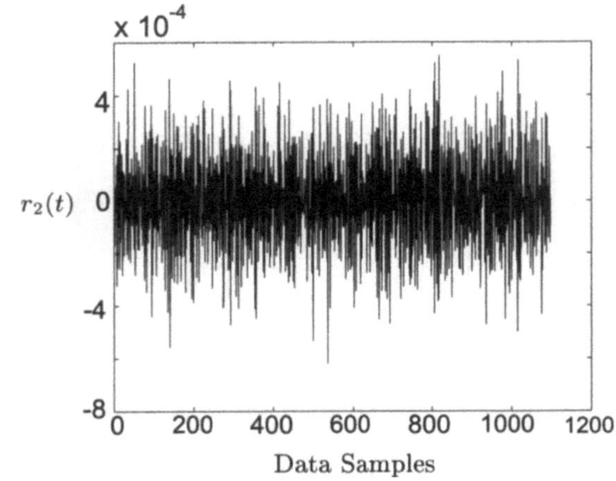

Fig. 5.73. The residual $r_2(t)$ computed by the observer with eigenstructure assignment procedure in the fault–free case.

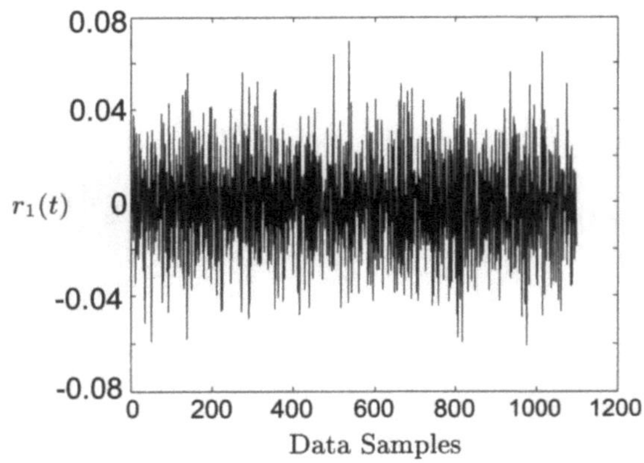

Fig. 5.74. The residual $r_1(t)$ for $y_1(t)$ from the Luenberger observers when a system fault occurs.

disturbance and fault effects are decoupled and residuals can be used as malfunction detectors. Moreover, fault occurrence is highlighted by an abrupt change in the residual signals.

Finally, it is worth noting how the results obtained by this robust approach state the enhancement of the proposed diagnosis method with respect to a classical scheme which uses Luenberger dynamic observer for residual generation.

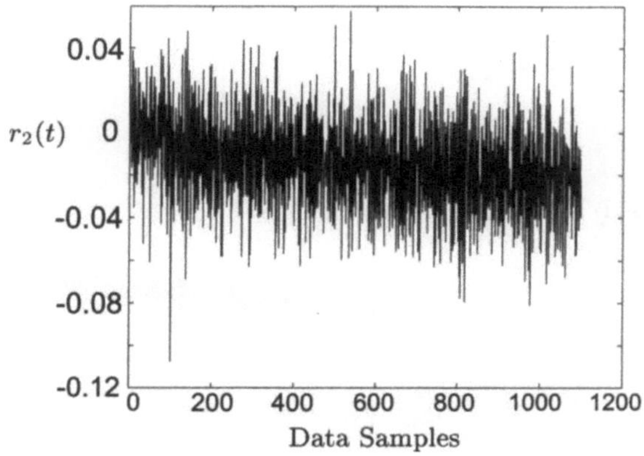

Fig. 5.75. The residual $r_2(t)$ for $y_2(t)$ from the Luenberger observers when a system fault occurs.

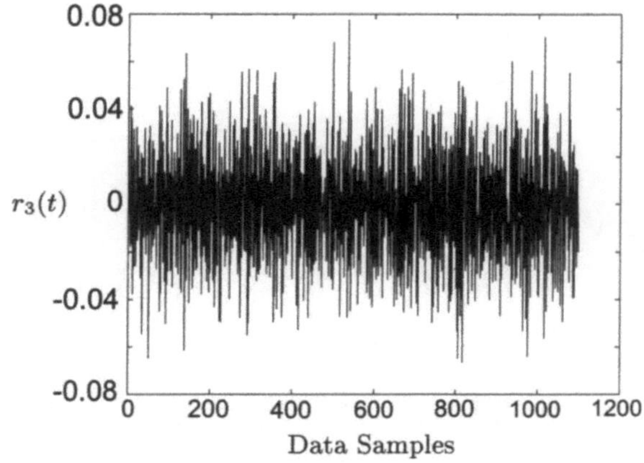

Fig. 5.76. The residual $r_3(t)$ for $y_3(t)$ from the Luenberger observers when a system fault occurs.

5.8 Robust Residual Generation Problem

One critical limitation of the model–based approach to fault diagnosis is a consequence of the fact that modelling uncertainty is inevitable and its effects are very variable. For complex systems such as a gas turbine plant, the effects of uncertainty can be significant, due to high non–linearity and operating point dependency. This chapter has focused on a robust fault detection method using both disturbance de–coupling residual generator and system identification techniques.

In order to design robust FDI schemes, we should have some description of modelling uncertainty. Furthermore, it is necessary to make sure that

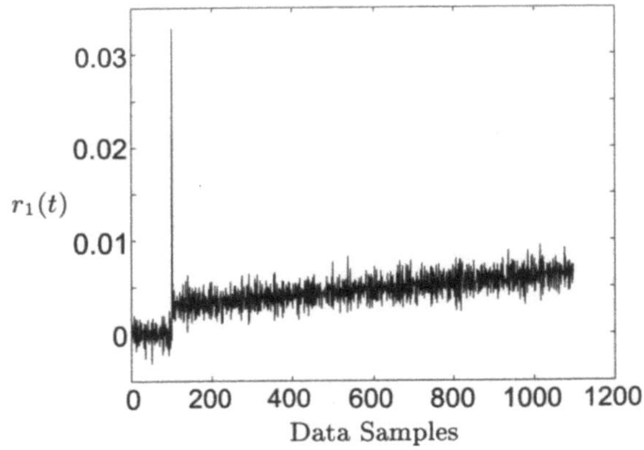

Fig. 5.77. The residual $r_1(t)$ from the observer with eigenstructure assignment in case of a system fault.

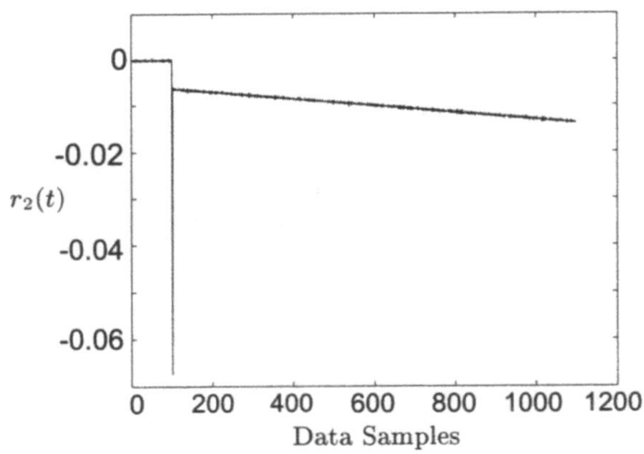

Fig. 5.78. The $r_2(t)$ residuals from the observer with eigenstructure assignment in case of a system fault.

this description can be handled in a straightforward and systematic manner. Modelling uncertainty can be accounted for using an additional term in the dynamic equation of the system; this additional term has a certain structure (*i.e.*, structured uncertainty).

Normally, it is assumed that the distribution of this additional term is known *a priori*. Based on this description, the disturbance de–coupling approach is used to design a robust FDI scheme. However for most real systems, the structure of the uncertainty is unknown.

This chapter has studied the methods for determining the structure of uncertainty. The main aim has been to bridge theoretical assumptions with practical reality. Two principle methods for determining disturbance distribution matrices have been presented. The first method is based upon direct determination and optimisation. This is a simple and direct approach which does not require real or simulated system input and output data. Its disadvantage is that it requires some *a priori* information about modelling uncertainty. However, this chapter has presented ways to determine disturbance distribution matrices for a wide range of possible situations. Hence, it is claimed that the method is general in application. The second method is the estimation and de–convolution method. One disadvantage is that it requires that the system have more than n (state dimension) independent measurements. However, for many fault diagnosis problems, *e.g.* the gas turbine system, there are usually a large number of measurements available and the dynamics of the system can be approximated by a relatively low order model. The method can be used for a number of fault diagnosis problems and, as real or simulated system input and output data are used, the results can be affected by the system inputs; different inputs may give arise different distribution matrices. This is a disadvantage of this estimation method.

It can be seen that the two methods have compromising properties. One can choose which method is more suitable for a particular problem. In this chapter, an example is used to illustrate the application of the method to several power systems. They are very complex processes and the non–linearities and modelling errors are inevitable. This presents a big challenge for achieving reliable FDI using model–based approaches. Excellent results have been obtained and these indicate the effectiveness of the method for detecting soft (small) faults.

This chapter has focused on the robustness problem of detecting faults but extensions to this study allow us to consider robust isolation of faults, based upon the same de–coupling principles. This has been the subject of a further study by the authors.

5.9 Summary

In this chapter, several simulated and real examples were presented in order to test the FDI techniques presented in Chapter 4.

Complete design procedures for the detection, isolation and identification of faults concerning actuators, components, input and output sensors of industrial processes have been described.

The fault diagnosis was performed using banks of dynamic observers and Unknown Input Observers or, when the measurement noises are not negligible, banks of Kalman filters and Unknown Input Kalman filters.

Single step and ramp faults on the actuators, components, input and output sensors and multiple faults on the output sensors were considered on the real and simulated processes.

The FDI methods exploited in the chapter do not require any physical knowledge of the processes under observation since the input–output links were obtained by means of identification methods.

Under this assumption, identification techniques recalled in Chapter 3 were applied in order to obtain suitable models of the processes under investigation.

The procedures were applied to different models of industrial gas turbines.

The results obtained by this approach indicate that the minimal detectable faults on the various sensors are of interest for the industrial diagnostic applications.

CHAPTER 6
CONCLUDING REMARKS

This monograph seeks to provide a deep view of the system identification problem for fault diagnosis with special regard to industrial applications. Methods are developed for designing efficient algorithms for model–based fault detection, isolation and identification. The main focus has been placed on the identification of a reliable model of the system under investigation, as it has been recognised that model–based fault detection performance, which also include false alarm rejection, is strictly related to the "quality" of the model and measurements exploited for fault diagnosis.

This achievement have been pursued by means of a number of intermediate stages discussed in the book, namely:

1. Analysis of existing strategies for model–based residual generation, such as unknown input observers, dynamic observers and Kalman filters.
2. Development of new theory and techniques for identifying a model of the monitored system from input-output data, even in presence of measurement noise.
3. Introduction of a new method for generating robust residuals using de–coupling techniques.
4. Application of the methods and technique to both real systems and accurate model of industrial processes.

It is important to note that, the results discussed are of a general nature and are applicable, not only to particular systems treated specifically in this book, but to a wide class of linear and non–linear dynamic systems.

In the following, the main topics and contributions presented in the monograph are summarised chapter by chapter.

Chapter 1 has presented an introduction to the fault diagnosis problem and outlined the structure of the book. Briefly, the international nomenclature concerning the FDI theory was recalled. Moreover, the chapter outlines developments in the field of fault detection and diagnosis during the decade 1991–2001. Therefore, by going through the relevant literature, the chapter recalled main FDI applications in order to understand the goals of the contributions and to compare the different approaches.

Chapter 2 has presented the basic principles and general framework for model–based FDI. The residual generation was identified as the essence of this framework and some basic definition concerning residual properties were given. This chapter has provided comments upon some commonly used residual generation approaches. Examples of their applicability have been discussed and suggestions for the selection of methods have also been given. The chapter concluded with a discussion on integrating different diagnostic methods for FDI in complex dynamic systems using neural networks and fuzzy models.

Chapter 3 has investigated the problem of identifying an accurate model of the monitored system in order to apply model-based FDI techniques. In the chapter, different procedures were presented for the identification of both linear and non–linear dynamic system from data affected by noise. Linear, piecewise affine and fuzzy models were also exploited.

Chapter 4 has given a development of unknown input observers for residual generation. This chapter presented the full–order Unknown Input Observer structure, its existence conditions and design procedure. Using its structure, the residuals can be also made to have directional properties. The degrees of design freedom can be exploited to satisfy disturbance decoupling conditions. Moreover, the design of classical dynamic observers in deterministic environment, Kalman filters and Kalman filters with unknown inputs, when measurements are affected by noises, was shown and applied for FDI goals. Actuator, component, input and output fault detection, isolation and identification schemes were finally discussed.

Chapter 5 has presented several simulated and real examples in order to test the FDI techniques developed in Chapter 4. Complete design procedures for the detection, isolation and identification of faults concerning actuators, components, input and output sensors of industrial processes were described. The fault diagnosis was performed by using unknown input and dynamic observers or, when the measurement noise signals are not negligible, Kalman filters are used. Step and ramp faults on the actuators, components, input and output sensors and multiple faults on the output sensors were considered on the real and simulated processes. The FDI methods exploited in the chapter do not require any physical knowledge of the processes under observation since the input–output links are obtained by means of identification methods.

It is believed that the problems addressed in this book which have not been fully studied before are important in process diagnosis, and we hope readers find the methods of approaching the problems both interesting and practical. We did out best to present both methodologies and practical applications in a homogeneous manner. In particular, industrial case studies have been proposed to illustrate how these methods can be successfully applied.

We also hope that the book will provide stimulus to researchers, since the field is still open to further development. Particularly, Section 6.1 outlines possible areas of deeper investigation.

6.1 Suggestions for Future Work

Model–based FDI has been studied for over 20 years, however it is still an open research domain and many problems are waiting to be solved. The material presented in this book has inevitably had to end before all the interesting topics for future FDI research could be fully explored. In the following sections the authors describe some important topics for further research.

6.1.1 Frequency Domain Residual Generation

As described through this monograph, there are many methods, such as Unknown Input Observers, eigenstructure assignment and robust parity relations, for eliminating or minimising disturbance and modelling error effects on residuals and hence for achieving robustness in FDI. However, these techniques were developed for ideal systems or with a special uncertainty structure, and then efforts have been made to include non–ideal or more general uncertainty.

In contrast, frequency domain design methods are designed to possess robustness properties. In particular, H_∞ optimisation has been developed from the very beginning with the understanding that no design goal of a system can be perfectly achieved without being compromised by an optimisation in the presence of uncertainty, hence this technique is very suitable for tackling uncertainty issues.

As described through this monograph, there are many methods, such as Unknown Input Observers, eigenstructure assignment and robust parity relations, for eliminating or minimising disturbance and modelling error effects on residuals and hence for achieving robustness in FDI. However, these techniques were developed for ideal systems or with a special uncertainty structure, and then efforts have been made to include non–ideal or more general uncertainty.

In contrast, frequency domain design methods are designed to possess robustness properties. In particular, H_∞ optimisation has been developed from the very beginning with the understanding that no design goal of a system can be perfectly achieved without being compromised by an optimisation in the presence of uncertainty, hence this technique is very suitable for tackling uncertainty issues.

Patton *et al.* [Patton *et al.*, 1986] first discussed the possibility of using frequency domain information to design FDI algorithms. The design of a residual generator in the frequency domain was firstly based on a frequency domain optimal observer and then by using the factorisation of the transfer function matrix of the monitored system. These methods were developed and later extended by Ding and Frank [Ding and Frank, 1990]. Some important modifications in robust FDI design were made by Gertler [Gertler, 1998] by using the factorisation–based H_∞ optimisation technique. The more elegant and advanced H_∞ optimisation methods are based mainly on the use of the Algebraic Riccati Equations (ARE). In particular, the robust FDI estimation problem was solved by using Riccati equation approach through the use of H_∞ and μ robust estimator synthesis methods [Chen and Patton, 1999]. These approaches were further extended to time–variant and non–linear systems.

The majority of studies discussed so far involve the use of a slightly modified H_∞ filter for residual generation. That is to say the design objective is to minimise the effect of disturbances and modelling errors on the estimation error and subsequently on the residual. The residual has to be remain sensitive to faults whilst the effect of disturbance has to be minimised. Hence, the essential idea is to reach an acceptable compromise between disturbance robustness and fault sensitivity. The final goal is to find an observer design which provides the maximum ratio between fault sensitivity and disturbance sensitivity

$$J = sup_{Q(s)} \frac{\| Q(s)G_f(s) \|_\infty}{\| Q(s)G_d(s) \|_\infty} \tag{6.1}$$

over a frequency range, $Q(s)G_f(s)$ being the transfer matrix between the residual and fault, whilst $Q(s)G_d(s)$ being the transfer matrix between the residual and disturbance.

Solutions for this optimisation problem were given and revised, in order to obtain robust FDI technique [Chen and Patton, 1999]. Unfortunately, it was shown that $\| Q(s)G_f(s) \|_\infty$ may be smaller than $\| Q(s)G_d(s) \|_\infty$ in certain frequency range even their ratio 6.1 has been maximised.

It should be pointed out that the transfer function matrix $G_d(s)$ can only be defined for disturbances, hence the technique presented can only deal with robustness against disturbance. The robust problem with respect to modelling errors has still not been solved. The only solution suggested is to calculate the residual bound and set and adaptive threshold.

Few progresses were made solving the robust FDI problem against modelling errors when μ synthesis with H_∞ optimisation is incorporated. Robust FDI design based on H_∞ optimisation and μ synthesis is still in its early development, even if some research is still needed. This could be a direction for future research which has great potential.

In connection with frequency domain, FDI techniques can exploit a different identification approach from the one presented in Chapter 3. For example, an identification method based on the frequency domain approach for Errors–In–Variable models and its application to the dynamic Frisch scheme estimation technique in still in development [Beghelli *et al.*, 1994a, Beghelli *et al.*, 1997]. Such a procedure can provide an accurate estimation of the transfer matrices $Q(s)$, $G_f(s)$ and $G_d(s)$ from input-output measurements affected by white, mutually uncorrelated and correlated noises.

This general method, using the frequency domain approach, facilitates a unique determination of both the characteristics of the noise affecting the data ($G_d(s)$) as well as the transfer matrices ($Q(s)$ and $G_f(s)$) of the process under investigation. A comparison between time-domain and frequency-domain approaches can be found in [Beghelli *et al.*, 1994b].

6.1.2 Adaptive Residual Generators

The system dynamics and parameters may vary or may be perturbed during the system operation. A fault diagnosis system designed for a system model corresponding to nominal system operation may not perform well when applied to the system with perturbed conditions.

To overcome this problem, instead of using complex non–linear models, a residual generator scheme using adaptive observers were proposed. The idea is to estimate and compensate system parameter variations. Figure 6.1 illustrates the basic principle of this approach. It can be applied to linear systems with parametric variations if stability and convergence conditions are satisfied.

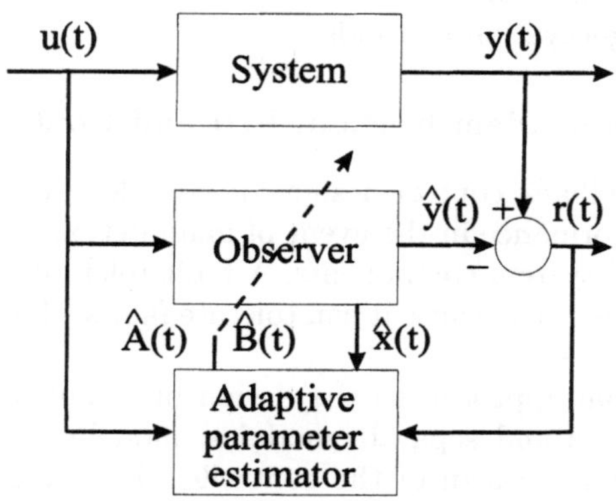

Fig. 6.1. Residual generator with adaptive observer.

Adaptive residual generation schemes for both linear and non–linear uncertain dynamic systems using adaptive observers were proposed in the literature [Patton *et al.*, 1989]. Unfortunately, the disadvantage of this approach is the complexity.

Chen and Patton [Chen and Patton, 1999] presented an alternative way to generate adaptive symptoms using a method to estimate the bias term in the residuals due to modelling errors, then compensate it adaptively. This technique decreases the effects of uncertainties on residuals. The approach to estimate such a bias term in residuals rather than computing modelling errors themselves avoids complicated estimation algorithms.

The state $\hat{x}(t)$, and parameters, $\hat{\boldsymbol{A}}(t)$, $\hat{\boldsymbol{B}}(t)$ simultaneous estimation algorithms presented in [Söderström and Stoica, 1987, Ljung, 1999] can be also used to generate adaptive residuals. With reference to Figure 6.1, observer parameters are linked by the relations

$$\left.\begin{array}{lcl} \hat{\boldsymbol{x}}(t+1) & = & \hat{\boldsymbol{A}}(t)\hat{\boldsymbol{x}}(t) + \hat{\boldsymbol{B}}(t)\boldsymbol{u}(t) \\ \hat{\boldsymbol{y}}(t) & = & C\hat{\boldsymbol{x}}(t). \end{array}\right\} \tag{6.2}$$

An adaptive residual generation algorithm normally involves both state and $\hat{x}(t)$ and parameter $\hat{\boldsymbol{A}}(t), \hat{\boldsymbol{B}}(t)$ estimation can be considered as a combination of observer and identification based FDI approaches. Hence, complementary advantages in both approaches can be gained.

For all adaptive methods, the main problem to be tackled is that fault effects may be compensated as well as modelling errors and parameter variations. This makes the detection for incipient faults almost impossible whilst for abrupt faults this can be acceptable. To overcome this problem, the effect of faults can be considered as a slow varying parameter which can be estimated along with parameters. Under the assumption that parameters and faults varying at different rates, two filters with different gains can be used. However, much research effort is still needed in the theory and application of adaptive residual generation methods.

6.1.3 Integration of Identification, FDI and Control

A conventional feedback control design for complex systems may result in unsatisfactory performance in the event of malfunction in input–output sensors, actuators and system components. A fault tolerant closed–loop control system is very attractive because it can tolerate faults whilst also maintaining desirable performance.

The conventional approach to the design of a fault-tolerant control includes different steps and separate modules: modelling or identification of the controlled system, design of the controller, FDI scheme and a method for reconfiguring the control system. Identification and design of the controller can be performed separately or using combined methods. Hence, the FDI and controller are linked through the reconfiguration module. The fundamental problem with such a system lies in the identification stage, in the independent design of the control and FDI modules. Significant interactions occurring among these modules can be neglected. There is therefore a need for a research study into the interactions between system identification, control design, the FDI stage [Diversi *et al.*, 2002] and the fault–tolerant control design strategy.

6.1.4 Fault Identification

Fault identification is the most important of all the fault diagnosis tasks. When a fault is estimated, detection and isolation can be easily achieved since the fault nature can improve the diagnosis process. However, the fault identification problem itself has not gained enough research attention.

Most fault diagnosis techniques, such as parameter identification, parity space and observer–based methods cannot be directly used to identify faults in sensors and actuators.

Very little research has been done to overcome the fault identification problem. The Kalman filter for statistical testing and fault identification was proposed in [Patton *et al.*, 1989]. However, the statistical testing methods can impose a high computational demand.

Recently, a fault identification scheme solving a system inversion problem was proposed [Simani *et al.*, 1998b, Chen and Patton, 1999, Simani *et al.*, 1999d, Simani and Patton, 2002b, Simani and Fantuzzi, 2002]. In the scheme depicted in Figure 6.2 fault identification is performed by estimating the non–linear relationship between residuals and fault magnitudes. This is possible because robust residuals should only contain fault information.

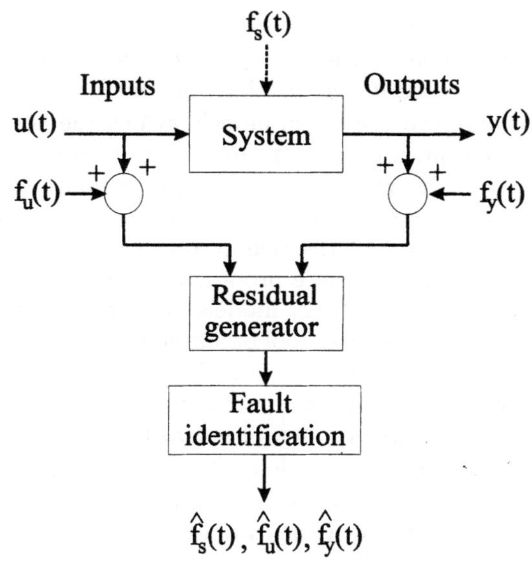

Fig. 6.2. Fault estimation scheme.

Such a non–linear function approximation and estimation can be performed by using neural networks or an inversion of the transfer matrix between residuals and faults.

6.1.5 Fault Diagnosis of Non–Linear Dynamic Systems

The central task in model-based fault detection is the residual generation. Most residual generation techniques are based on linear system models. For non–linear systems, the traditional approach is to linearise the model around the system operating point. However, for systems with high non–linearity and a wide dynamic operating range, the linearised approach fails to give satisfactory results.

One solution is to use a large number of linearised models corresponding to a range of operating points. This means that a large number of FDI schemes corresponding to each operating points is needed.

Hence, it is important to study residual generation techniques which tackle non–linear dynamic systems directly. There are some research studies on the residual generation of non–linear dynamic systems, for example using non–linear observers [Frank *et al.*, 2000, De Persis and Isidori, 2001, Chen and Patton, 1999]. There have been some attempts to use non–linear observers to solve non–linear system FDI problem [De Persis and Isidori, 2001, Chen and Patton, 1999], *e.g.* non–linear unknown input observers, including adaptive observers and sliding mode observers. If the class of nonlinearities can be restricted, observers for bilinear systems were also proposed [Chen and Patton, 1999].

On the other hand, the analytical models, which the non–linear observer approaches are based on, are not easy to obtain in practice. Sometimes, it is impossible to model the system using an explicit mathematical model. To overcome this problem, it is desirable to find a universal approximate model which can be used to represent the real system with an arbitrary degree of accuracy. Different approaches were proposed and they are currently under investigation: neural networks, fuzzy models and hybrid models.

Neural networks are a powerful tool of handling non–linear problems. One of the most important advantages of neural networks is their ability to implement non–linear transformations for functional approximation problems [Schilling *et al.*, 2001]. Therefore, neural networks can be used in a number of ways to tackle fault diagnosis problems for non–linear dynamic systems. In early publications, they were mainly exploited as fault classifier with steady state processes, whereas, recently, neural networks have been used as residual generators and for modelling non–linear dynamic systems for FDI purposes [Chen and Patton, 1999] [Simani *et al.*, 1998b, Simani *et al.*, 1999d, Simani, 2000a, Simani and Patton, 2002b, Simani and Fantuzzi, 2002].

Fuzzy models can be used both as residual classifier [Sneider and Frank, 1996] and as non–linear system parametric model [Ying, 1994]. In the second case, the main idea is to build an FDI scheme based on fuzzy observers. Estimated outputs and residuals are computed as fuzzy fusion of local observer output and residuals. The main problem of this approach concerns the stability of the global observer. A linear matrix inequality method was proposed by Patton [Chen and Patton, 1999] using

Lyapunov theorem, but this solution can be quite conservative, so more researches still open on the way.

Hybrid models can describe the behaviour of any non–linear dynamic process if they are described as a composition of several local affine models selected according to the process operating conditions [Hyb, 1998]. Instead of exploiting complicated non–linear models obtained by modelling techniques, it is possible to describe the plant by a collection of affine models. Such a compound system requires the identification of the local models from data. Several works [Rovatti *et al.*, 1998b, Simani *et al.*, 1999c, Simani *et al.*, 1999b, Fantuzzi *et al.*, 2002] addressed a method for the identification and the optimal selection of the local affine models from a sequence of noisy measurements acquired from the process. Application of these results to model–based fault diagnosis is another research area worth of mention.

REFERENCES

[IFI, 1983] (1983). *Reliable computing and fault tolerance*, meeting in Como, Italy. IFIP working group 10.4.

[RAM, 1988] (1988). Reliability, Availability and maintainability Dictionary. ASQC Quality press. Milwaukee.

[Mat, 1990] (1990). *MATLAB User's Guide*. MathWorks Inc. South Natick, MA, U.S.A.

[Hyb, 1998] (1998). Special issue on hybrid control systems. IEEE Trans. on Automatic Control, vol. 43, n. 4.

[Akaike, 1974] Akaike, H. (1974). A new look at the statistical model identification. *IEEE Trans. Automatic Control*, 19(6):716–723.

[Alexandru *et al.*, 2000] Alexandru, M., Combastel, C., and Gentil, S. (2000). Diagnostic decision using recurrent neural network. In *Proc. of the 4th IFAC Symposium on Fault Detection, Supervision and Safety for Technical Processes*, volume 1, pages 410–415, Budapest, Hungary.

[Appleby *et al.*, 1991] Appleby, B., Dowdle, J., and Vander Velde, W. (1991). Robust estimator design using μ synthesis. In *Proc. of the 30th Conf on Decision & Control*, pages 640–644, Brighton, UK.

[Ayoubi, 1995] Ayoubi, M. (1995). Neuro–Fuzzy Structure for Rule Generation and Application in the Fault Diagnosis of Technical Processes. In *Proc. of the American Control Conference, ACC'95*, pages 2757–2761, Washington, USA.

[Babu and Murty, 1994] Babu, G. and Murty, M. (1994). Clustering with evolution strategies. *Pattern Recognition*, 27(2):321–329.

[Babuška, 1998] Babuška, R. (1998). *Fuzzy Modeling for Control*. Kluwer Academic Publishers.

[Babuška, 2000] Babuška, R. (2000). *Fuzzy Modelling and Identification Toolbox*. Control Engineering Laboratory, Faculty of Information Technology and Systems, Delft University of Technology, Delft, The Netherlands, version 3.1 edition. (Available at http://lcewww.et.tudelft.nl/ ˜babuska).

[Babuška *et al.*, 1997] Babuška, R., Keizer, R., and Verhaegen, M. (1997). Identification of nonlinear dynamic systemsas a composition of local linear parametric or state space models. In *Proc. of SYSID'97*, Fukuoka, Japan.

[Babuška and Verbruggen, 1995] Babuška, R. and Verbruggen, H. B. (1995). Identification of composite linear models via fuzzy clustering. In *Proc. 3rd ECC'95*, pages 1207–1212, Rome, Italy.

[Backer, 1995] Backer, E. (1995). *Computer–assisted reasoning in cluster analysis*. Prentice Hall, New York.

[Bakiotis *et al.*, 1979] Bakiotis, C., Raymond, J., and Rault, A. (1979). Parameter identification and discriminant analysis for jet engine mechanical state diagnosis. In *IEEE Conference on Decision and Control*, Fort Lauderdale.

[Banks and Kathur, 1989] Banks, S. and Kathur, S. (1989). Structure and control of piecewise linear system. *Int. J. of Control*, 50:346–358.

[Basseville, 1988] Basseville, M. (1988). Detecting changes in signals and systems: A survey. *Automatica*, 24(3):309–326.

[Basseville, 1997] Basseville, M. (1997). Information criteria for residual generation and fault detection and isolation. *Automatica*, (33):783–803.

[Basseville and Benveniste, 1986] Basseville, M. and Benveniste, A. (1986). Detection of abrupt changes in signals and and dynamical systems. In *Lecture Notes in Control and Information Sciences*, volume 77, London. Springer–Verlag.

[Basseville and Nikiforov, 1993] Basseville, M. and Nikiforov, I. V. (1993). *Detection of Abrupt Changes: Theory and Application.* Prentice–Hall Inc.

[Beard, 1971] Beard, R. V. (1971). Failure accomodation in linear systems through self-reorganisation. Technical Report MVT-71-1, Man Vehicle Lab., Cambridge, Mass.

[Beghelli *et al.*, 1994a] Beghelli, S., Castaldi, P., Guidorzi, R. P., and Soverini, U. (1994a). A comparison between different model selection criteria in Frisch scheme identification. *Systems Science Journal*, 20(1):77–84. Wroclaw, Polonia.

[Beghelli *et al.*, 1994b] Beghelli, S., Castaldi, P., and Soverini, U. (1994b). Dynamic Frisch scheme identification: time and frequency domain approaches. In *IFAC'94*. 10th IFAC Symposium on System Identification.

[Beghelli *et al.*, 1990] Beghelli, S., Guidorzi, R. P., and Soverini, U. (1990). The Frisch scheme in dynamic system identification. *Automatica*, 26(1):171–176.

[Beghelli *et al.*, 1997] Beghelli, S., Guidorzi, R. P., and Soverini, U. (1997). A frequencial approach to the dynamic Frisch scheme identification. In *ECC'97*, Brussels, Belgium. 4th European Control Conference.

[Beghelli and Soverini, 1992] Beghelli, S. and Soverini, U. (1992). Identification of linear relations from noisy data: Geometrical aspects. *System and Control Letters*, 18(5):339–346.

[Bemporad and Morari, 1999] Bemporad, A. and Morari, M. (1999). Control od systems integrating logic, dynamics, and constraints. *Automatica*, 35(3):407–428.

[Benvenuti *et al.*, 1993] Benvenuti, E., Bettocchi, R., Cantore, G., and Negri di Montenegro, G. Spina, P. R. (1993). Gas Turbine Cycle Modelling Oriented to Component Performance Evaluation from Limited Design or Test Data. In *Proceeding of 7th ASME COGEN–TURBO*, pages 327–337, Bournemouth, UK.

[Bettocchi *et al.*, 1996] Bettocchi, R., Spina, P. R., and Fabbri, F. (1996). Dynamic Modelling of Single–Shaft Gas Turbine. In *ASME Paper 96–GT–332*, pages 1–9.

[Bezdek, 1980] Bezdek, J. (1980). A convergence theorem for the fuzzy isodata clustering algorighms. *IEEE Trans. Pattern Anal. Machine Intell.*, PAMI–2(1):1–8.

[Bezdek, 1981] Bezdek, J. (1981). *Pattern recognition with fuzzy objective function.* Plenum Press, New York.

[Bezdek *et al.*, 1987] Bezdek, J., Hathaway, R., Howard, R., Wilson, C., and Windham, M. (1987). Local convergence analysis of a grouped variable version of coordinate descendent. *Journal of Optimization Theory and Applications*, 54(3):471–477.

[Bezdek and Pal, 1992] Bezdek, J. and Pal, S. (1992). *Fuzzy models for pattern recognition.* IEEE Press, New York.

[Billings and Voon, 1983a] Billings, S. and Voon, W. (1983a). Structure detection and model validity test in the identification of nonlinear systems. *IEE Proceedings*, 130D(4):193–199.

[Billings and Voon, 1983b] Billings, S. and Voon, W. (1983b). Structure detection and model validity tests in the identification of nonlinear systems. *IEE Proc.*, 130(4):193–200.

[Billings and Voon, 1986] Billings, S. A. and Voon, W. (1986). Correlation based model validity test for non–linear models. *International Journal of Control*, 44(1):235–244.

[Blotenberg, 1993] Blotenberg, W. (1993). A model for the dynamic simulation of a two–shaft industrial gas turbine with dry low NOx combustor. In *ASME*, number 93–GT–355, pages 1–11. ASME.

[Boyd *et al.*, 1994] Boyd, S., Ghaoui, L., Feron, E., and Balakrishnan, V. (1994). *Linear Matrix Inequalities in System and Control Theory*. SIAM, Philadelphia.

[Brown and Harris, 1994a] Brown, M. and Harris, C. (1994a). *Neurofuzzy adaptive modelling and control*. Prentice Hall.

[Brown and Harris, 1994b] Brown, M. and Harris, C. (1994b). *Neurofuzzy Adaptive Modelling and Control*. Prentice Hall.

[Calado *et al.*, 2001] Calado, J., Korbicz, J., Patan, K., Patton, R., and Sá da Costa, J. (2001). Soft computing approaches to fault diagnosis for dynamic systems. *European Journal of Control*, 7(2–3):248–286.

[Carpenter and Grossberg, 1987] Carpenter, G. and Grossberg, S. (1987). A massively parallel architecture for a self-organizing neural pattern recognition machine. *Computer Vision, Graphics and Image Processing*, 37:54–115.

[Castaldi and Soverini, 1998] Castaldi, P. and Soverini, U. (1998). Identification of errors-in-variables models and optimal output reconstruction. In Beghi, A., Finesso, L., and Picci, G., editors, *Proc. of the MNST'98 Symposium*, pages 727–730, Padova, Italy. Il Poligrafo.

[Chang and Hsu, 1995] Chang, S. K. and Hsu, P. L. (1995). A novel design for the unknown input fault detection observer. *Control Theory and Advanced Technology*, 10(4).

[Chen and Patton, 1999] Chen, J. and Patton, R. J. (1999). *Robust Model–Based Fault Diagnosis for Dynamic Systems*. Kluwer Academic.

[Chen and Patton, 2000] Chen, J. and Patton, R. J. (2000). Standard H_∞ filter formulation of robust fault detection. In Edelmayer, A. M., editor, *SAFEPROCESS2000, 4th IFAC Symposium on Fault Detection, Supervision and Safety for Technical Processes*, volume 1, pages 256–261, Budapest, Hungary. IFAC 2000, IFAC 2000.

[Chen and Patton, 2001] Chen, J. and Patton, R. J. (2001). Fault–tolerant control systems design using the linear matrix inequality method. In *European Control Conference, ECC'01*, pages 1993–1998, Porto, Portugal.

[Chen *et al.*, 1996a] Chen, J., Patton, R. J., and Liu, G. P. (1996a). Optimal residual design for fault–diagnosis using multiobjective optimisation and genetic algorithms. *Int. J. Sys. Sci.*, 27(6):567–576.

[Chen *et al.*, 1993] Chen, J., Patton, R. J., and Zhang, H. Y. (1993). A multi–criteria optimization approach to the design of robust fault detection algorithm. In *Proc. of Int. Conf. on Fault Diagnosis: TOOLDIAG'93*, Toulouse, France.

[Chen *et al.*, 1996b] Chen, J., Patton, R. J., and Zhang, H. Y. (1996b). Design of unknown input observer and robust fault detection filters. *Int. J. Control*, 63(1):85–105.

[Chen *et al.*, 1990a] Chen, S., Billings, A. S., Cowan, C. F. N., and Grant, P. M. (1990a). Practical identification of NARMAX models using radial basis function. *Int. J. Control*, 52:1327–1350.

[Chen and Billings, 1989] Chen, S. and Billings, S. (1989). Representation of non–linear systems: the NARMAX model. *Int. J. Control*, 49:1013–1032.

[Chen *et al.*, 1990b] Chen, S., Billings, S., Cowan, C. F., and Grant, P. (1990b). Practical identification of NARMAX model using radial basis functions. *Int. J. Control*, 52:1327–1350.

[Chen *et al.*, 1991] Chen, S., Cowan, C., and Grant, P. (1991). Orthogonal least squares learning algorithm for radial basis function networks. *IEEE Trans. Neural Networks*, 2(2):302–309.

[Chen *et al.*, 1997] Chen, Z., Patton, R. J., and Chen, J. (1997). Robust fault-tolerant system synthesis via LMI. In *Proc. of IFAC Symposium on Fault Detection, Supervision and Safety for Technical Processes: SAFEPROCESS'97*, volume 1, pages 347–352, The University of Hull, UK.

[Chiang *et al.*, 2001] Chiang, L. H., Russel, E. L., and Braatz, R. D. (2001). *Fault Detection Diagnosis in Industrial Systems.* Advanced Textbooks in Control and Signal Processing. Springer–Verlag London Limited, London, Great Britain.

[Chow and Willsky, 1980] Chow, E. Y. and Willsky, A. S. (1980). Issue in the development of a general algorithm for reliable failure detection. In *Proc. of the 19th Conf. on Decision & Control*, Albuquerque, NM.

[Chow and Willsky, 1984] Chow, E. Y. and Willsky, A. S. (1984). Analytical redundancy and the design of robust detection systems. *IEEE Trans. Automatic Control*, 29(7):603–614.

[Chowdhury and Aravena, 1998] Chowdhury, F. N. and Aravena, J. L. (1998). A modular methodology for fast fault detection and classification in power systems. *IEEE Trans. of Control System Technology*, 6(5).

[Chung and Speyer, 1998] Chung, W. H. and Speyer, J. L. (1998). A game theoretic fault detection filter. *IEEE Trean. on Automatic Control*, 43(2):143–161.

[Clark, 1978] Clark, R. N. (1978). Instrument fault detection. *IEEE Trans. Aero. & Electronic Systems*, 14(3).

[Clark, 1989] Clark, R. N. (1989). *Fault Diagnosis in Dynamic Systems: Theory and Application*, chapter 2, pages 21–45. Prentice Hall.

[Cottle, 1982] Cottle, R. (1982). Minimal triangulation of the 4-cube. *Discrete Mathematics*, 40:25–29.

[Daly *et al.*, 1979] Daly, K. C., Gai, E., and Harrison, J. V. (1979). Generalized likelihood test for FDI in redundancy sensor configurations. *J. of Guidance, Control & Dynamics*, 2(1):9–17.

[Davis, 1991] Davis, L. (1991). *Handbook of Genetic Algorithms.* Van Nostrand Reinhold, New York.

[de Boor, 1978] de Boor, C. (1978). *A practical guide to splines.* Springer–Verlag, New York.

[De Persis and Isidori, 2001] De Persis, C. and Isidori, A. (2001). A geometric approach to non–linear fault detection and isolation. *IEEE Transactions on Automatic Control*, 45(6):853–865.

[Delmaire *et al.*, 1999] Delmaire, G., Cassar, P., Staroswiecki, M., and Christophe, C. (1999). Comparison of multivariable identification and parity space techniques for FDI purpose in M.I.M.O. systems. In *ECC'99*, Karlsruhe, Germany.

[Demuth and BealeDemuth, 1997] Demuth, H. and BealeDemuth, M. (1997). *Neural Network Toolbox For Use with MATLAB.* The MathWorks Inc., Version 3.0 edition. South Natick, MA, U.S.A.

[DeSarbo, 1982] DeSarbo, W. (1982). Gennclus: New models for general nonhierarchical clustering analysis. *Psychometrika*, 47(4):449–476.

[Dexter and Benouarets, 1997] Dexter, A. L. and Benouarets, M. (1997). Model–based fault diagnosis using fuzzy matching. *IEEE Trans. on Sys. Man. and Cyber. Part A: Sys. & Humans*, 27(5):673–682.

[Dietz *et al.*, 1989] Dietz, W. E., Kiech, E. L., and Ali, M. (1989). Jet and rocket engine fault diagnosis in real time. *J. of Neural Network Computing*, 1:5–18.

[Ding *et al.*, 1999] Ding, S. X., Jeinsch, T., Ding, E. L., Zhou, D., and Wang, G. (1999). Application of Observer–Based FDI Schemes to the Three Tank System. In *European Control Conference, ECC'99*, Karlsruhe, Germany.

[Ding *et al.*, 2000] Ding, S. X., Jeinsch, T., Frank, P. M., and Dind, E. L. (2000). A unified approach to the optimisation of fault detection systems. *Int. J. of Adaptive Control and Signal Processing*, 14(7):725–745.

[Ding and Frank, 1990] Ding, X. and Frank, P. M. (1990). Fault detection via factorization approach. *Syst. Contr. Lett.*, 14(5):431–436.

[Ding and Frank, 1991] Ding, X. and Frank, P. M. (1991). Frequency domain approach and threshold selector for robust model–based fault detection and isolation. In *Preprint of IFAC/IMACS Symposium SAFEPROCESS'91*, volume 1, pages 307–312. Baden–Baden.

[Diversi and Guidorzi, 1998] Diversi, R. and Guidorzi, R. P. (1998). Filtering–oriented identification of multivariable errors-in-variables models. In Beghi, A., Finesso, L., and Picci, G., editors, *Proc. of the MNST'98 Symposium*, pages 775–778, Padova, Italy. Il Poligrafo.

[Diversi *et al.*, 2002] Diversi, R., Simani, S., and Soverini, U. (2002). Robust residual generation for dynamic processes using de–coupling technique. In *CCA'02. Proc. of the Conference on Control Applications*, Glasgow, Scotland. IEEE Control Systems Society.

[Drag and Patton, 2001] Drag, G. R. and Patton, R. J. (2001). Robust fault detection using Luenberger–type unknown input observers: a parametric approach. *Int. J. Systems Science*, 32(4).

[Duan *et al.*, 2002] Duan, G., How, D., and Patton, R. (2002). Robust Fault Detection in Descriptor Systems via Generalised Unknown Input Observers. *Int. J. Systems Science.*

[Duda and Hart, 1973] Duda, R. and Hart, P. (1973). *Pattern classification and scene analysis.* John Wiley & Sons, New York.

[Dunn, 1974] Dunn, J. (1974). A fuzzy relative of the ISODATA process and its use in detecting compact well–separated clusters. *International Journal of Cybernetics and Systems*, 3(3):32–57.

[Edelmayer *et al.*, 1997] Edelmayer, A., Bokor, J., and Keviczky, L. (1997). A scaled L_2 optimisation approach for improving sensitivity of $H_i nfty$ detection filters for LTV systems. In Bányász, C., editor, *Preprints of the 2nd IFAC Symp. on Robust Control Design: RECOND97*, pages 543–548, Budapest, Hungary.

[Edwards and Spurgeon, 1994] Edwards, C. and Spurgeon, S. (1994). On the development of discontinuous observers. *International Journal of Control*, 59(1):1211–1229.

[Edwards *et al.*, 2000] Edwards, C., Spurgeon, S. K., and Patton, R. J. (2000). Sliding mode observers for fault detection and isolation. *Automatica*, 36(1):541–553.

[Emami-Naeini *et al.*, 1988] Emami-Naeini, A., Akhter, M., and Rock, M. (1988). Effect of model uncertainty on failure detection: the threshold selector. *IEEE Trans. on Automatic Control*, 33(2).

[Fantuzzi and Rovatti, 1996] Fantuzzi, C. and Rovatti, R. (1996). On the approximation capabilities of the homogeneous Takagi–Sugeno model. *Proceedings of the Fifth IEEE International Conference on Fuzzy Systems*, pages 1067–1072.

[Fantuzzi *et al.*, 1998] Fantuzzi, C., Rovatti, R., Simani, S., and Beghelli, S. (1998). Fuzzy modeling with noisy data. In *EUFIT'98*, volume 3, pages 1615–1619, Aachen, Germany. The 6th European Congress on Intelligent Techniques and Soft Computing.

[Fantuzzi and Simani, 2002] Fantuzzi, C. and Simani, S. (2002). Parametric identification in robust fault detection. In *IFAC'02*, Balcelona, Spain. 15th IFAC World Congress on Automatic Control. Invited paper, accepted.

[Fantuzzi *et al.*, 2001a] Fantuzzi, C., Simani, S., and Beghelli, S. (2001a). Parameter identification for eigenstructure assignment in robust fault detection. In *ECC'01*, pages 149–154, Porto, Portugal. European Control Conference 2001.

[Fantuzzi *et al.*, 2001b] Fantuzzi, C., Simani, S., and Beghelli, S. (2001b). Robust fault diagnosis of dynamic processes using parametric identification with eigenstructure assignment approach. In CSS, I., editor, *CDC'01*, pages 155–160, Orlando, Florida, U.S.A. 2001, 40th IEEE Conference on Decision and Control.

[Fantuzzi *et al.*, 2002] Fantuzzi, C., Simani, S., Beghelli, S., and Rovatti, R. (2002). Identification of piecewise affine models in noisy environment. *International Journal of Control*. accepted.

[Filbert and Metzger, 1982] Filbert, D. and Metzger, K. (1982). Quality test of systems by parameter estimation. In *9th IMEKO Congress*, Berlin.

[Frank, 1990] Frank, P. M. (1990). Fault diagnosis in dynamic systems using analytical and knowledge based redundancy: A survey of some new results. *Automatica*, 26(3):459–474.

[Frank, 1993] Frank, P. M. (1993). Advances in observer-based fault diagnosis. *Proc. TOOLDIAG'93 Conference*. CERT, Toulose (F).

[Frank, 1994a] Frank, P. M. (1994a). Application of fuzzy logic process supervision and fault diagnosis. In *SAFEPROCESS'94: Preprints of the IFAC Symposium on Fault Detection, Supervision and Safety for Technical Processes*, volume 2, pages 531–538, Espoo, Finland.

[Frank, 1994b] Frank, P. M. (1994b). Enhancement of robustness on observer-based fault detection. *International Journal of Control*, 59(4):955–983.

[Frank *et al.*, 2000] Frank, P. M., Ding, S. X., and Köpper-Seliger, B. (2000). Current Developments in the Theory of FDI. In *SAFEPROCESS'00: Preprints of the IFAC Symposium on Fault Detection, Supervision and Safety for Technical Processes*, volume 1, pages 16–27, Budapest, Hungary.

[Frank and Ding, 1997] Frank, P. M. and Ding, X. (1997). Survey of robust residual generation and evaluation methods in observer-based fault detection system. *Journal of Process Control*, 7(6):403–424.

[Friedmann, 1991] Friedmann, J. (1991). Multivariable adaptive regression Splines. *The Annal of Statistics*, pages 1–141.

[Frisch, 1934] Frisch, R. (1934). *Statistical Confluenece Analysis by Means of Complete Regression Systems*. University of Oslo, Economic Institute, publication n. 5 edition.

[Funahashi, 1989] Funahashi, K. (1989). On the approximate realization of continous mappings by neural networks. *Neural Networks*, 2:183–192.

[Füssel *et al.*, 1997] Füssel, D., Ballé, P., and Isermann, R. (1997). Closed loop fault diagnosis based on a non-linear process model and automatic fuzzy rule generation. In *Proc. of IFAC Symposium on Fault Detection, Supervision and Safety for Technical Process SAFEPROCESS'97*, The University of Hull, UK.

[Geiger, 1982] Geiger, G. (1982). Monitoring of an elecrical driven pump using continuous-time parameter estimation models. In Pergamon Press, editor, *6th IFAC Symposium on Identification and Parameter Estimation*, Washington.

[Gertler, 1988] Gertler, J. (1988). Survey of model-based failure detection and isolation in complex plants. *IEEE Control System Magazine*, pages 3–11.

[Gertler, 1991] Gertler, J. (1991). Generating directional residuals with dynamic parity equations. *Proc. IFAC/IMACS Symp. SAFEPROCESS'91*. Baden Baden (G).

[Gertler, 1995] Gertler, J. (1995). Diagnosing Parametric Faults: from Parameter Estimation to Parity Relations. In *ACC'95*, pages 1615–1620, Seatle, Washingtown.

[Gertler, 1998] Gertler, J. (1998). *Fault Detection and Diagnosis in Engineering Systems.* Marcel Dekker, New York.

[Gertler and Monajemy, 1993] Gertler, J. and Monajemy, R. (1993). Generating directional residuals with dynamic parity equations. *Proc. of the 12th IFAC World Congress*, 7:505–510. Sydney.

[Gertler and Singer, 1990] Gertler, J. and Singer, D. (1990). A new structural framework for parity equation–based failure detection and isolation. *Automatica*, 26(2):381–388.

[Goldberg, 1989] Goldberg, D. E. (1989). *Genetic Algorithms in Search, Optimisation and Machine Learning.* Addison Wesley Publishing Company.

[Guidorzi, 1975] Guidorzi, R. P. (1975). Canonical Structures in the Identification. *Automatica*, 11:361–374.

[Guidorzi, 1981] Guidorzi, R. P. (1981). Invariants and canonical forms for system structural and parametric identification. *Automatica*, (17):117–133.

[Guidorzi, 1996] Guidorzi, R. P. (1996). On the use of minimal parametrisations in multivariable ARMAX identification. In *IFAC'96*, pages 31–36, S. Francisco, USA. 13th IFAC World Congress.

[Guidorzi *et al.*, 1982] Guidorzi, R. P., Losito, P., and Muratori, T. (1982). The range error test in the structural identification of linear multivariable systems. *IEEE Transactions on Automatic Control*, 27(5):1044–1054.

[Gustafson and Kessel, 1979] Gustafson, D. E. and Kessel, W. C. (1979). Fuzzy clustering with a variance covariance matrix. In *Proc. IEEE CDC'79*, pages 161–166, San Diego, CA, USA.

[Hadamard, 1964] Hadamard, J. (1964). *La theorie des equations aus derivees partielles.* Editions Scientifiques, Pekin.

[Haiman, 1991] Haiman, M. (1991). A simple and relatively efficient triangulation of the n-cube. *Discrete Computational Geometry*, 6:287–289.

[Hathaway and Bezdek, 1983] Hathaway, R. and Bezdek, J. (1983). Switching regression models and fuzzy clustering. *IEEE Trans. of Fuzzy Systems*, 1(3).

[Hermans and Zarrop, 1996] Hermans, F. and Zarrop, M. (1996). Sliding mode observers for robust sensor monitoring. In *Proceedings 13th IFAC World Congress*, pages 211–216, San Francisco, USA.

[Himmelblau, 1978] Himmelblau, D. M. (1978). Fault Diagnosis in Chemical and Petrochemical Processes. *Elsevier*. Amsterdam.

[Himmelblau *et al.*, 1991] Himmelblau, D. M., Barker, R. W., and Suewatanakul, W. (1991). Fault classification with the aid of artificial neural networks. In *IFAC/IMACS Symposium SAFEPROCESS '91*, volume 2, pages 369–373, Baden Baden, Germany.

[Höfling and Pfeufer, 1994] Höfling, T. and Pfeufer, T. (1994). Detection of additive and multiplicative faults - Parity space vs. parameter estimation. In *Proc. IFAC SAFEPROCESS Symposium '94*, Espoo, Finland.

[Hohmann, 1977] Hohmann, H. (1977). *Automatic monitoring and failure diagnosis for machine tools.* Dissertation, T. H. Darmstadt, Germany. in German.

[Hoskins and Himmelblau, 1988] Hoskins, J. C. and Himmelblau, D. M. (1988). Artificial neural network models of knowledge representation in chemical engineering. *Comp. chem. Engng*, 12:881–890.

[Hou and Patton, 1996] Hou, M. and Patton, R. J. (1996). An LMI approach to H_-/H_∞ fault detection observers. In *CONTROL'96*, pages 305–310, University of Exter, UK. IEE.

[Hou and Patton, 1997] Hou, M. and Patton, R. J. (1997). An H_∞/H_- approach to the design of robust fault diagnosis observers based upon LMI optimisation. In *Proceedings of the 4th European Control Conference, ECC'97*, Brussels.

[Hunt *et al.*, 1992a] Hunt, J., Lee, M., and Price, C. (1992a). An introduction to qualitative model-based reasoning. In *Preprints IFAC/IFIP/IMACS Int. Sym. on Artificial Intelligence in Real-Time Control*, Delft (The Nederland).

[Hunt *et al.*, 1992b] Hunt, K., Sbarbaro, D., Zbikowki, R., and Gawthrop, P. (1992b). Neural networks for control system: a survey. *IEEE Trans. Neural Networks*, 28:1083–1112.

[Isermann, 1984] Isermann, R. (1984). Process fault detection based on modeling and estimation methods: A survey. *Automatica*, 20(4):387–404.

[Isermann, 1992] Isermann, R. (1992). Estimation of physical parameters for dynamic processes with application to an industrial robot. *Int. J. of Control*, 55:1287–1298.

[Isermann, 1993] Isermann, R. (1993). Fault diagnosis via parameter estimation and knowledge processing. *Automatica*, 29(4):815–835.

[Isermann, 1994a] Isermann, R. (1994a). Integration of fault detection and diagnosis methods. In *Proc. IFAC SAFEPROCESS Symposium '94*, Espoo, Finland.

[Isermann, 1994b] Isermann, R. (1994b). *Supervision and fault diagnosis.* VDI-Verlag, Düsseldorf. In German.

[Isermann, 1997] Isermann, R. (1997). Supervision, fault detection and fault diagnosis methods: an introduction. *Control Engineering Practice*, 5(5):639–652.

[Isermann, 1998] Isermann, R. (1998). On fuzzy logic applications for automatic control, supervision and fault diagnosis. *IEEE Trans. on Sys. Man. and Cyber. Part A: Sys. & Humans*, 28(2):221–235.

[Isermann and Ballé, 1997] Isermann, R. and Ballé, P. (1997). Trends in the application of model-based fault detection and diagnosis of technical processes. *Control Engineering Practice*, 5(5):709–719.

[Isermann and Freyermuth, 1992] Isermann, R. and Freyermuth, B., editors (1992). *Fault Detection, Supervision and Safety for Technical Processes*, volume 6 of *IFAC Symposia Series*. SAFEPROCESS'91, Pergamon Press.

[Isermann and Füssel, 1999] Isermann, R. and Füssel, D. (1999). Knowledge-based fault detection and diagnosis systems. Tutorial workshop in ECC 1999, Karlsruhe,Germany.

[Jackson, 1991] Jackson, J. E. (1991). *A user's guide to principal components.* Wiley-Interscience, N.J.

[Jager, 1995] Jager, R. (1995). *Fuzzy logic in control.* PhD thesis, Delft University of Technology, Delft, The Netherlands.

[Jain and Dubes, 1988] Jain, A. and Dubes, R. (1988). *Algorithms for clustering data.* Englewood Cliffs: Prentice Hall.

[Jang, 1993] Jang, J. (1993). ANFIS: Adaptive network based fuzzy inference system. *IEEE Transactions on Systems, Man., & Cybernetics*, 23(3):665–684.

[Jang, 1994] Jang, J. (1994). Structure determination in fuzzy modelling: a fuzzy CART approach. In *Proc. of IEEE International Conf. on Fuzzy Systems.*

[Jang and Sur, 1995] Jang, J. and Sur, R. (1995). Neuro–fuzzy modeling and control. *Proc. IEEE*, 83(3):378–405.

[Jazwinski, 1970] Jazwinski, A. H. (1970). *Stochastic processes and filtering theory.* Academic Press, New York.

[Johansen and Foss, 1993] Johansen, T. and Foss, B. (1993). Constructing NARMAX models using ARMAX models. *Int. J. Control*, 58(5):1125–1153.

[Johansen, 1996] Johansen, T. A. (1996). Robust identification of takagi-sugeno-kang fuzzy models using regularization. In *Fifth IEEE International Conference on Fuzzy Systems*, New Orleans, USA.

[Jones, 1973] Jones, H. L. (1973). *Failure detection in linear systems.* PhD thesis, Dept. of Aeronautics, M.I.T., Cambridge, Mass.

[Juditsky *et al.*, 1995] Juditsky, A., Hjalmarsson, H., Beneviste, A., Delyon, B. Ljung, L., Sjöberg, J., and Zhang, Q. (1995). Nonlinear black–box modelling in system identification: a mathematical foundation. *Automatica*, 31(12):1691–1724.

[Kalman, 1982a] Kalman, R. E. (1982a). Identification from real data. In Hazewinkel, M. and Rinnoy Kan, A. H. G., editors, *Current Developments in the Interface: Economics, Econometrics, Mathematics,*, pages 161–196. D. Reidel, Dordrecht, The Netherlands.

[Kalman, 1982b] Kalman, R. E. (1982b). System Identification from Noisy Data. In Bednarek, A. R. and Cesari, L., editors, *Dynamical System II*, pages 135–164. Academic Press, New York.

[Kalman, 1984] Kalman, R. E. (1984). Identification of noisy systems. In *50th Anniversary Symp. Steklov Institute of Mathematics, U.S.S.R. Academy of Sciences*, Moskva.

[Kalman, 1990] Kalman, R. E. ((Springer-Verlag, Berlin, 1990)). Nine lectures on identification. *Lecture Notes on Economics and Mathematical System.*

[Kavuri and Venkatasubramanian, 1994] Kavuri, S. N. and Venkatasubramanian, V. (1994). Neural network decomposition strategies for large-scale fault diagnosis. *Int. J. of Control*, 59(3):767–792.

[Klir and Yuan, 1995] Klir, G. J. and Yuan, B. (1995). *Fuzzy Sets and Fuzzy Logic: Theory and Applications.* Prentice Hall.

[Korbicz *et al.*, 1999] Korbicz, J., Patan, K., and Obuchowicz, A. (1999). Dynamic neural network for process modelling in fault detection and isolation systems. *Applied Mathematics and Computer Science*, 9(2):519–546. Technical University of Zielona Gora, Poland.

[Korbicz and Obuchowitcz, 1999] Korbicz, J. anf Patan, K. and Obuchowitcz, K. (1999). Dynamic neural networks for process modelling in fault detection and isolation systems. *Journal of Applied Mathematics and Computer Science*, 9(3):519–546.

[Koscielny *et al.*, 1999] Koscielny, J., Syfert, M., and Bartys, M. (1999). Fuzzy–logic fault diagnosis of industrial process actuators. *Journal of Applied Mathematics and Computer Science*, 9(3):637–652.

[Kosko, 1994] Kosko, B. (1994). Fuzzy systems as universal approximators. *IEEE Transactions on Computers*, 43:1329–1333.

[Kramer, 1987] Kramer, M. A. (1987). Malfunction diagnosis using quantitative models with non-boolean reasoning in expert systems. *AIChE*, (33):130–140.

[Krishnapuram and Freg, 1992] Krishnapuram, R. and Freg, C. (1992). Fitting an unknown number of lines and planes to image data through compatible cluster merging. *Pattern Recognition*, 25(4):385–400.

[Lee *et al.*, 1994] Lee, Y., Hwang, C., and Shih, Y. (1994). A combined approach to fuzzy model identification. *IEEE Trans. Sys. Man, Cybern.*, 24(5):736–744.

[Leonard and Kramer, 1991] Leonard, J. A. and Kramer, M. A. (1991). Radial basis function for calssifying process faults. *IEEE Control System Magazine*, 11(3):31–38.

[Leontaritis and Billings, 1985b] Leontaritis, I. and Billings, S. A. (1985b). Input-output parametric models for non–linear systems part II: stochastic non-linear systems. *Int. J. Control*, 41(2):329–344.

[Leontaritis and Billings, 1985a] Leontaritis, I. and Billings, S. A. (1985a). Input-output parametric models for non-linear systems part I: deterministic non-linear systems. *Int. J. Control*, 41(2):303–328.

[Leshno *et al.*, 1993] Leshno, M., Lin, V. Y., Pinkus, A., and Shocken, S. (1993). Multilayer feedforward networks with a non polynomial activation function can approximate any function. *Neural Networks*, 6:861–867.

[Liu and Patton, 1998] Liu, G. P. and Patton, R. J. (1998). *Eigenstructure Assignment for Control System Design.* John Wiley & Sons, England.

[Ljung, 1999] Ljung, L. (1999). *System Identification: Theory for the User.* Prentice Hall, Englewood Cliffs, N.J., second edition.

[Lou *et al.*, 1986] Lou, X., Willsky, A., and Verghese, G. (1986). Optimal robust redudancy relations for failure detection in uncertainty systems. *Automatica*, 22(3):333–344.

[Luenberger, 1971] Luenberger, D. G. (1971). An introduction to observers. *IEEE Transactions on Automatic Control*, AC-16(6):596–602.

[Luenberger, 1979] Luenberger, D. G. (1979). *Introduction to Dynamic System: Theory, Models and Application.* John Wiley and Son, New York.

[Mamdani, 1976] Mamdani, E. (1976). Advances in the linguistic synthesis of fuzzy controllers. *Int. J. Man–Machine Studies*, 8:669–678.

[Mamdani and Assilian, 1995] Mamdani, E. and Assilian, S. (1995). An experiment in linguistic synthesis with fuzzy logic controller. *Int. J. Man–Machine Studies*, 7(1):1–13.

[Mangoubi *et al.*, 1992] Mangoubi, R., Appleby, B. D., and Farrell, J. R. (1992). Robust estimation in fault detection. In *Proc. of the 31st Conf. on Decision & Control*, pages 2317–2322, Tucson, AZ, USA.

[Mara, 1976] Mara, P. (1976). Triangulation for the cube. *Journal of Combinatorial Theory (A)*, 20:170–177.

[Marcu and Mirea, 1997] Marcu, T. and Mirea, L. (1997). Robust detection and isolation of process faults using neural networks. *IEEE Control System Magazine*, pages 72–79.

[Marcu *et al.*, 1999] Marcu, T., Mirea, L., and Frank, P. (1999). Development of dynamic neural networks with application to observer–based fault detection and isolation. *Journal of Applied Mathematics and Computer Science*, 9(3):547–570.

[Massoumnia *et al.*, 1989] Massoumnia, M., Verghese, G. C., and Willsky, A. S. (1989). Failure detection and identification. *IEEE Trans. Automat. Contr.*, 34:316–321.

[Massoumnia, 1986] Massoumnia, M. A. (1986). *A geometric appoach to failure detection and identification in linear systems.* PhD thesis, Massachusetts Institute of Technology, Massachusetts, USA.

[MathWorks, 1998] MathWorks (1998). *Neural Network Toolbox: User's Guide.* MathWorks Inc. South Natick, MA, U.S.A.

[McDuff and Simpson, 1990] McDuff, R. J. and Simpson, P. K. (1990). An adaptive reasonance diagnostic system. *J. of Neural Network Computing*, (2):19–29.

[McGraw and Harbisson-Briggs, 1989] McGraw, K. and Harbisson-Briggs, K. (1989). *Knowledge Acquisition: Principles and Guidelines.* Englewood Cliffs: Prentice Hall.

[Meneganti *et al.*, 1998] Meneganti, M., Saviello, F., and Tagliaferri, R. (1998). Fuzzy neural networks for classification and detection of anomalies. *IEEE Trans. on Neural Networks*, 9(5):848–861.

[Morozov, 1984] Morozov, V. (1984). *Methods for Solving Incorrectly Posed Problems.* Springer, Berlin.

[Murray-Smith and Johansen, 1997] Murray-Smith, R. and Johansen, T. A. (1997). *Multiple model approaches to nonlinear modelling and control.* Taylor & Frencis, London, UK.

[Napolitano *et al.*, 1998] Napolitano, M. R., Widon, D. A., Casanova, J. L., Innocenti, M., and Silvestri, G. (1998). Kalman filters and neural–networks schemes for sensor validation in flight control system. *IEEE Trans. on Control System Technology*, 6(5):596–611.

[Nauck and Kruse, 1998] Nauck, D. and Kruse, R. (1998). Nefclass – a soft computing tool to build readable fuzzy classifiers. *BT Technol. Journal*, 16(3).

[Nelles, 2001] Nelles, O. (2001). *Nonlinear System Identification*. Springer–Verlag Berlin Heidelberg, Germany.

[Nelles and Isermann, 1996] Nelles, O. and Isermann, R. (1996). Basis function networks for interpolation of local linear models. In *Proc. of the 35th IEEE Conference on Decision and Control*, volume 4, pages 470–475, Kobe, Japan.

[Niemann and Stoustrup, 1996] Niemann, H. and Stoustrup, J. (1996). Filter design for failure detection and isolation in the presence of modelling errors and disturbances. In *Proc. of the 35th IEEE Conf. on Decision and Contr.*, pages 1155–1160, Kobe, Japan.

[Norton, 1986] Norton, J. (1986). *An Introduction to Identification*. Academic Press, London.

[Palade *et al.*, 2002] Palade, V., Patton, R. J., Uppal, F. J., Quevedo, J., and Daley, S. (2002). Fault diagnosis of an industrial gas turbine using neuro–fuzzy methods. In *IFAC'02*, Balcelona, Spain. 15th IFAC World Congress on Automatic Control.

[Patton and Chen, 1993] Patton, R. and Chen, J. (1993). Optimal selection of unknown input distribution matrix in the design of robust observers for fault diagnosis. *Automatica*, 29(4):837–841.

[Patton and Chen, 1994a] Patton, R. and Chen, J. (1994a). A review of parity space approaches to fault diagnosis for aerospace systems. *AIAA Journal of Guidance, Control & Dynamics*, 17(2):278–285.

[Patton *et al.*, 1992] Patton, R., Chen, J., and Zhang, H. (1992). Modelling methods for improving robustness in fault diagnosis of jet engine system. In *31*-st *IEEE Conference on Decision and Control*, pages 2330–2335.

[Patton *et al.*, 1999a] Patton, R., Lopez-Toribio, C., and Uppal, F. (1999a). Artificial Intelligence Approaches to fault diagnosis for dynamic systems. *Journal of Applied Mathematics and Computer Science*, 9(3):471–518.

[Patton *et al.*, 1986] Patton, R., Willcox, S., and Winter, J. (1986). A parameter insensitive technique for aircraft sensor fault diagnosis using eigenstructure assignment and analytical redundancy. In *Proc. of the AIAA Conference on Guidance, Navigation & Control*, number 86–2029–CP, Williamsburg, VA.

[Patton, 1999] Patton, R. J. (1999). Preface to the Papers from the 3rd IFAC Symposium SAFEPROCESS'97. *Control Engineering Practice*, 7(1):201–202.

[Patton and Chen, 1991a] Patton, R. J. and Chen, J. (1991a). A review of parity space approaches to fault diagnosis. In *IFAC Symposium SAFEPROCESS '91*, Baden-Baden.

[Patton and Chen, 1991b] Patton, R. J. and Chen, J. (1991b). Robust fault detection using eigenstructure assignment: a tutorial consideration and some new results. *30*–th *IEEE Conference on Decision and Control*, pages 2242–2247.

[Patton and Chen, 1994b] Patton, R. J. and Chen, J. (1994b). A review of parity space approaches to fault diagnosis for aerospace systems. *AIAA Journal of Guidance, Control & Dynamics*, 17(2):278–285.

[Patton and Chen, 1994c] Patton, R. J. and Chen, J. (1994c). A review of parity space approaches to fault diagnosis for aerospace systems. *AIAA J. of Guidance, Contr. & Dynamics*, 17(2):278–285.

[Patton and Chen, 1997] Patton, R. J. and Chen, J. (1997). Observer–based fault detection and isolation: Robustness and applications. *Control Eng. Practice*, 5(5):671–682.

[Patton and Chen, 2000] Patton, R. J. and Chen, J. (2000). On eigenstructure assignment for robust fault diagnosis. *Int. J. of Robust & Non–Linear Control*, 10(9).

[Patton *et al.*, 1989] Patton, R. J., Frank, P. M., and Clark, R. N., editors (1989). *Fault Diagnosis in Dynamic Systems, Theory and Application.* Control Engineering Series. Prentice Hall, London.

[Patton *et al.*, 2000] Patton, R. J., Frank, P. M., and Clark, R. N., editors (2000). *Issues of Fault Diagnosis for Dynamic Systems.* Springer–Verlag, London Limited.

[Patton and Hou, 1997] Patton, R. J. and Hou, M. (1997). h_∞ estimation and robust fault detection. In *Proc. of the 1997 European Control Conference*, Brussels, Belgium. ECC'97. (CD-ROM).

[Patton and Hou, 1998] Patton, R. J. and Hou, M. (1998). Design of fault detection and isolation observers: a matrix pencil approach. *Automatica*, 34:1135–1140.

[Patton *et al.*, 2001a] Patton, R. J., Lopez-Toribio, C. J., and Simani, S. (2001a). Robust fault diagnosis in a chemical process using multiple model identification. In CSS, I., editor, *CDC'01*, pages 149–154, Orlando, Florida, U.S.A. 2001, 40th IEEE Conference on Decision and Control.

[Patton *et al.*, 2001b] Patton, R. J., Lopez-Toribio, C. J., Simani, S., Morris, J., Martin, E., and Zhang, J. (2001b). Actuator fault diagnosis in a continuous stirred tank reactor using identification techniques. In *ECC'01*, pages 2729–2734, Porto, Portugal. European Control Conference 2001.

[Patton *et al.*, 1999b] Patton, R. J., Lopez-Toribio, C. J., and Uppal, F. I. (1999b). Artificial Intelligence Approaches to Fault Diagnosis. *Applied Mathematics and Computer Science*, 9(3):471–518.

[Pau, 1981] Pau, L. F. (1981). *Failure Diagnosis and Performance Control.* Marcel Dekker, New York.

[Pettit and Wellstead, 1995] Pettit, N. and Wellstead, P. (1995). Analyzing piecewise linear dynamical systems. *IEEE Control System*, pages 43–50.

[Potter and Suman, 1977] Potter, I. E. and Suman, M. C. (1977). Thresholdless redundancy management with array of skewed instruments. Technical Report AGARDOGRAPH-224, AGARD, Integrity in Electronic Flight Control Systems.

[Priestly, 1988] Priestly, M. (1988). *Non–linear non–stationary time series analysis.* Academic press.

[Rault *et al.*, 1971] Rault, A., Richalet, A., Barbot, A., and Sergenton, J. P. (1971). Identification and modelling of a jet engine. In *IFAC Symposium od Digital Simulation of Continuous Processes*, Gejör.

[Ray and Luck, 1991] Ray, A. and Luck, R. (1991). An introduction to sensor signal validation in redundant measurement systems. *IEEE Contr. Syst. Mag.*, 11(2):44–49.

[Rich and Venkatasubramanian, 1987] Rich, S. H. and Venkatasubramanian, V. (1987). Model-based reasoning in diagnostic expert system for chemical process plant. *Comp. chem. Engng*, 11:111–122.

[Rissanen, 1978] Rissanen, J. (1978). Modelling by shortest data description. *Automatica*, (14):465–471.

[Rovatti, 1996] Rovatti, R. (1996). Takagi–sugeno models as approximators in sobolev norms: the siso case. In *Fifth IEEE International Conference on Fuzzy Systems*, New Orleans (Louisiana).

[Rovatti *et al.*, 1998a] Rovatti, R., Borgatti, M., and Guerrieri, R. (1998a). A geometric approach to maximum-speed n-dimensional linear interpolation in rectangular grids. *IEEE Transactions on Computers*, 47:894–899.

[Rovatti *et al.*, 2000] Rovatti, R., Fantuzzi, C., and Simani, S. (2000). High–speed DSP–based implementation of piecewise–affine and piecewise–quadratic fuzzy systems. *The Signal Processing Journal*, 80(6):951–963.

[Rovatti *et al.*, 1998b] Rovatti, R., Fantuzzi, C., Simani, S., and Beghelli, S. (1998b). Parameter Identification for Piecewise Linear Model with Weakly Varying Noise. In *CDC'98*, volume 4, pages 4488–4489, Tampa, Florida. 1998 IEEE Conference on Decision and Control.

[Sadrnia *et al.*, 1997] Sadrnia, M. A., Chen, J., and Patton, R. J. (1997). Robust H_∞/μ observer–based residual generation for fault diagnosis. In Pergamon, ., editor, *Proc. of the IFAC Symp. on Fault Detection, Supervision and Safety for Technical Processes: SAFEPROCESS'97*, pages 155–162, Univ. of Hull, UK.

[Sallee, 1982] Sallee, J. (1982). A triangulation of the n-cube. *Discrete Mathematics*, 40:81–86.

[Sallee, 1984] Sallee, J. (1984). Middle-cut triangulations of the n-cube. *SIAM Journal on Algebraic and Discrete Methods*, 5:407–419.

[Sauter *et al.*, 1997] Sauter, D., Rambeaux, F., and Hamelin, F. (1997). Robust fault diagnosis in a H_∞ setting. In Pergamon, ., editor, *Proc. of the IFAC Symp. on Fault Detection, Supervision and Safety for Technical Processes: SAFEPROCESS'97*, pages 867–874, Univ. of Hull, UK.

[Schilling *et al.*, 2001] Schilling, R., Carroll, J.J., J., and Al-Ajlouni, A. (2001). Approximation of non–linear systems with radial basis function neural networks. *IEEE Transactions on Neural Networks*, 12(1):1–15.

[Seber and Wild, 1989] Seber, G. and Wild, C. (1989). *Nonlinear regression*. John Wiley & Sons, New York.

[Setnes and Kaymak, 1998] Setnes, M. and Kaymak, U. (1998). Extended fuzzy c–means with volume prototypes and cluster merging. In *Proceedings EUFIT'98*, volume 3, pages 1360–1364, Aachen, Germany.

[Shann and Fu, 1995] Shann, J. and Fu, H. (1995). A fuzzy neural network for rule acquiring on fuzzy control systems. *Fuzzy Sets and Systems*, 71(1):345–357.

[Shapiro, 1977] Shapiro, A. H. (1977). *The Dynamics and Thermodynamics of Compressible Fluid Flow.* John Wiley and Sons;, London.

[Siebert and Isermann, 1976] Siebert, H. and Isermann, R. (1976). Fault diagnosis via on-line correlation analysis. Technical Report 25-3, VDI/VDE Darmstadt, Germany. In German.

[Simani, 1999a] Simani, S. (1999a). Fuzzy multiple inference identification and its application to fault diagnosis of industrial processes. In *ISAS'99/SCI'99*, volume 7, pages 185–191, Orlando, FL, USA. The Fifth Conference of the ISAS (Information Systems Analysis and Synthesis)/The Third Conference of the SCI (Systemics, Cybernetics and Informatics).

[Simani, 1999b] Simani, S. (1999b). Sensor fault diagnosis of a power plant: an approach based on state estimation techniques. In Mastorakis, N. E., editor, *IMACS-IEEE'99*, volume Recent Advances in Signal Processing and Communications, pages 274–281, Athens. International Conference on Computer Engineering in System Applications, World Scientific Engineering Society.

[Simani, 2000a] Simani, S. (2000a). Fault Diagnosis of a Power Plant at Different Operating Points using Neural Networks. In *SAFEPROCESS2000*, volume 1, pages 192–196, Budapest, Hungary. 4th Symposium on Fault Detection Supervision and Safety for Technical Processes. Invited session.

[Simani, 2000b] Simani, S. (2000b). Multi Model Based Fault Diagnosis of a Sugar Cane Crushing Process. In *SAFEPROCESS2000*, volume 2, pages 657–662, Budapest, Hungary. 4th Symposium on Fault Detection Supervision and Safety for Technical Processes.

[Simani and Fantuzzi, 2000] Simani, S. and Fantuzzi, C. (2000). Fault diagnosis in power plant using neural networks. *International Journal of Information Sciences*, 127(3–4):125–136. Special Issue: Applications to Intelligent Manufacturing and Fault Diagnosis: PART 1 - Fault Diagnosis.

[Simani and Fantuzzi, 2002] Simani, S. and Fantuzzi, C. (2002). Neural networks for fault diagnosis and identification of industrial processes. In *ESANN'02*, pages 489–494, Bruges, Belgium. Proc. of the 10th European Symposium on Artificial Neural Networks. Invited paper.

[Simani *et al.*, 1999a] Simani, S., Fantuzzi, C., and Beghelli, S. (1999a). Improved observer for sensor fault diagnosis of a power plant. In *MED99. The 7th IEEE Mediterranean Conference on Control & Automation*, pages 826–834, Haifa, Israel.

[Simani *et al.*, 2000a] Simani, S., Fantuzzi, C., and Beghelli, S. (2000a). Diagnosis techniques for sensor faults of industrial processes. *IEEE Transactions on Control Systems Technology*, 8(5):848–855.

[Simani *et al.*, 2002] Simani, S., Fantuzzi, C., and Patton, R. (2002). Identification and fault diagnosis of a simulated model of an industrial gas turbine. *IEEE Transactions on Control Systems Technology*. (under revision).

[Simani *et al.*, 1998a] Simani, S., Fantuzzi, C., Rovatti, R., and Beghelli, S. (1998a). Noise rejection in parameters identification for piecewise linear fuzzy models. In *WCCI'98, FUZZ-IEEE'98*, pages 378–383, Ancorage, Alaska. 1998 IEEE International Conference on Fuzzy Systems.

[Simani *et al.*, 1999b] Simani, S., Fantuzzi, C., Rovatti, R., and Beghelli, S. (1999b). Non–linear algebraic system identification via piecewise affine models in stochastic environment. In *CDC'99*, pages 1083–1088, Phoenix, AZ, U.S.A. 1999 IEEE Conference on Decision and Control.

[Simani *et al.*, 1999c] Simani, S., Fantuzzi, C., Rovatti, R., and Beghelli, S. (1999c). Parameter identification for piecewise linear fuzzy models in noisy environment. *International Journal of Approximate Reasoning*, 1–2(22):149–167.

[Simani *et al.*, 1998b] Simani, S., Fantuzzi, C., and Spina, P. R. (1998b). Application of a neural network in gas turbine control sensor fault detection. In *CCA'98*, volume 1, pages 182–186, Trieste, Italy. 1998 IEEE Conference on Control Applications.

[Simani *et al.*, 1999d] Simani, S., Marangon, F., and Fantuzzi, C. (1999d). Fault diagnosis in a power plant using artificial neural networks: analysis and comparison. In *ECC'99*, pages 1–6, Karlsruhe, Germany. European Control Conference 1999.

[Simani and Patton, 1999] Simani, S. and Patton, R. J. (1999). Identification and fault diagnosis of a simulated model of an industrial gas turbine. Technical Report 1, Department of Electronic Engineering at the University of Hull, Hull, U.K.

[Simani and Patton, 2002a] Simani, S. and Patton, R. J. (2002a). Model–based data–driven approaches to robust fault diagnosis in chemical processes. In *IFAC'02*, Balcelona, Spain. 15th IFAC World Congress on Automatic Control. Invited paper.

[Simani and Patton, 2002b] Simani, S. and Patton, R. J. (2002b). Neural networks for fault diagnosis of industrial plants at different working points. In *ESANN'02*, pages 495–500, Bruges, Belgium. Proc. of the 10th European Symposium on Artificial Neural Networks. Invited paper.

[Simani *et al.*, 2000b] Simani, S., Patton, R. J., Daley, S., and Pike, A. (2000b). Fault diagnosis of a simulated model of an industrial gas turbine prototype using identification techniques. In *SAFEPROCESS2000*, volume 1, pages 518–524, Budapest, Hungary. 4th Symposium on Fault Detection Supervision and Safety for Technical Processes.

[Simani *et al.*, 2000c] Simani, S., Patton, R. J., Daley, S., and Pike, A. (2000c). Identification and fault diagnosis of an industrial gas turbine prototype model.

In CSS, I., editor, *CDC'00*, pages 2615–2620, Sydney, Australia. 2000, 39th IEEE Conference on Decision and Control.

[Simani and Spina, 1998] Simani, S. and Spina, P. R. (1998). Kalman filtering to enhance the gas turbine control sensor fault detection. In *6th IEEE Med '98*, pages 443–450, Alghero, Sardinia, Italy. The 6th IEEE Mediterranean Conference on Control and Automation.

[Simani *et al.*, 1998c] Simani, S., Spina, P. R., Beghelli, S., Bettocchi, R., and Fantuzzi, C. (1998c). Fault detection and isolation based on dynamic observers applied to gas turbine control sensors. In *ASME TURBO EXPO LAND, SEA & AIR '98*, number 98-GT-158 in ASME, pages 1–11, Stockholm, Sweden. The 43rd ASME Gas Turbine and Aeroengine Congress, Exposition and Users Symposium, STOCKHOLM INTERNATIONAL FAIR.

[Sjöberg *et al.*, 1995] Sjöberg, J., Zhang, Q., Ljung, L., Beneviste, A., Delyon, B., Glorennec, P.-Y., Hjalmarsson, H., and Juditsky, A. (1995). Nonlinear black–box modelling in system identification: a unified overview. *Automatica*, 31(12):1691–1724.

[Skeppstedt *et al.*, 1992] Skeppstedt, A., Ljung, L., and Millnert, M. (1992). Construction of composite models from observed data. *Int. J. Control*, 55:141–152.

[Slotine *et al.*, 1987] Slotine, J., Hedrick, J., and Misawa, E. (1987). On sliding observers for nonlinear systems. *Transactions of the ASME: Journal of Dynamic Systems, Measurement and Control*, 109:245–252.

[Sneider and Frank, 1996] Sneider, H. and Frank, P. (1996). Observer–based supervision and fault detection in robots using nonlinear and fuzzy logic residual evaluation. *IEEE Transactions on Control Systems Technology*, 4(3):274–282.

[Söderström and Stoica, 1987] Söderström, T. and Stoica, P. (1987). *System Identification.* Prentice Hall, Englewood Cliffs, N.J.

[Sontag, 1981] Sontag, E. (1981). Nonlinear regulation: The piecewise linear approach. *IEEE Trans. on Automatic Control*, 26:346–358.

[Speyer, 1999] Speyer, J. L. (1999). Residual sensitive fault detection filters. In *MED'99*, pages 835–851, Haifa, Israel.

[Sreedhar *et al.*, 1993] Sreedhar, R., Fernandez, B., and Masada, G. (1993). Robust fault detection in nonlinear systems using sliding mode observers. In *Proceedings of the IEEE Conference on Control Applications*, pages 715–721.

[Stoustrup and Niemann, 1998] Stoustrup, I. and Niemann, H. (1998). Fault detection for nonlinear systems - a standard problem approach. In *Proc. of the 37th IEEE Conf. on Decision & Control*, pages 96–101, Tampa, Florida, USA.

[Stoustrup *et al.*, 1997] Stoustrup, J., Grimble, M. J., and Niemann, H. (1997). Design of integrated systems for the control and detection of actuator/sensor faults. *Sensor Review*, 17(2):138–149.

[Strang and Fix, 1973] Strang, G. and Fix, G. (1973). *An Analysis of the Finite Element Method.* Prentice-Hall.

[Sugeno and Kang, 1988] Sugeno, M. and Kang, G. (1988). Structure identification of fuzzy model. *Fuzzy Set and Systems*, 28:15–33.

[Tachibana and Furuhashi, 1994] Tachibana, K. and Furuhashi, T. (1994). A hierarchical fuzzy modelling method using genetic algorithm for identification of concise submodels. In *Proc. of 2nd Int. Conference on Knowledge–Based Intelligent Electronic Systems*, Adelaide, Australia.

[Takagi and Sugeno, 1985] Takagi, T. and Sugeno, M. (1985). Fuzzy identification of systems and its application to modeling and control. *IEEE Transaction on System, Man and Cybernetics*, SMC-15(1):116–132.

[Tan and Edwards, 2001] Tan, C. P. and Edwards, C. (2001). An LMI approach for designing sliding mode observers for fault detection and isolation. In *European Control Conference, ECC'01*, pages 481–486, Porto, Portugal.

[The MathWorks Inc, 1990] The MathWorks Inc (1990). *MATLAB User's Guide.* The MathWorks, Inc, Natick, Massachusetts, USA.

[The MathWorks Inc, 1991] The MathWorks Inc (1991). *SIMULINK User's Guide.* Mathworks Inc, Natick, Massachusetts, USA.

[Tikhonov and Arsenin, 1977] Tikhonov, A. and Arsenin, V. (1977). *Solution of Ill-posed Problems.* Winston and Wiley, Washington.

[Tou and Gonzalez, 1974] Tou, J. T. and Gonzalez, R. C. (1974). *Pattern recognition principles.* Addison–Wesley Publishing.

[Ulieru and Isermann, 1993] Ulieru, M. and Isermann, R. (1993). Design of fuzzy-logic based diagnostic model for technical process. *Fuzzy Set and Systems*, 58(3):249–271.

[Uppal and Patton, 2002] Uppal, F. J. and Patton, R. J. (2002). Fault diagnosis of an electro–pneumatic valve actuator using neural networks with fuzzy capabilities. In *ESANN'02*, Bruges, Belgium. Proc. of the 10th European Symposium on Artificial Neural Networks. Invited paper.

[Uppal *et al.*, 2002] Uppal, F. J., Patton, R. J., and Palade, V. (2002). Neuro-fuzzy based fault diagnosis applied to an electro–pneumatic valve. In *IFAC'02*, Balcelona, Spain. 15th IFAC World Congress on Automatic Control.

[Utkin, 1977] Utkin, V. (1977). Variable structure with sliding modes. *IEEE Trans. AC*, 22:212–222.

[Utkin, 1992] Utkin, V. (1992). *Sliding modes in control and optimisation.* Springer–Verlag, Berlin, 3rd edition.

[van Huffel and Vandewalle, 1991] van Huffel, S. and Vandewalle, J. (1991). The Total Least Squares Problem: Computational Aspects and Analysis. *Frontiers in Applied Mathematics.* Philadelphia, USA.

[Venkatasubramanian and Chan, 1989] Venkatasubramanian, V. and Chan, K. (1989). A neural network methodology for process fault diagnosis. *AIChE J.*, (35):1993–2002.

[Verhaegen and Dewilde, 1992] Verhaegen, M. and Dewilde, P. (1992). Subspace model identification. Part I: the output error state space model identification class of algorithms. *International Journal of Control*, 56(1):1187–1210.

[Walcott and Żak, 1988] Walcott, B. and Żak, S. (1988). Combined observer-controller synthesis for uncertain dynamical systems with applications. *IEEE Transactions on Systems, Man. and Cybernetics*, 18:88–104.

[Wang, 1992] Wang, L. (1992). Fuzzy systems are universal approximators. In *Proc. first IEEE Int. Conf. on Fuzzy Syst.*, S. Diego (CA).

[Wang, 1995] Wang, L.-X. (1995). The design and analysis of fuzzy identifiers of nonlinear dynamic systems. *IEEE Transaction on Automatic Control*, 40(1):11–23.

[Wang *et al.*, 1975] Wang, S. H., Davison, E. J., and Dorato, P. (1975). Observing the state of systems with unmeasurable disturbance. *IEEE Trans. on Automatic Control*, 20:716–717.

[Watanabe and Himmelblau, 1982] Watanabe, K. and Himmelblau, D. M. (1982). Instrument fault detection in systems with uncertainties. *Int. J. System Sci.*, 13(2):137–158.

[Weerasinghe *et al.*, 1998] Weerasinghe, M., Gomm, J., and Williams, D. (1998). Neural network for fault diagnosis of a nuclear fuel processing plant at different operating points. *Control Engineering Practice*, 6:281–289.

[Werbos, 1990] Werbos, P. J. (1990). Backpropagation through time: what it does and how to do it. *Proc. IEEE*, 78(10):1550–1560.

[Widrow and Lehr, 1990] Widrow, B. and Lehr, M. A. (1990). 30 years of adaptive neural networks: Perceptron, madaline, and backpropagation. *Proc. IEEE*, 78(9):1415–1442.

[Willsky, 1976] Willsky, A. S. (1976). A survey of design methods for failure detection in dynamic systems. *Automatica*, 12(6):601–611.

[Wu and Harris, 1996] Wu, Z. Q. and Harris, C. J. (1996). Neuro–fuzzy modelling and state estimation. In *IEEE Medit. Symp. on Control and Automation: Circuits, Systems and Computers '96*, pages 603–610, Hellenic Naval Academy, Piraeus, Greece.

[Wünnenberg, 1990] Wünnenberg, J. (1990). *Observer–based fault detection in dynamic systems.* PhD thesis, University of Duisburg, Duisburg, Germany.

[Wünnenberg and Frank, 1987] Wünnenberg, J. and Frank, P. M. (1987). Sensor fault detection via robust observer. In *System Fault Diagnosis, Reliability, and Related Knowledge-Based Approaches*, volume 1, pages 147–160. S. Tzafestas et al edition.

[Wünnenberg and Frank, 1990] Wünnenberg, J. and Frank, P. M. (1990). Robust observer–based detection for linear and non–linear systems with application to robot. In *Proc. of IMACS Annals on Computing & Applied Mathematics MIM-S^2:90*, Brussels.

[Xie and Soh, 1994] Xie, L. and Soh, Y. C. (1994). Robust Kalman filtering for uncertain systems. *Systems and Control Letters*, 22:123–129.

[Xie *et al.*, 1994] Xie, L., Soh, Y. C., and de Souza, C. E. (1994). Robust Kalman filtering for uncertain discrete-time systems. *IEEE Transaction on Automatic Control*, 39:1310–1314.

[Ying, 1994] Ying, H. (1994). Sufficient conditions on general fuzzy systems as function approximators. *Automatica*, 30:521–525.

[Yu *et al.*, 1999] Yu, D., Gomm, J., and Williams, D. (1999). Sensor fault diagnosis in a chemical process via RBF Neural Networks. *Control Engineering Practice*, 7:49–55.

[Zeng and Singh, 1996] Zeng, X.-J. and Singh, M. (1996). Approximation accuracy analysis of fuzzy systems as function approximators. *IEEE Transactions on Fuzzy Systems*, 4:44–63.

[Zhang and Morris, 1996] Zhang, J. and Morris, J. (1996). Process modeling and fault diagnosis using fuzzy neural networks. *Fuzzy Sets and Systems*, 79(1):127–140.

[Zhou *et al.*, 1996] Zhou, K., Doyle, J. C., and Glover, K. (1996). *Robust and Optimal Control.* Prentice Hall, New Jersey.

INDEX

H_∞ filter
– residual generation, 254
H_∞ optimisation, 253
– residual generation, 254
μ synthesis
– residual generation, 254
χ^2 innovation test, 184

Adaptive residual generation, 255
Analytical redundancy, 169

Correlation test, 184
Cumulative sum algorithm, 184

Dedicated observer scheme, 123
Disturbance, 4
Disturbance de–coupling, 131
Disturbance distribution matrix, 133, 248
– estimation, 132, 136, 137
– identification, 139
– optimisation, 139
Double–shaft gas turbine
– description, 199
– disturbance de–coupling, 209
– fuzzy model, 210
– fuzzy residual generation, 211
– identification, 199, 201
– Kalman filter, 208
– minimal detectable faults, 214
– Output observer, 207
– Pont–sur–Sambre, 199
– UIO, 203
Dynamic observer, 117
– bank, 122
– FDI, 176

Eigenstructure assignment, 118, 132
Eigenvalue assignment, 118
Error, 3

Failure, 3
Fantuzzi-Simani-IFAC:2002, 139
Fault, 3
– abrupt, 5
– additive, 5
– incipient, 5
– multiplicative, 5
Fault Detection
– H_∞ methods, 253
Fault detection, 4
– H_∞ methods, 47, 49
– active robustness, 50
– disturbance decoupling, 46
– in dynamic systems, 19
– input sensor, 125
– model uncertainty, 8
– model–based, 7, 19
– neural networks, 53
– output sensor, 124
– passive robustness, 50
– performace index, 49
– redudancy methods, 5
– robust methods, 9
Fault diagnisis, 4
Fault diagnosis
– non–linear system, 258
Fault identification, 4, 116, 143, 144, 256
– pattern recognition, 53
Fault isolation, 4, 124, 131
– input sensor, 123
– output sensor, 122
Fault location, 22
Fault model, 21, 26
– actuator fault, 25
– multiplicative fault, 25
– sensor noise, 25
– sensors fault, 24
– state–spece model, 26
– transfer function model, 27

Fault signature, 123
Fault tolerant control, 256
FDI
– integration, 256
– model–based, 20
Frequency domain, 253
– design, 253
– residual generation, 253
Fuzzy clustering, 95
– c–Menas algorithm, 97
– Gustafson–Kessel algorithm, 98
– product–space, 100, 105
Fuzzy model
– antecented fuzzy sets, 92, 107
– consequent crisp functions, 92, 93, 109
– defuzzification, 94
– Takagi–Sugeno, 142
Fuzzy models
– residual generation, 142

Gas turbine, 157
– description, 158
– diagram, 163
– double–shaft, 199
– identification, 168
– model prototype, 214
– modelling, 158, 160
– SIMULINK scheme, 161
– single–shaft, 169
– single–shaft model, 171
Gas turbine prototype, 214
– description, 215, 216
– disturbance de–coupling, 243
– eigenstructure assignment, 242
– fault description, 221
– fault isolation, 235
– identification, 215
– Kalman filter, 233
– minimal detectable faults, 239
– output observer, 220
– robust residual generation, 243
Generalised observer scheme, 123

Hankel matrix, 65
Hybrid model, 259

IFAC, 1
Innovation test, 184
Input sensor
– fault detection, 125
– fault isolation, 123

Kalman filter, 130
– bank, 122
– design, 130
– parameter estimation, 142
– residual, 184

Low rank approximation, 135
Luenberger observer, 116

Malfunction, 3
Model reduction, 133
Monitoring, 4
Multi Layer Perceptron, 145
Multiple model approach, 142

Neural network, 145
– back–propagation, 146
– FDI, 143, 149
– multiple operating points, 147
– supervised, 146
Neuro–Fuzzy, 54, 150
– ARMA model, 154
– B–spline, 56
– FDI, 151
– hierarchical networks, 56
– Mamdani model, 152
– residual evaluation, 155
– residual generation, 57, 152, 154
– structure identification, 57
– Sugeno–type, 55, 152
Non–linear observer
– fault diagnosis, 258
Non–linear system
– fault diagnosis, 258
– fault identification, 143
– hybrid model, 259
– linearisation, 258
– modelling, 142, 143, 150, 258
– Neuro–Fuzzy, 150
– residual generation, 142, 150, 152, 154
– sliding mode observer, 127

Observer, 117
– eigenvalues, 117
Output observer, 116, 122
Output sensor
– fault detection, 124
– fault isolation, 122

Parameter estimation
– equation error methods, 32
– output error mehtods, 34
parameter estimation
– Kalman filter, 142

Parity equations, 40
Parity relation, 253
Pole placement, 118
Pont–sur–Sambre, 199

Radial Basis Function, 145
Regression
– non–linear, 103
Residual, 4, 6, 115
– generation, 115
– robustness, 131, 254
– sensitivity, 117
Residual evaluation
– Fuzzy threshold, 57
Residual generation
– observer–based approach, 35
Residual analysis, 20, 44
– fuzzy decision-making, 52
– residual evaluation, 21
– residual generation, 21
– with statistical methods, 44
Residual evaluation
– Neuro–Fuzzy, 155
Residual generation, 28, 30
– H_∞ filter, 254
– adaptive, 255
– adaptive threshold, 50
– bank of observers, 38
– comparing with threshold, 31
– factorisation method, 253
– frequency domain, 253
– fuzzy model, 142
– Kalman filters approach, 37
– MIMO processes, 38
– neural network, 144
– Neuro–Fuzzy, 152, 154
– neuro–fuzzy, 58
– output observers, 39
– parameter estimation, 141
– techinques, 31
– via parameter estimation, 32
– with parity equations, 42
Robust residual generation, 116, 131, 247

SAFEPROCESS, 1
Single–shaft gas turbine
– fuzzy identification, 189
– fuzzy residual generation, 189
– Kalman filter, 183
– Kalman filter residual, 185
– minimum detectable fault, 186
– multiple working conditions, 196, 197
– multiple–model, 190
– neural network, 191
– output observer, 178
– sensor FDI, 176
– thresholds, 177
– UIO, 177
Single–shaft gast turbine
– sensor fault identification, 192
Singleton model, 94
Singular value decomposition, 135
Sliding Mode Observer, 127
– design, 128
– structure, 129
Supervision, 4
Symptom, 4
System identification, 11, 61
– affine systems, 61, 82
– Frisch scheme, 61
–– algebraic case, 68
–– dynamic case, 70
–– MIMO case, 73
– fuzzy systems, 90
– homogenous Takagi Sugeno fuzzy models, 93
– Takagi Sugeno fuzzy models, 92
System model
– ARX, 63
– Error in Variable (EIV), 25
– error in variable (EIV), 62, 63, 83
– fuzzy systems, 52, 89
– fuzzy systems structure estimation, 102
– hybrid, 75
– linear systems, 24, 62
– non linear systems, 52
– non–linear ARX, 104
– non–linear systems, 75
– piecewise affine, 75
– state–space realisation, 64

Uncertainty
– bounded, 135
– parameter, 134
– structured, 248
– unstructured, 134
Uncertainty unstructured, 135
Unknow Input Observer
– design procedure, 125
Unknown Input Kalman Filter, 123, 130, 131
Unknown Input Observer, 119, 123
– de–coupling, 120
– design procedure, 122

– existence conditions, 121
– full–order, 120
– structure, 120

MIX
Papier aus verantwortungsvollen Quellen
Paper from responsible sources
FSC® C105338

If you have any concerns about our products,
you can contact us on
ProductSafety@springernature.com

In case Publisher is established outside the EU,
the EU authorized representative is:
Springer Nature Customer Service Center GmbH
Europaplatz 3, 69115 Heidelberg, Germany

Printed by Libri Plureos GmbH
in Hamburg, Germany